A Practical Guide to Aerial Photography with an Introduction to Surveying

A Practical Guide to Aerial Photography with an Introduction to Surveying

John A. Ciciarelli
Pennsylvania State University

VAN NOSTRAND REINHOLD
New York

Softcover reprint of the hardcover 1st edition 1991
Library of Congress Catalog Card Number 90-20481
ISBN-13: 978-1-4684-6505-1 e-ISBN-13: 978-1-4684-6503-7
DOI: 10.1007/978-1-4684-6503-7

Manufactured in the United States of America

Published by Van Nostrand Reinhold
115 Fifth Avenue
New York, New York 10003

Chapman and Hall
2-6 Boundary Row
London, SE1 8HN

Thomas Nelson Australia
102 Dodds Street
South Melbourne 3205
Victoria, Australia

Nelson Canada
1120 Birchmount Road
Scarborough, Ontario M1K 5G4, Canada

16 15 14 13 12 11 10 9 8 7 6 5 4 3 2 1

Library of Congress Cataloging-in-Publication Data

Ciciarelli, John A., 1942–
A practical guide to aerial photography: with an introduction to surveying
John A. Ciciarelli.
p. cm.
Includes bibliographical references and index.
ISBN-13: 978-1-4684-6505-1
1. Photography, Aerial. 2. Aerial photogrammetry. I. Title.
TR810.C55 1991
778.3′5—dc20 90-20481
CIP

Contents

Preface

The aim of this book and the rationale behind writing it are twofold, and are born of experience in working in the field with geologists, builders, landscapers, and people in the construction and building trades in general. This author, in his capacity as a geologist, air photo interpreter, and surveyor, has been asked to consult on very simple tasks that should be a part of the general knowledge of nearly everyone in the building trades. There is a definite and clear need for a short, simple, well-illustrated, and concise book that treats problems encountered every day by people in the field, and for training such people. This is such a book.

The primary aim of the book, then, is not to write about new knowledge in the areas of photogrammetry or mathematics, or to discuss the latest procedures in using complex electrooptical surveying devices, but to give the common, everyday worker in the field (and there are many of them) some assistance and guidance.

The book is divided into five major sections. The first section, i.e., the first three chapters, deals with the basic mathematics needed to function as a surveyor and as a user of aerial photography. It should be mentioned that only those aspects of the mathematics that are germane to the general subject material of the book are discussed. Moreover, this is accomplished with a minimum of theoretical and derivational presentations so that the reader does not need an extensive background in mathematics to be able to use and apply the equations and formulae. For example, understanding simple trigonometry for squaring up foundations, computing areas of regular and irregular plots, and computing volumes for earth work often not well understood. These are discussed in some detail.

Chapter 3 is also a good example of this pragmatic approach. It deals with accuracy and precision and how to obtain both with instruments in common and everyday use by the small construction or landscape contractor. Complex, expensive instruments are not always needed to obtain satisfactory accuracy. This chapter, then, instructs the reader in a straightforward manner how to squeeze the maximum accuracy and precision from simple and inexpensive equipment.

The second section (Chapters 4–6) provides information needed for a person in the field or the owner of a small construction company with limited resources to be able to use aerial photography. Aerial photography, a very powerful tool that can be used in many projects, is almost always overlooked in small construction projects. The simple procedures of sketching, triangulation, true-north determinations, etc. on an aerial photograph can be quite useful and should be understood. Moreover, many workers in the field have little understanding of how to interpret an air photo in terms of slope stability, rock fractures, or soils—all important to the builder. Because of these needs, these topics are dealt with as well.

These chapters show not only how to use existing conventional aerial photography, but also how to obtain photography tailored to specific needs. For example, discussed are techniques to calibrate your own hand-held camera to take your own photography, how to accurately communicate to an aerial photographic contractor what your needs are, and how to use aerial photographs for simple and elementary interpretation. Another extremely useful section treats the subject of transferring data points from an aerial photograph to a map in their spatially correct positions.

There is also a need for understanding the basics in common surveying, layout, and mapping. What is needed is not so much theory or by-the-book field techniques, but rather ways to extract greater accuracy and precision from the common equipment that a small construction enterprise is likely to have; there are some interesting and little-used techniques which should be in the repertoire of the worker in the field. In view of this, the last section included in the text encompasses three chapters on surveying. These are not intended to cover the entire gamut of surveying tasks, but again to give a person working in the field, with limited experience and resources, a basic understanding of the most common elements of surveying. Covered are items such as taping, turning angles with a transit, layout work, cut-and-fill concepts, and topographic mapping. As with the work on aerial photography, these topics are presented in a useful and pragmatic fashion without the extra baggage of theoretical and background discussions.

The final two parts of the book are contained in the appendices. The abundant material found there is more than just trigonometric and unit conversion tables, although this type of information is there. Two major topics in these sections can be of great assistance to the reader. First, the section titled ''Additional Layout Techniques with a Transit'' gives many examples of procedures that can be extremely helpful in the field. The second aid is the computer programs which enable the user to solve area and volume problems with the speed, accuracy, and convenience afforded by a personal computer.

The disk is available free of charge to instructors using the book. It can be obtained by writing to the Engineering Editor, Van Nostrand Reinhold, 115 Fifth Avenue, New York, NY 10003.

The second aim of this book is to provide a good textbook which will furnish the fundamental requirements for students who might someday work as surveyors, landscape contractors, building contractors, or *any* field practitioner and for students wishing to continue their studies in photogrammetry and/or surveying. This book also offers basics that are essential to more advanced work. Whether one has many years of experience or is a rank novice, it is hoped that the book will furnish some new and useful insights and perspectives into the fascinating and very practical topics of aerial photography and surveying.

Acknowledgments

Many colleagues have contributed time, energy, and expertise to make this volume a reality. It is impossible to acknowledge every person involved and everyone who assisted me in this effort. However, I am indebted to the following for their important and direct contributions; I would like to thank this group for their work in reviewing the manuscript at various stages along its route to completion: William "Pete" Blackwood, Blackwood Construction Company; Bill Fussetti, Michael Baker Corporation; Tom Seybert, Department of Civil Engineering, Penn State University; Melvin Podwysocki, U.S. Geological Survey; Larry Souleret, Souleret Engineering; and Robert Clark, U.S. Geological Survey.

Thanks are extended to Janet Colaluca, June Golletti, and Norine Kirk, all of Penn State, for their work in typing both the manuscript and all the correspondence with Van Nostrand Reinhold. For assistance in obtaining needed documents and data for various phases of the project, I would like to thank Ralph Viola, Aliquippa and Southern Railroad (Ret.) and Rocco DeMaiolo of Penn State. For guidance in preparing the photographs, and for invaluable help, suggestions, and encouragement from the beginning to the end of the project, special thanks and acknowledgments go to Jim Luce, also of Penn State University.

Extra thanks and indebtedness are extended to Ted Stahl of Wallover and Mitchell Architects, who did an outstanding job in drafting the many line drawings, diagrams, and maps. I would also like to state my appreciation for his patience while working with constantly changing deadlines and requirements.

I reserve my highest commendation and testimonial, and my most heartfelt gratitude, for Tom Ciciarelli, U.S. Geological Survey, for his skillful work in designing and writing the computer programs that are an integral part of this book. Without his knowledge, expertise, and experience in this area, the book would be lacking an important and very useful dimension.

Finally, I would like to thank my family (my wife Kathy, and children Greg, Jill, and Matt) for their encouragement, help, and understanding during this entire project. At the same time, I apologize to them for the disruption in our daily routine and for the constantly paper-littered dining room table.

A Practical Guide to Aerial Photography with an Introduction to Surveying

1
Trigonometry

ANGLES AND THEIR MEASUREMENT

Because almost all measurements made on aerial photographs and those made in the field by surveyors and builders are lines and angles, an understanding of the portion of mathematics dealing with these elements is essential for our work. Trigonometry, which is derived from two Greek words, *trigon* (triangle) and *metric* (measurement), and which is the only link between straight-line measurements and angular measurements, satisfies our requirements. No attempt is being made here however to understand trigonometry in its entirety. We will examine only some very basic concepts germane to practical aerial photography and surveying.

Because the mutual relationship between sides of triangles, sides of complex polygons, property boundaries, building corners, road intersections, and many other features are defined in terms of angles, we will begin with their study. When line *OB* is rotated about point *O*, it is said to describe an angle (see Fig. 1-1). The beginning position, *OA*, is called the *initial side*, while the final position, *OB*, is called the *terminal side*, and the point of rotation is called the *vertex* of the angle. The angle generated in Fig. 1-1 can be designated either as angle *AOB* or simply as angle *O*.

The amount of rotation about point *O* is measured in degrees. It should be mentioned here that for accurate work, one-degree increments are usually not small enough. Therefore, keep in mind that one degree can be subdivided into 60 minutes of arc, and a minute can be further subdivided into 60 seconds of arc. Remember, too, that the minute and second designations refer to parts of a degree and *not* time. As an example of this notation, the equalities in Table 1-1 are noted.

As you might suspect there are different categories, classifications, or types of angles. These are given names, and should be kept in mind. Fig. 1-2 illustrates four types of angles that can be generated by rotating line *OB* about point *O*. An *acute angle* is an angle of less than 90°, an *obtuse angle* is one which is more than 90°, a *straight angle* is 180°, and a *right angle* contains 90°.

During your work you may be called upon to read many different varieties of maps, charts, plans, field notes, etc. drawn by different people under various

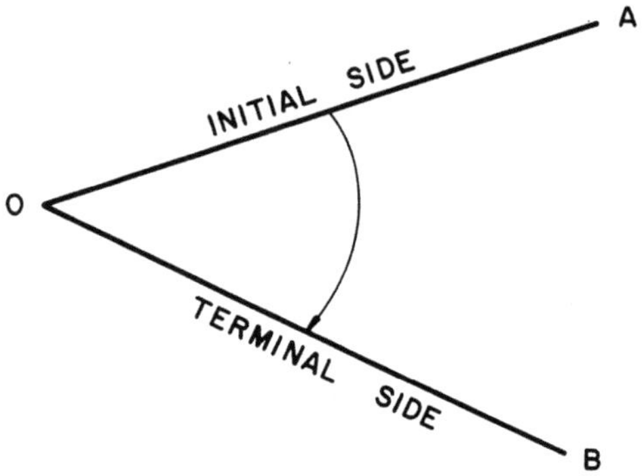

Fig. 1-1. Parts of an angle.

Table 1-1.

$1\frac{1}{4}°$	=	1°15′
$1\frac{1}{2}°$	=	1°30′
$1\frac{3}{4}°$	=	1°45′
1°30.25′	=	1°30′15″
1°30.50′	=	1°30′30″
1°30.75′	=	1°30′45″

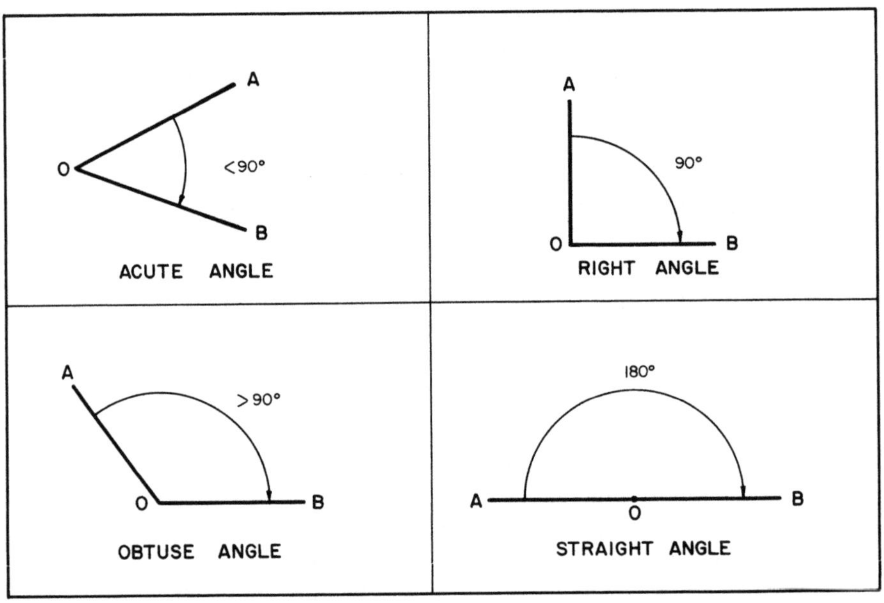

Fig. 1-2. Various types of angles.

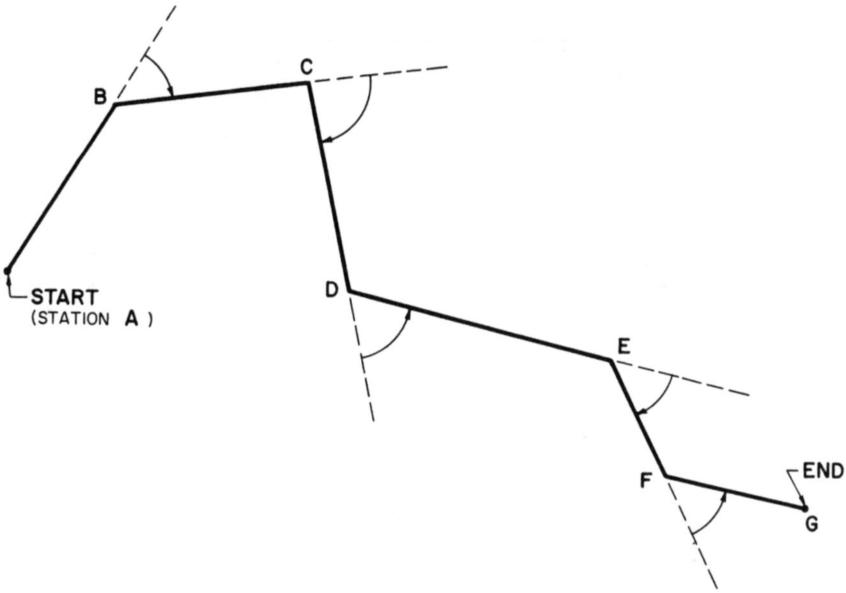

Fig. 1-3. Measurement of deflection angles (A to G is an *open traverse*).

conditions, with some perhaps drawn many years ago. Therefore, it is important to understand a few of the more common ways that angles and the directions of lines are measured and recorded.

First, deflection angles are frequently used when doing route surveys or when working with open traverses, i.e., traverses that do not return to their starting points. A *deflection angle* is defined as a change in the direction of a course at a station (see Fig. 1-3). It should be obvious from Fig. 1-3 that unless the direction of the deflection (left or right) is properly noted and recorded a serious error may result.

Second, in working with triangles, other more complex polygons, and with closed traverses (those that return to their starting points), it is convenient to use what are called interior or included angles. An *interior angle* is formed within a closed figure between the adjacent sides (see Fig. 1-4). Note that interior angles may be acute, right, or obtuse angles.

The orientation or direction of property lines recorded on deeds and other official documents are almost always shown as bearings. The *bearing* of a line is simply its direction measured from either north or south, whichever is closer (see Fig. 1-5). Thus all bearing angles will be 90° or less. Care should be taken when working with bearings to determine whether or not they are measured from true north (and south) or from magnetic north (and south). A complete explanation of the differences between the two is given in Appendix F.

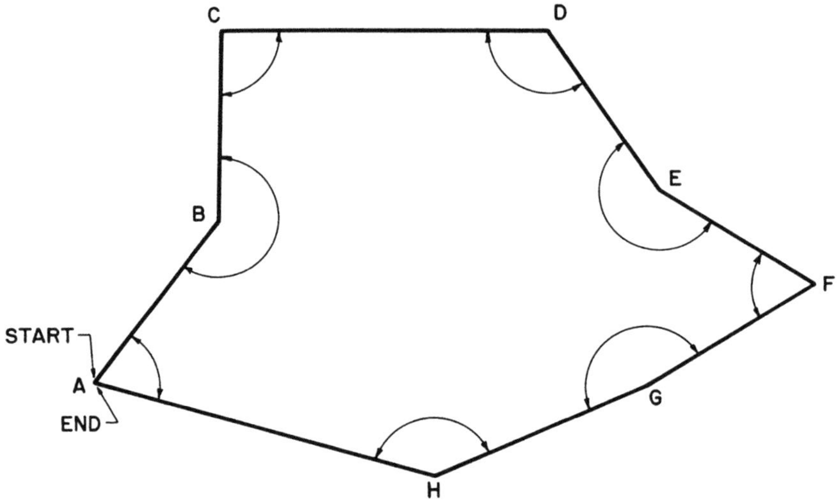

Fig. 1-4. Measurement of interior angles (*ABCDEFGHA* is a *closed traverse*).

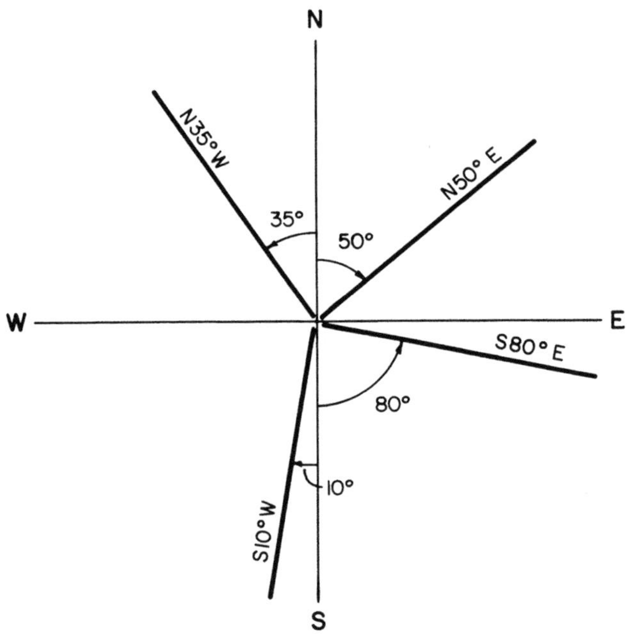

Fig. 1-5. Measurement of bearing angles.

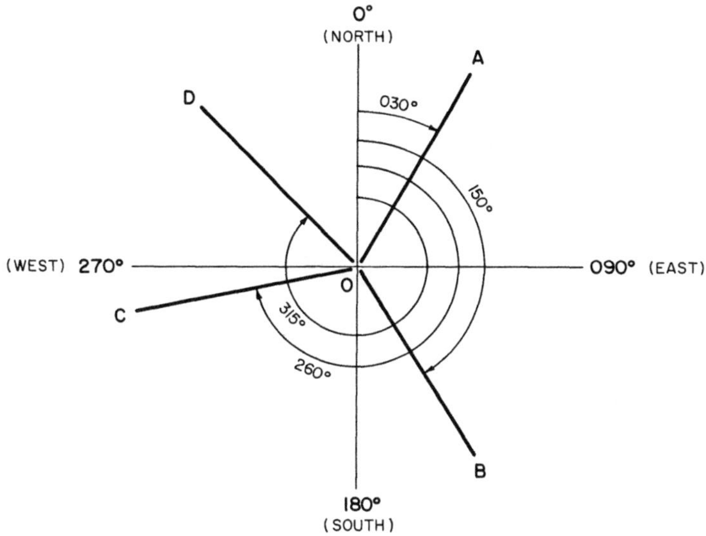

Fig. 1-6. Measurement of azimuth.

Another method of recording the direction of a line with respect to a reference is referred to as the azimuth method. An *azimuth* gives the direction of a line measured in degrees in a clockwise direction usually from true north (see Fig. 1-6). Using this system for recording and measuring line directions, a line may have an azimuth between 0° and 360°.

It is important to note that when bearings and azimuths are stated, it is understood that they are the directions of lines from a point of origin to some terminal point. Thus, if the direction is reversed, then the terms *back azimuth* and *back bearing* are employed. To illustrate this, Figs. 1-7 and 1-8 show the relationship between azimuth and back azimuths and bearings and back bearings, respectively. Fig. 1-7 shows line *AB* with an azimuth of 065°, while *BA* has a back azimuth of 245°. Fig. 1-8 shows line *AB* with a bearing of N70°W, while line *BA* has a back bearing of S70°E.

It is a simple matter from the preceeding figures and explanation to note that a back azimuth of a line can be determined from its forward azimuth both conceptually and arithmetically. If the azimuth of a line is greater than 180°, subtract 180°, and if the azimuth of a line is less than 180°, add 180°. In the event bearings are used, back bearings can be found simply by changing the letter designation, e.g., N to S or S to N *and* changing E to W or W to E.

It should be evident at this point that it is important for you to become very familiar with converting from one type of angular notation to another, i.e., bearings to azimuths and vice-versa. You should be at ease with converting

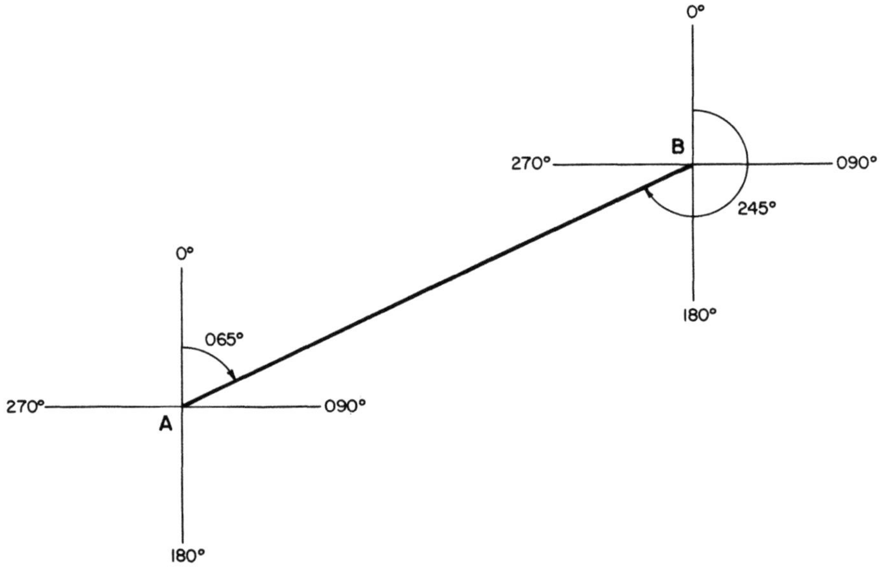

Fig. 1-7. Azimuth and back azimuth.

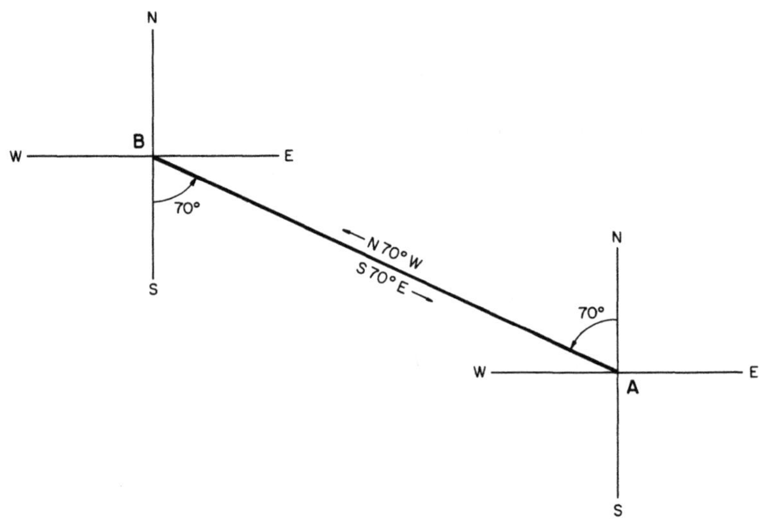

Fig. 1-8. Bearing and back bearing.

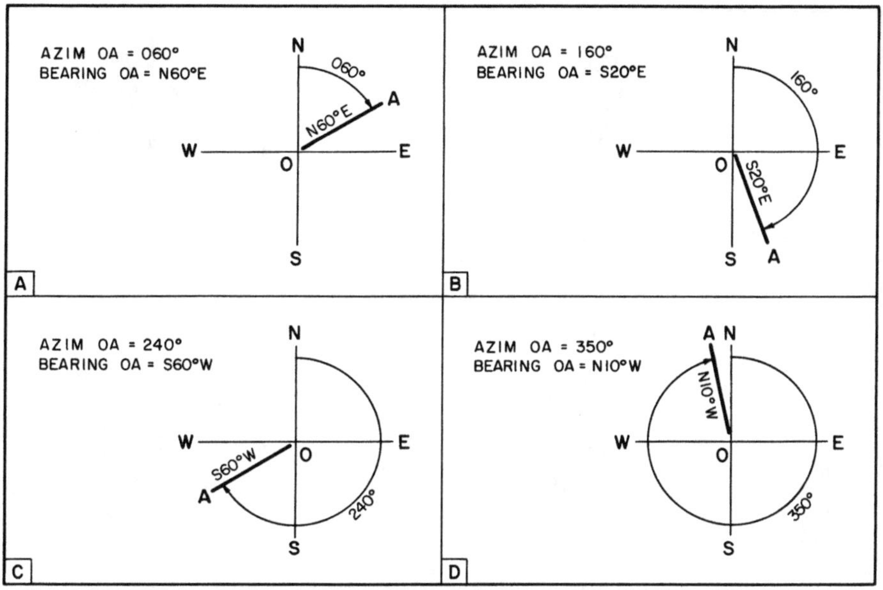

Fig. 1-9. Azimuth versus bearing.

bearings and azimuths of adjacent property lines to interior angles (see Figs. 1-9 to 1-11). They will help you to visualize the computations needed for this. Study these carefully.

TRIGONOMETRIC FUNCTIONS

As was mentioned at the beginning of this chapter trigonometry bears the only links between angular measurements and straight line measurements. It is appropriate then at this point to examine these relationships. They are not difficult to understand, and they can be quite useful in our work in aerial photography and surveying for they allow us to easily solve many difficult problems.

Referring to Fig. 1-12, note that side *a* is the same length in all of the triangles. Note, too, that as side *b* is made longer, angle *A* decreases. By the same procedure, if side a increases, and side *b* remains constant, angle *A* then increases. Also observe that as angle *A* increases angle *B*, its *complimentary angle*, decreases and vice-versa. These are a few of the relationships that exist between the side and angles of right triangles. (A *right triangle* is one in which an angle is a 90° angle.) These relationships permit us to say in general that the size of the angles in a right triangle depends on the ratios between the lengths of any two of the sides. Because these relationships (or ratios) are the foundation of trigonometry, they have been formalized and given names.

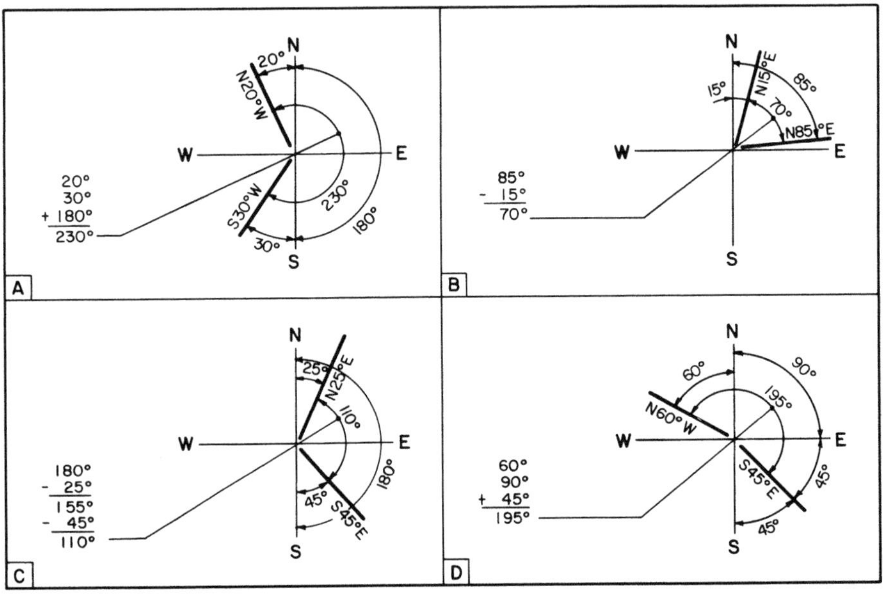

Fig. 1-10. Determining interior angles between two given bearings.

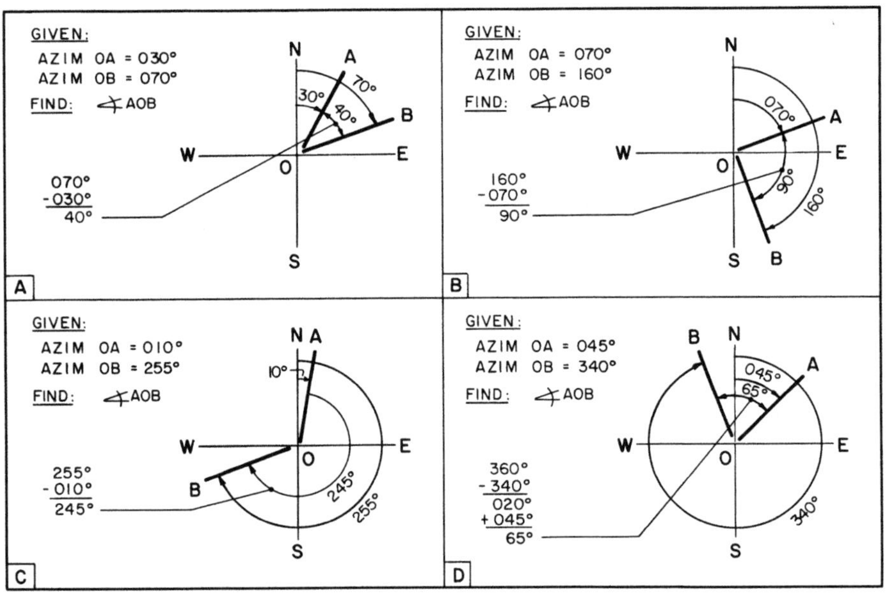

Fig. 1-11. Determining interior angles from given azimuths.

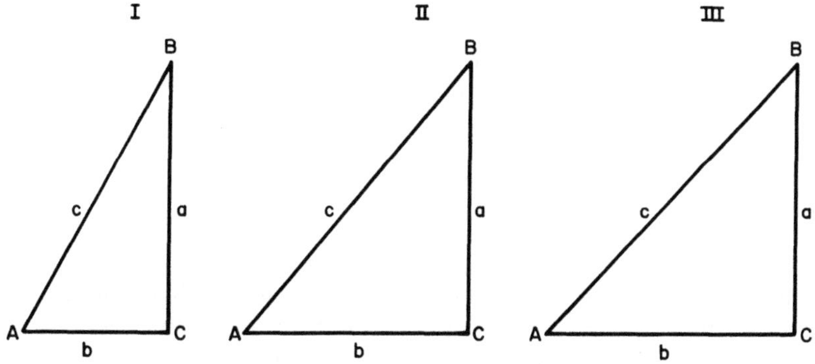

Fig. 1-12. Relationships between angles of a right triangle and the length of its sides.

By taking two of the sides of a right triangle at a time, six ratios called *trigonometric functions* can be expressed. For example, the length of the side opposite angle A divided by the length of the hypotenuse (the side opposite of the right angle) is called the sine of angle A. The term sine is abbreviated to "sin" though is pronounced as if the "e" were still on the end. Using the nomenclature of the triangles in Fig. 1-12, all six functions are given in Table 1-2.

With these six functions it is possible to *solve*, i.e., determine all the sides and angles of *any* right triangle, provided that at least one side and one of the

Table 1-2.

Function Name (general)		Side Name (general)		Abbreviation of Function (labeled angle)		Labeled Side
Sine	=	$\frac{\text{Opposite side}}{\text{Hypotenuse}}$	=	$\sin A$	=	$\frac{a}{c}$
Cosine	=	$\frac{\text{Adjacent side}}{\text{Hypotenuse}}$	=	$\cos A$	=	$\frac{b}{c}$
Tangent	=	$\frac{\text{Opposite side}}{\text{Adjacent side}}$	=	$\tan A$	=	$\frac{a}{b}$
Cosecant	=	$\frac{\text{Hypotenuse}}{\text{Opposite side}}$	=	$\csc A$	=	$\frac{c}{a}$
Secant	=	$\frac{\text{Hypotenuse}}{\text{Adjacent side}}$	=	$\sec A$	=	$\frac{c}{b}$
Cotangent	=	$\frac{\text{Adjacent side}}{\text{Opposite side}}$	=	$\cot A$	=	$\frac{b}{a}$

Table 1-3.

$\sin B$	=	$\cos A$
$\cos B$	=	$\sin A$
$\tan B$	=	$\cot A$
$\cot B$	=	$\tan A$
$\sec B$	=	$\csc A$
$\csc B$	=	$\sec A$

acute angles is known, *or* two sides are known. Because the functions are ratios, and because no absolute values are placed on them, they can be used to solve problems of astronomical proportions or problems on the molecular scale. Because there are only six possible combinations, and because they are so useful, it would be very helpful to memorize them. Other trigonometric relationships to be introduced later in this chapter will simply be extensions of these functions.

Note that the acute angle opposite angle A, angle B, is its *complimentary angle*, i.e., they add up to 90°. The functions of complimentary angles can be seen in Table 1-3. The pairs of functions, i.e., sin/cos, tan/cot, sec/csc, are called *cofunctions*. From this it can be stated that the function of an angle is equal to the corresponding cofunction of the complimentary angle. As the functions and corresponding cofunction are equal, there is no advantage in using one versus the other. As a general statement, choose the function that best fits the problem and data at hand. We will try a few examples later in this chapter.

USING TABLES OF TRIGONOMETRIC FUNCTIONS

Suppose you need to know one of the angles of a right triangle and you have the dimensions of the hypotenuse (50 ft) and the side adjacent to angle A (40 ft). From Table 1-2 you can see that the ratio determined by the adjacent side/hypotenuse is $\cos A$. Thus

$$\cos A = \frac{40 \text{ ft}}{50 \text{ ft}} = \frac{\text{Adjacent side}}{\text{Hypotenuse}} = 0.8000 \tag{1-1}$$

What now needs to be determined is the angle that the ratio 0.8000 represents. For this we need a *Table of Trigonometric Functions* (see Appendix A, Tables 1-1 and 1-2). Such a table is a tabulation of computed values of the sines, cosines, tangents, etc., for all acute angles. If you scan the table in the cosine column, you will discover that 0.8000 is closest to 0.80003. This number represents an angle of 36°52′, which is the angle we want. So we were able to determine one of the angles of a right triangle by utilizing the length of two of the sides *and* the Table of Trigonometric Functions. Of course, it is also possible to use the tables in reverse. For example, if we know the value of an angle in degrees, its function (sin, cos, tan, etc.) can readily be obtained.

It should be noted here that tables of trigonometric functions vary in accuracy. Some only show three significant digits, others four, with some showing as many as eight. A table showing four significant digits, for example, is referred to as a four-place table. Many electronic pocket calculators have trigonometric functions already programmed into them, as do most home microcomputers; where these are available, the actual printed tables are not needed.

SOLUTION OF RIGHT TRIANGLES

From the above discussion it should be obvious that solving right triangles is an extremely important and useful procedure, so two examples are shown here for practice. The first example involves finding the two acute angles when two sides are known.

Figure 1-13 shows a right triangle in which the lengths of two of the sides are given. We must determine the value of the two acute angles:

$$\begin{aligned} \tan A &= \frac{a}{b} = \frac{68.24'}{106.50'} \\ &= 0.64075 \qquad (1\text{-}2) \\ A &= 32°39' \end{aligned}$$

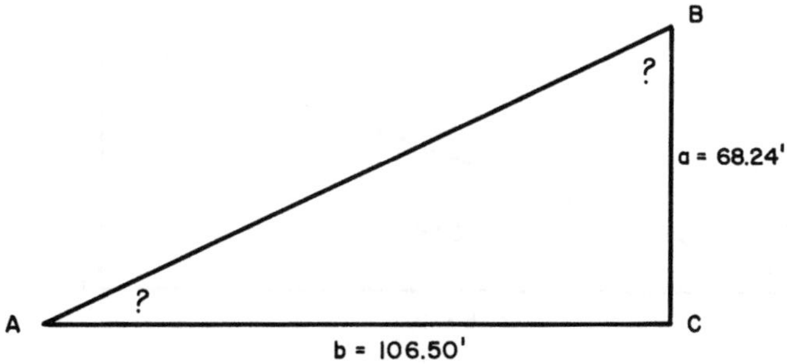

Fig. 1-13. Right triangle given two sides and no angles.

Because the three angles of a triangle add up to 180°, angle B can be determined by Eq. (1-3):

$$\begin{aligned} 180° &= \measuredangle A + \measuredangle B + \measuredangle C \\ \measuredangle B &= 180° - \measuredangle C - \measuredangle A \\ &= 180° - 90°00 - 32°39' \\ &= 57°21' \end{aligned} \tag{1-3}$$

In a second example, the remaining sides of a right triangle need to be determined while only one side and one of the acute angles is known (see Fig. 1-14):

$$\begin{aligned} \cos A &= \frac{b}{c} \\ .94646 &= \frac{b}{236.55} \\ 223.88' &= b \end{aligned} \tag{1-4}$$

Using Eq. (1-3), angle B is 71°10′.

$$\begin{aligned} \sin A &= \frac{a}{b} \\ .32282 &= \frac{a}{236.55} \\ 76.36' &= a \end{aligned} \tag{1-5}$$

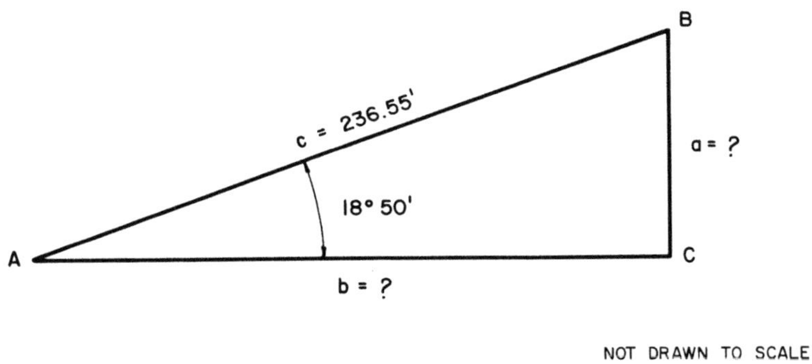

Fig. 1-14. Right triangle given one acute angle and one side.

You probably noticed that in the second example other approaches could have been used to determine the sides and angles. For example, once side b was calculated, the tangent of angle A could have been used to determine side a, but if an error were made in calculating b, it would have been carried on to a. Therefore, it is a good idea to determine *unknown* parts of a triangle from *given* or *known* data rather than from calculated data.

SOLUTION OF OBLIQUE TRIANGLES

In surveying, landscaping, and aerial photography work it is not always possible or practical to work with right triangles. Frequently situations call for the use of oblique triangles. An *oblique triangle* is one that has no right angles. It is possible to solve oblique triangles just as we did with right triangles, again provided three of the six parts (of which one must be a side) are known.

The above provision can be expressed in only four ways (in most trigonometry texts they are commonly referred to as *cases*):

Case I	1 side and 2 angles are known
Case II	2 sides and the angle opposite one of them are known
Case III	2 sides and the included angle are known
Case IV	3 sides are known

The formulas for solving each of the above cases are given by the *Law of Sines* and the *Law of Cosines* and are discussed below. Be aware that they can also be used to solve right triangles, but are usually not employed for this purpose because they can be unnecessarily cumbersome.

The *Law of Sines* states that in any triangle the sides are proportional to the sines of the opposite angles:

$$\frac{a}{\sin A} = \frac{b}{\sin B} = \frac{c}{\sin C} \tag{1-6}$$

Expanding this relationship for simplicity, we get:

$$\begin{aligned} \frac{a}{\sin A} &= \frac{b}{\sin B} \\ \frac{b}{\sin B} &= \frac{c}{\sin C} \\ \frac{c}{\sin C} &= \frac{a}{\sin A} \end{aligned} \tag{1-7}$$

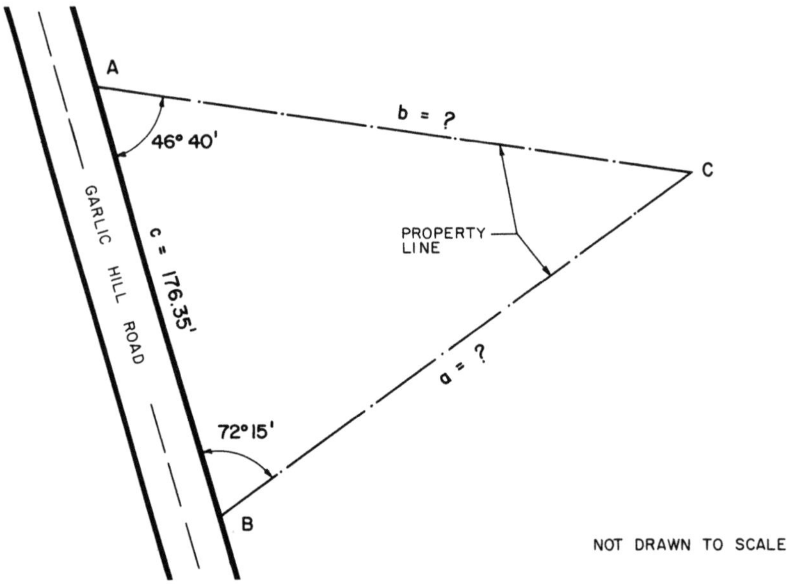

Fig. 1-15. Oblique triangle given one side and two angles.

The following example will illustrate the usefulness of the Law of Sines. Using Fig. 1-15, suppose a triangular-shaped lot has a frontage (the front boundary line that abuts the street) of 176.35 ft and the other two sides make angles of 46°40′ and 72°15′ with the street. What are the lengths of the other two property lines?

In the first step in the solution, angle C is calculated to be 61°05′ by Eq. (1-3). Once computed, using Eq. (1-8) we can determine side a:

$$\frac{C}{\sin C} = \frac{a}{\sin A}$$

$$a = \frac{C \sin A}{\sin C} = \frac{(176.35')(0.72737)}{(0.87532)} \tag{1-8}$$

$$= 146.54'$$

The remaining side b can be ascertained in the same fashion, this time using Eq. (1-9):

$$\frac{b}{\sin B} = \frac{c}{\sin C}$$

$$b = \frac{C \sin B}{\sin C} = \frac{(176.35')(0.95240)}{(0.87532)} \tag{1-9}$$

$$= 191.88'$$

Many triangles can be solved using the Law of Sines, though not all. Yet another relationship exists among the sides and angles of an oblique triangle to treat the cases not covered by the Law of Sines.

The *Law of Cosines* states that in any triangle the square of any side equals the sum of the squares of the other two sides minus twice the product of these two sides and the cosine of the included angle:

$$a^2 = b^2 + c^2 - 2bc \cos A \tag{1-10}$$

By solving for cos A we get:

$$\cos A = \frac{b^2 + c^2 - a^2}{2bc} \tag{1-10a}$$

and by cyclically interchanging the sides and angles, we therefore obtain the following relationships:

$$b^2 = c^2 + a^2 - 2ca \cos B \tag{1-11}$$

$$\cos B = \frac{c^2 + a^2 - b^2}{2ca} \tag{1-11a}$$

$$c^2 = a^2 + b^2 - 2ab \cos C \tag{1-12}$$

$$\cos C = \frac{a^2 + b^2 - c^2}{2ab} \tag{1-12a}$$

As an illustrative example of the Law of Cosines, suppose from station *A* you sight the opposite shores of a small lake at distances indicated in Fig. 1-16, and you measure angle *A* at 39°27′. How far is it across the lake from *B* to *C*? Using Eq. (1-10) here, side *a* (distance from *B* to *C*) can be determined:

$$a^2 = b^2 + c^2 - 2bc \cos A$$

$$= (5.28.25)^2 + (436.33)^2 - 2(528.25)(436.33)(0.77218)$$

$$= 113{,}470.35$$

$$a = 336.85'$$

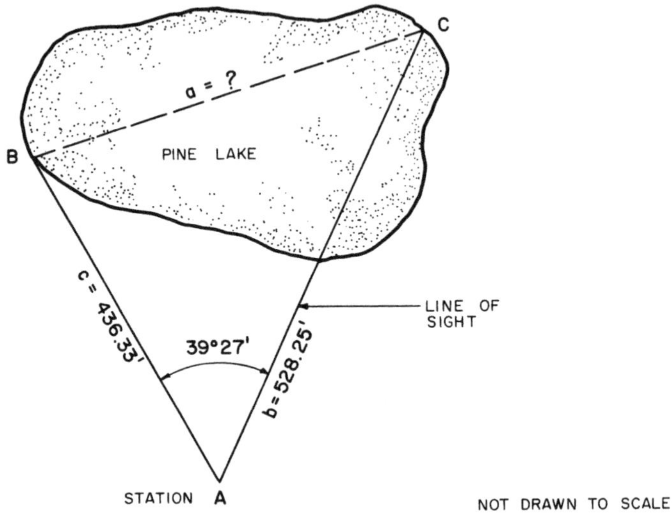

Fig. 1-16. Oblique triangle given two sides and the included angle.

With some study of Eqs. (1-6) through (1-12a), it can be noted that all oblique triangles can be solved by employing one or more of these relationships. A complete listing of equations is given in Appendix C, Table C-2.

"3,4,5" TRIANGLES

Because of the *Pythagorean Theorem*, which states that the sum of the squares of the lengths of the sides (legs) of a right triangle equals the square of the length of the hypotenuse, a few short cuts can be taken on occasion. The "3,4,5" triangle (sometimes called the magic triangle) shown in Fig. 1-17a is the simplest expression of this relationship.

Although all right triangles are an expression of the Phythagorean Theorem, only *some* have sides and a hypotenuse that are whole numbers. For example, a right triangle whose legs are 4 units and 5 units in length will not have a hypotenuse 6 units in length, but a hypotenuse of 6.4 units. However, a right triangle whose legs are 6 and 8 units in length will have a hypotenuse of 10 units. (Note these are multiples of 3, 4, and 5.) Actually *any* triangle whose dimensions are *multiples* of 3, 4, and 5 will have legs and a hypotenuse related to each other by whole numbers. Another integer triangle has legs 5, 12, and 13 units in length (see Figure 1-17b). A list of a number of these integer triangles is given in Table 1-4.

These are very useful in the building and landscaping trades as they can be used to "square up" walls, patios, decks, building layouts, etc. With a little

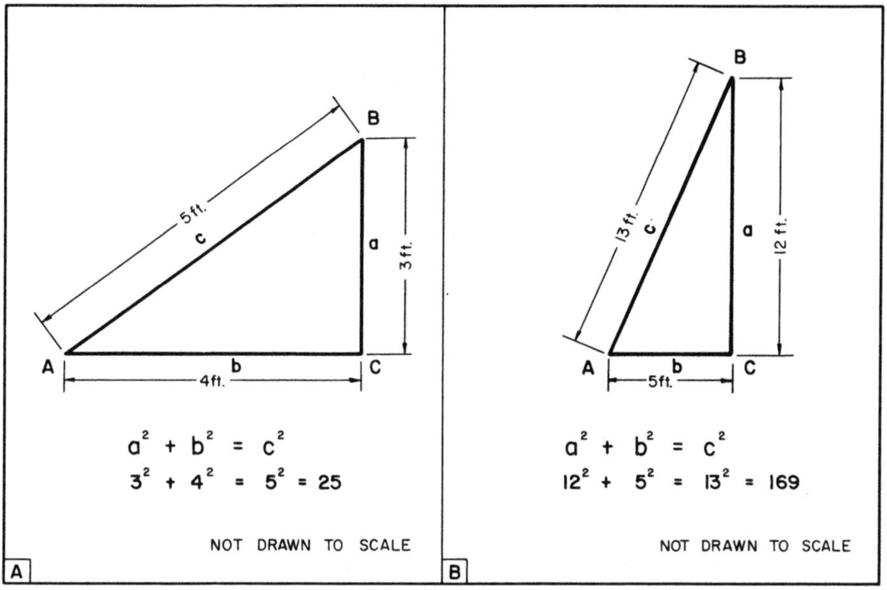

Fig. 1-17. "3,4,5" and "5,12,13" magic triangles.

Table 1-4.

b	*a*	*c*	*b*	*a*	*c*
3	4	5	8	15	17
5	12	13	12	35	37
7	24	25	65	72	97
16	30	34	28	45	53

practice it is possible to mentally solve for the sides of triangles that possess these ratios. What are the angles of "3,4,5" and "5,12,13" triangles?

COMPUTER SOLUTION OF TRIANGLES

While the above treatment presents the background needed to understand how triangles can be solved given certain parts, the actual calculation, of course, still needs to be done. With pocket and desktop calculators speed and accuracy are greatly increased over paper and pencil, not to mention convenience.

The microcomputer can make these calculations even more convenient. By following the simple instructions given in Appendix B, data from field notes, maps, scale drawings, aerial photographs, etc., can be directly entered into the computer with results displayed graphically on a monitor and/or via a printer.

2

Some Practical Area and Volume Considerations

AREAS OF REGULAR FIGURES

In the last chapter we discussed ways to solve triangles, which can be extremely useful in our work. However, in actual field practice and actual working situations much more complex figures are frequently encountered and must be understood. A very common problem that confronts photointerpreters, landscapers, builders, etc., is to ascertain the area of a given plot of ground. If the plot is a typical rectangular urban lot, it is a simple matter to calculate area. However, frequently one encounters many-sided and irregularly shaped polygonal areas that seem difficult to solve. Sometimes even areas that have no straight boundaries have to be analyzed. For example, suppose a landscaper needs to know the number of square feet in a project area that is bounded by a winding driveway and two streams. How can the area be determined so that the proper amount of materials can be ordered for the job? This chapter will aid in problems of this type.

We will start with simple shapes and then move to more complex ones. There are several ways to determine the areas of a triangle depending upon what parts are known. We will use the nomenclature in Fig. 2-1 and Eq. (2-1) through (2-3) for simple arbitrary triangles standing alone:

$$\text{Area } (A) = \tfrac{1}{2} \text{ length of base} \times (\text{length of corresponding altitude}) \quad (2\text{-}1)$$

$$A = \tfrac{1}{2}\, ab \sin C \quad (2\text{-}2)$$

$$A = \sqrt{s(s-a)(s-b)(s-c)} \quad (2\text{-}3)$$

where $S = \frac{1}{2}(a + b + c)$.

Four other equations can also be employed to determine areas of triangles provided the radius of an inscribed circle or the radius of a circumscribed circle is known (see Fig. 2-2 and Eqs. (2-4) through (2-7)). These techniques will

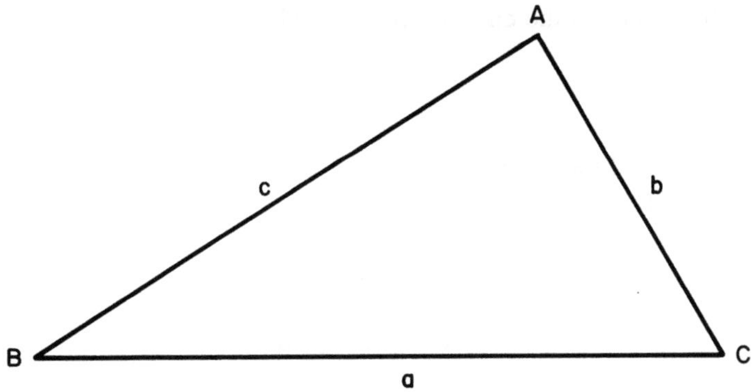

Fig. 2-1. Sides and angles of an arbitrary triangle.

seldom be used in actual field practice, but they are available and are included here for the sake of completeness and in the event they may be encountered sometime.

$$A = \frac{abc}{4R} \tag{2-4}$$

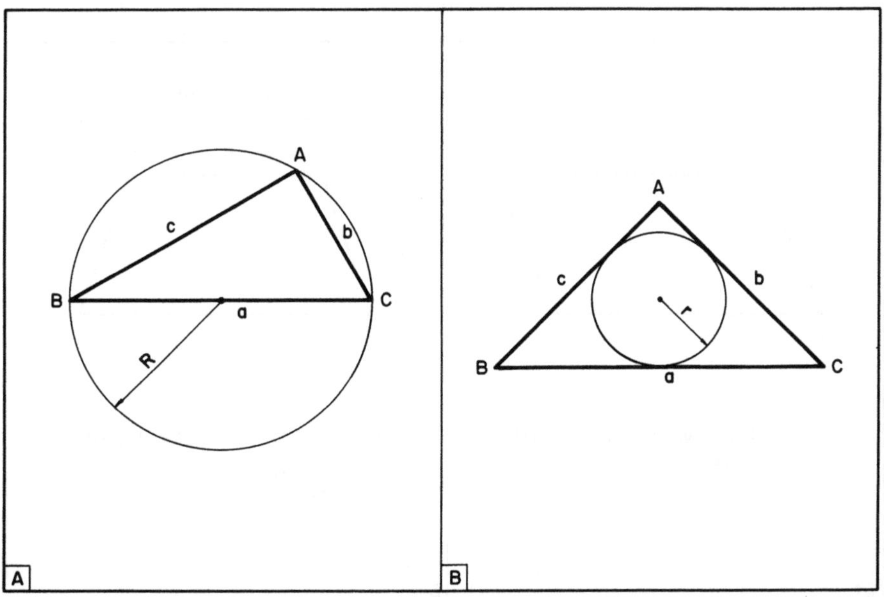

Fig. 2-2. Triangle *ABC* with circumscribing and inscribing circles.

where R is the radius of the circumscribing circle.

$$A = rs \tag{2-5}$$

where r = radius of the inscribing circle
$s = \frac{1}{2}(a + b + c)$.

$$A = 2r^2 \sin A \sin B \sin C \tag{2-6}$$

$$A = r^2 \cot \frac{A}{2} \cot \frac{B}{2} \cot \frac{C}{2} \tag{2-7}$$

As we move to determining the areas of figures with four or more sides, we will first consider those that are regular, i.e., those that exhibit symmetry. Far and away the most common figures encountered in the building trades, landscaping, etc., are rectangles and squares. Their areas are ascertained simply by multiplying their length by their width. For area calculation of more complex, though still regular four-sided figures, such as parallelograms and trapezoids, see Fig. 2-3 and Eqs. (2-8) and (2-9).

$$A = hb \qquad \text{(for parallelograms)} \tag{2-8}$$

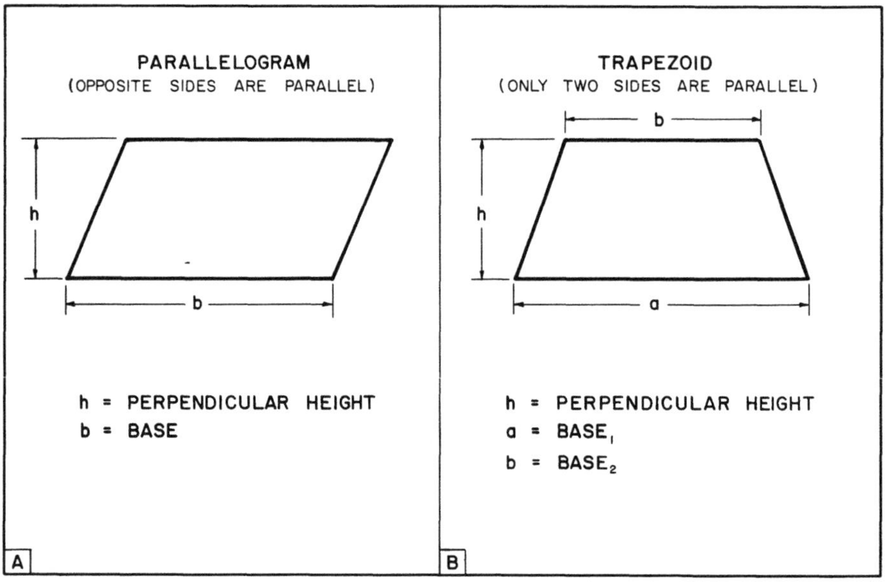

Fig. 2-3. Nomenclature of a parallelogram and trapezoid.

where h = perpendicular height
b = base.

$$A = \frac{(a+b)h}{2} \quad \text{(for trapezoids)} \tag{2-9}$$

where $a = \text{base}_2$.

Occasionally, when one needs to work with even more complex regular polygons, general formulas can be employed, depending upon what parts of the figure are known; see Fig. 2-4 and Eqs. (2-10) and (2-11):

$$A = \frac{nsr}{2} \tag{2-10}$$

where n = number of sides
s = length of a side
r = distance from center to side midpoints.

$$A = \frac{ns}{2}\sqrt{R^2 - \frac{s^2}{4}} \tag{2-11}$$

where R is the distance from the center to the vertices of adjoining sides.

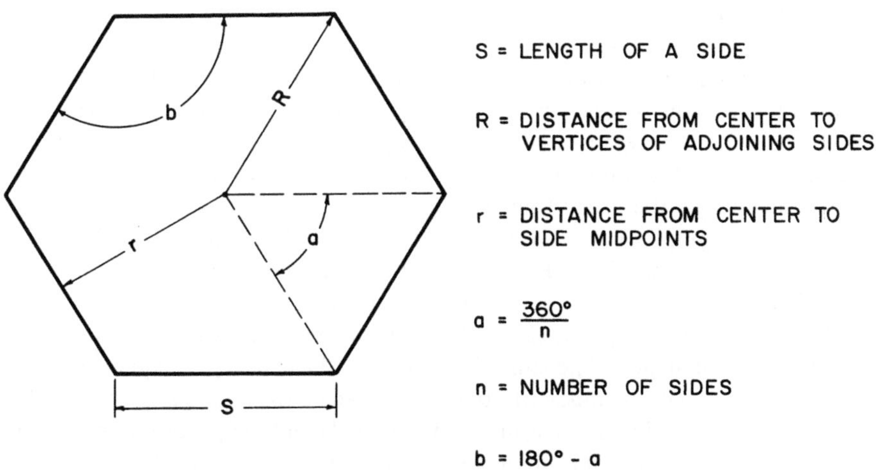

Fig. 2-4. Nomenclature of a regular *N*-sided polygon.

Table 2-1.

A	No. of Sides	4	5	6	7	8	9	10	11	12
B	Area when side = 1.00	1.000	1.721	2.598	3.634	4.828	6.181	7.694	9.366	11.196

Please note that as a short method, if a side of an *N*-sided polygon is known, multiply the square of the known side of the figure in question by the appropriate value indicated in Row B of Table 2-1.

For example, suppose a landscaper needs to determine the area of an octagon-shaped planting in the middle of a cul-de-sac. The sides of the octagon are 40 feet in length. In Table 2-1, Row B, find 4.828 under the "8" representing the eight sides of the octagon:

$$4.838 \times (40 \text{ ft})^2 = 7724.8 \text{ sq. ft}$$

As the number of sides increases, of course, the figure approaches a circle. For most practical situations, and in most field measurements, it might be easier to use the formula for the area of a circle, particularly when the number of sides reaches ten or twelve.

AREAS OF IRREGULAR FIGURES

To this point in working with regular figures, standard formulas could be used to calculate areas. With irregular figures, which vary infinitely in their shape and number of sides, no standard equations exist. The task, however, is not impossible. There are several options open, if it is possible and practical to return to the field or go back to the aerial photographs to make some additional measurements. An irregular *N*-sided polygon can be subdivided by a series of *diagonals* so as to generate or construct triangles whose areas can be determined and summed to obtain the area of the total figure. For example, suppose you need to determine the area of an irregularly shaped plot of ground viewed on an aerial photograph. You have already determined the value of *AB*, *BC*, *CD*, *DF*, and *FA* (see Fig. 2-5). From these data alone you cannot determine the area. However, simply by making two additional measurements, i.e., the lengths of *BF* and *CF*, and using Eq. (2-3), the areas of triangles *ABF*, *BCF*, and *DCF* can easily be calculated. The sum of the areas of these triangles equals the area of plot *ABCDF*. The accuracy of your results can be examined and checked by drawing and measuring a *different* set of diagonals and the procedure repeated.

Please be aware that the unknown figure does not necessarily have to be divided into triangles as in the above example. It can be divided into *any* com-

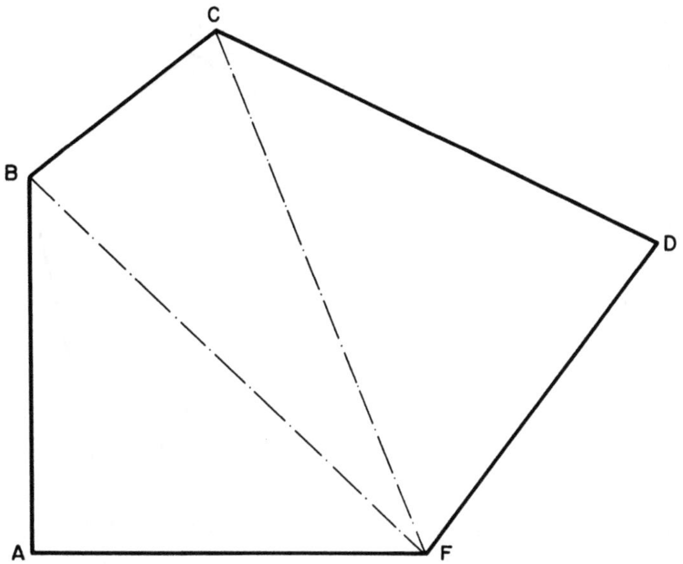

Fig. 2-5. Determining the area of *ABCDF* by using diagonals *BF* and *CF*.

bination of regular geometric shapes (rectangles, trapezoids, parallelograms, etc.) and the same procedure followed.

A high degree of accuracy can be maintained if these additional diagonal lengths are the result of direct field measurements or are directly taken from aerial photographs. Depending upon the accuracy required, the diagonal dimensions may also be determined from scale drawings or maps. A complete and detailed treatment of accuracy and precision will be presented in Chapter 3.

Areas of irregular plots can also be determined by other methods. Most advanced textbooks on surveying discuss the use of rectangular coordinates. Because rectangular coordinates uniquely define the location of points (corners of polygons) with respect to some arbitrarily chosen reference, their areas can be easily calculated.

Essentially what is done is algebraically adding the areas of the trapezoids formed by projecting lines from corners onto reference coordinate axes that are oriented north/south and east/west. As an example, suppose we want to determine the area of a plot bounded by *ABCDF* in Fig. 2-6. Eq. (2-12) nicely illustrates the crux of the method, while Eq. (2-13) illustrates the use of coordinate data for each corner as well as the calculation procedure:

$$\text{Area } ABCDF = \text{Area } BCcb + \text{Area } CDdc - \text{Area } DFfd - \text{Area } FAaf - \text{Area } ABba \tag{2-12}$$

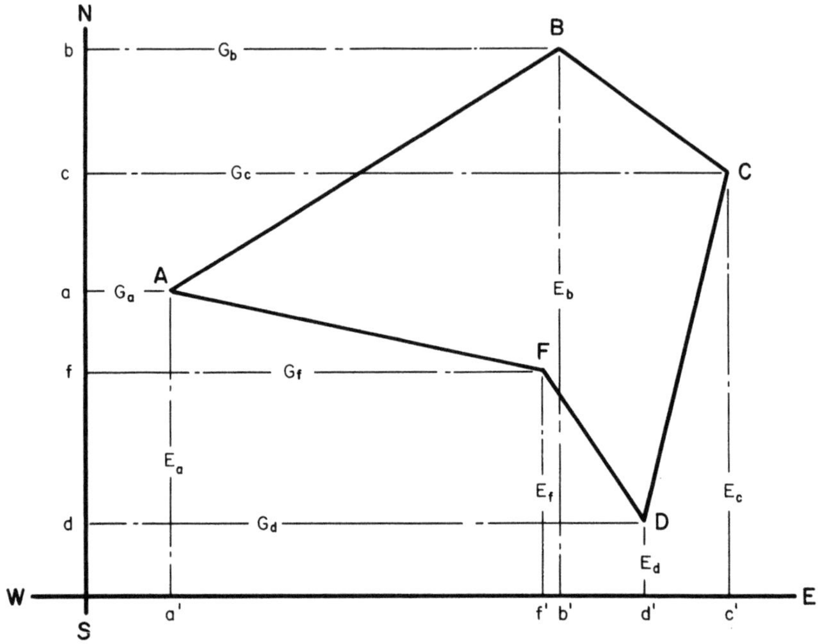

Fig. 2-6. Calculation of the area of *ABCDF* using the coordinates of the corners.

$$
\begin{aligned}
A = {} & \tfrac{1}{2}(G_b + G_c)(E_b - E_c) + \tfrac{1}{2}(G_c + G_d)(E_c - E_d) \\
& - \tfrac{1}{2}(G_d - G_c)(E_f - E_d) \\
& - \tfrac{1}{2}(F_f + G_a)(E_a - E_f) - \tfrac{1}{2}(G_a + G_b)(E_b - E_a) \qquad (2\text{-}13)
\end{aligned}
$$

Because this method can be tedious and laborious, especially if the area in question has many sides, care must be taken to ensure the accuracy of your data and of your calculation. Area calculation by computer as discussed in a later section will greatly lighten the burden of calculation and ensure accuracy.

AREAS OF CIRCLES AND CIRCULAR DIMENSIONS

We now consider circular features. Some property lines contain circular segments, and many landscaping projects contain circular or circle-like plots that need to be accurately measured. Just as we were able to summarize the area calculation of regular polygons in Table 2-1, regular circles and circle-related measurements are summarized in Figs. 2-7 through 2-11. Note that several dif-

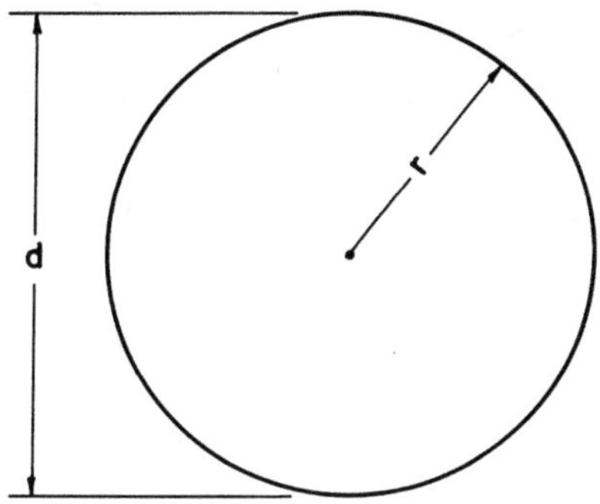

AREA $A = \pi r^2$

$A = \dfrac{d^2 \pi}{4}$

$A = 0.7854\, d^2$

CIRCUMFERENCE $C = 2\pi r$

$C = \pi d$

Fig. 2-7. Area and circumference of a circle.

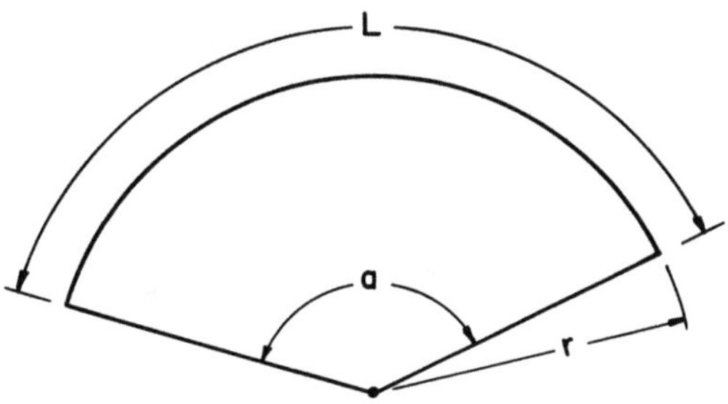

AREA $A = \dfrac{\pi r^2 a}{360°}$

$A = \dfrac{rL}{2}$

$A = 0.008727 a r^2$

$L = \dfrac{r a \pi}{180}$

$L = 0.01745 ra$

$r = \dfrac{2A}{L}$ $\qquad L = \dfrac{2A}{r}$

$r = \dfrac{57.296L}{a}$ $\qquad a = \dfrac{57.296L}{r}$

Fig. 2-8. Calculating the area and parts of a sector.

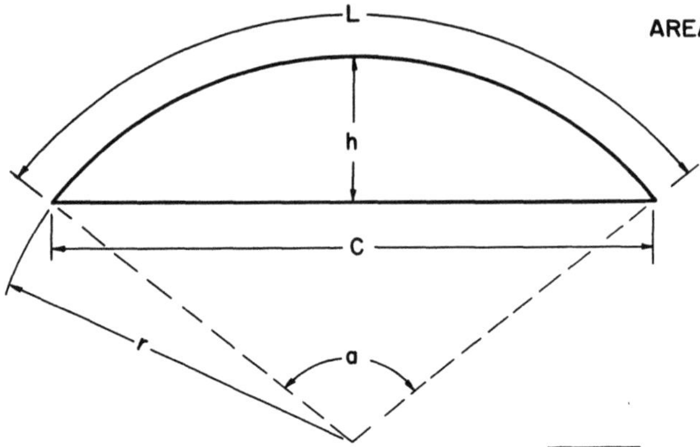

$$\text{AREA } A = \frac{rL - C(r-h)}{2}$$

$$C = 2\sqrt{h(2r-h)}$$

$$r = \frac{C^2 + 4h^2}{8h}$$

$$r = \frac{57.296L}{a}$$

$$L = 0.01745ra$$

$$h = r\,{}^1\!/_2\sqrt{4r^2 - C^2}$$

$$L = \sqrt{C^2 + 5.875h^2}$$

$$h = r(1 - \cos {}^a\!/_2)$$

$$a = \frac{57.296L}{r}$$

Fig. 2-9. Calculation of the area and parts of a circle segment.

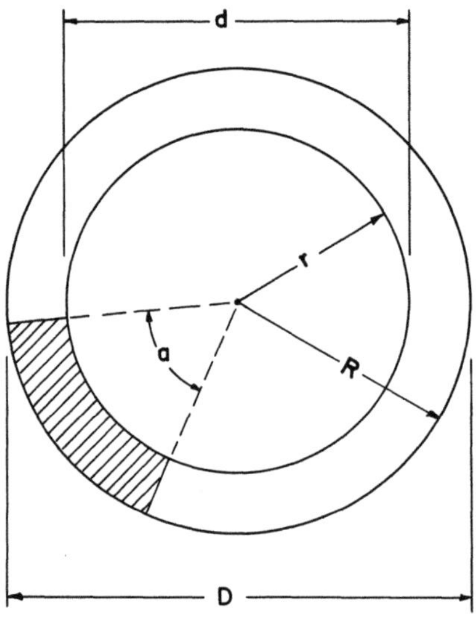

$$\text{AREA (ENTIRE RING) } A = (R^2 - r^2)\pi$$

$$\text{AREA (ENTIRE RING) } A = \frac{(D^2 - d^2)\pi}{4}$$

$$\text{AREA (SHADED AREA) } A = \frac{(R^2 - r^2)\pi a}{360°}$$

Fig. 2-10. Area calculation for a circular ring and part of a circular ring.

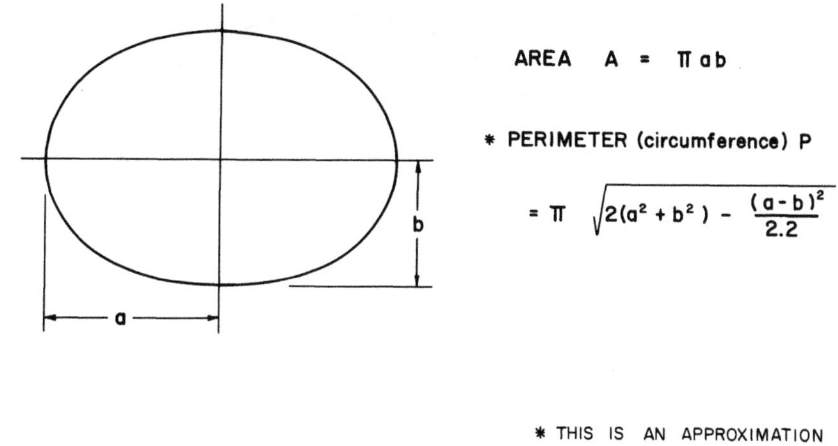

Fig. 2-11. Calculation of the area and the perimeter (circumference) of an ellipse.

ferent methods are presented for area calculation. Note, too, that formulas for other parts of circles, segments, chords, etc., are also given. Experience has shown that they all can be quite helpful in the field and in subsequent office work.

AREAS WITH CURVED OR IRREGULAR BOUNDARIES

In actual practice one frequently encounters plots that exhibit one or more boundaries that are portions of circles (see Fig. 2-12). In calculating the area of the plot in Fig. 2-12, it would be convenient to subdivide it into two parts. One part is defined by the polygon *ABCDF*, the area of which can be determined by one of the methods discussed above. The other portion, i.e., the area bounded by *AB* and the curved border, is a segment of a circle (see Fig. 2-9). The sum of the two areas is, of course, the total area. In some instances it might be necessary or convenient to use a sector rather than a segment of a circle, though the general procedure would be the same.

Frequently a project area may have an irregular boundary. For example, Fig. 2-13 shows an area that is bounded on one side by meandering stream *AB*. In a situation such as this, the usual procedure for area calculation calls for constructing a traverse from *F* to *G*. Then along *FG* at *regular intervals W*, construct perpendicular offsets *h* to the curved boundary. Please notice that this can be done either in the field, on an accurate map, aerial photograph, or scale drawing. If possible, and to increase accuracy, the offsets should be drawn at breaks or changes in the curves, as is shown in Fig. 2-13. In using this proce-

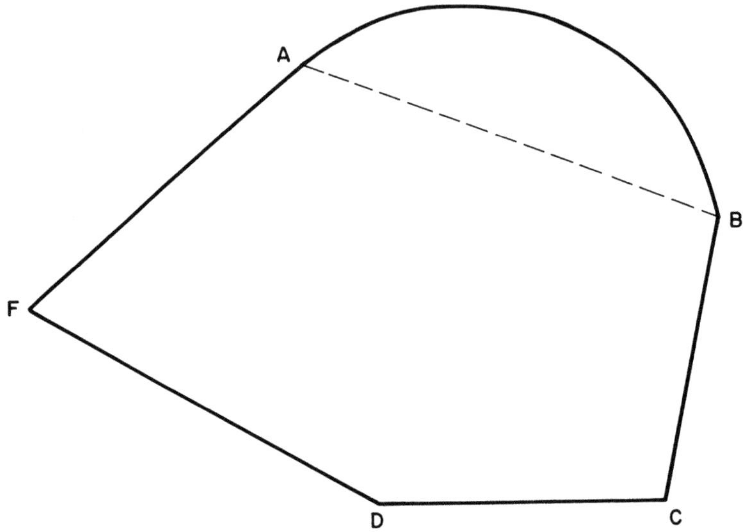

Fig. 2-12. Tract bounded by an arc of a circle *AB*.

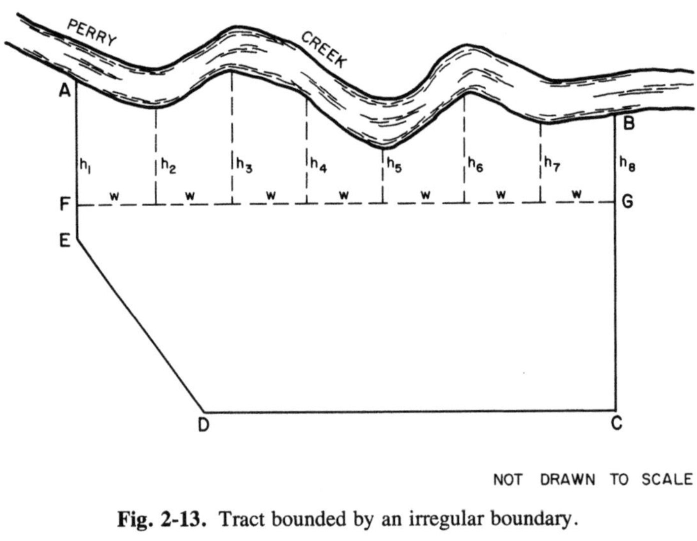

Fig. 2-13. Tract bounded by an irregular boundary.

dure it should be noted that the larger the number of offsets the more accurate the results. The area of the tract bounded by *ABGF* is given by:

$$\text{Area } A = W\left(\frac{h_1 + h_n}{2} + h_2 + h_3 + \ldots + h_{n-1}\right) \quad (2\text{-}14)$$

where n = number of offsets
h = length of offsets

As an illustration of the procedure, suppose the following data have been measured for tract *ABGF* in Fig. 2-13:

$$h_1 = 50.25' \quad h_5 = 35.25' \quad w = 25.00'$$

$$h_2 = 45.00' \quad h_6 = 45.90'$$

$$h_3 = 54.55' \quad h_7 = 41.10'$$

$$h_4 = 46.25' \quad h_8 = 43.55'$$

$$A = 25.00'\left(\frac{50.25' + 43.55'}{2} + 45.00' + 54.55' + 46.25' + 35.25' + 45.90' + 41.10'\right)$$

$$= 7873.75 \text{ ft}^2$$

The area of *ABGF* would be combined with the area of *EFGCD* to obtain the total area.

Three methods will be presented here that will enable the builder or landscaper to easily calculate the areas of plots that have *no* straight sides and are completely irregular. Situations may include estimating the area of a small pond, the area encompassed by a closed contour line on a topographic map, the area of an irregularly shaped patio, etc.

The first method is probably the least accurate of the three and should be employed only as a first or "rough" estimate, *or* in projects where great accuracy is not required. It is restricted to making measurements in the office using scale drawings or maps. Data needed can be provided by the scale of the map, aerial photograph or drawing, and by translucent graph paper or a quad-ruled overlay. (The squares may be drawn directly on the map if desired.) See Fig. 2-14. By counting squares and estimating parts of squares encompassed by the boundary of the plot, reasonably accurate results can be obtained.

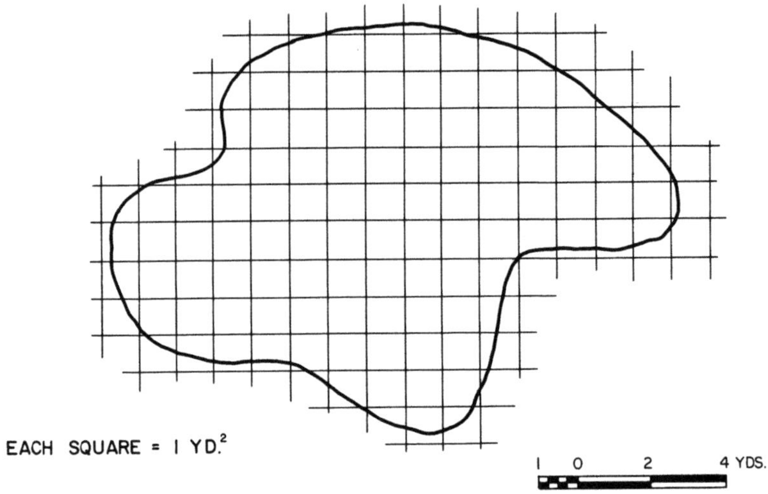

Fig. 2-14. Area of an irregular plot by counting squares.

The second procedure must also be carried out on a map or scale drawing. It calls for an instrument called a *polar planimeter* (see Fig. 2-15). A planimeter is a mechanical device used for measurement of irregular areas. It consists of two arms *OB* and *BT* that are hinged at *B*. As the tracing point *T* is moved *carefully* around the curve, a small wheel *W* records the motion. The wheel is marked with graduations and a vernier scale to record its exact number of rev-

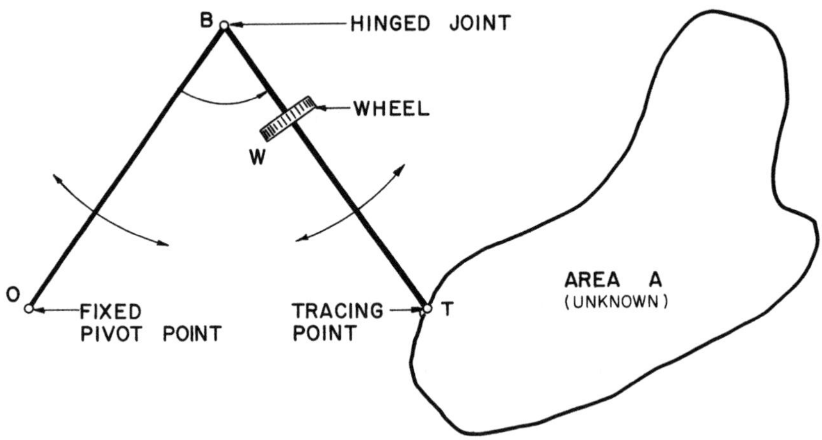

Fig. 2-15. Schematic diagram of a polar planimeter.

olutions which then can be converted to the area of the plot by Eq. (2-15):

$$(\text{Area})\ A = cn \tag{2-15}$$

where n = final reading on the wheel minus initial reading
c = length of $BT \times$ diameter of the wheel.

Step-by-step instructions are supplied with the instruments which vary with the manufacturer and with the model. Because each model is slightly different in design and operation, it would not be practical to discuss each of them here.

Our third method of irregular area computation is probably the most useful and the most accurate. It can be done either in the field, i.e., using actual field-measured data, or it can be done on maps and scale drawings. Here we are applying what is known as *Simpson's Rule*. The mathematical assumptions and derivations germane to its application here can be found in most advanced texts on surveying and is beyond the practical thrust of this book. Applying Simpson's Rule to area calculation involves four steps:

1. Divide the irregular area into an *even* number of N parallel strips of equal width W. Fig. 2-16 has been divided into 8 strips of all equal width.
2. Label the sides of the strips h_0, h_1, h_2, h_3, up to h_n
3. Measure the lengths of $h_0, h_1, h_2, h_3, \ldots, h_n$
4. Substitute the values of $h_0, h_1, h_2, h_3, \ldots, h_n$ into Eq. (2-16):

$$\text{Area } A = \frac{w}{3}\left[(h_0 + h_n) + 4(h_1 + h_3 + \ldots + h_{n-1}) + 2(h_2 + h_4 + \ldots + h_{n-2})\right] \tag{2-16}$$

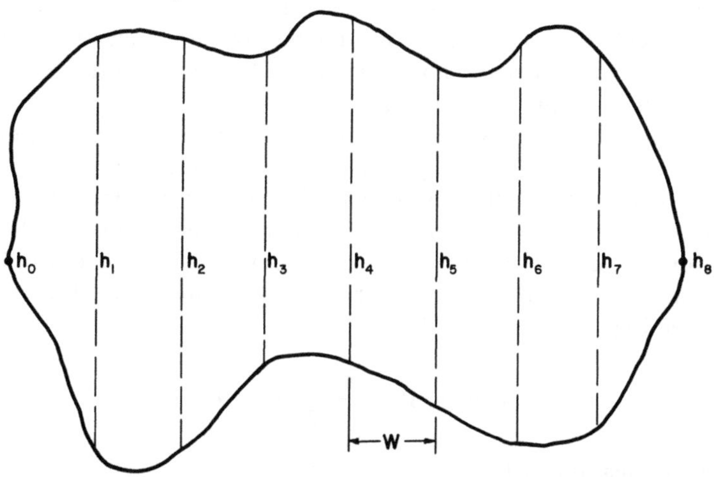

Fig. 2-16. An irregular plot divided into an even number of strips of width W for application of Simpson's Rule.

Using Fig. 2-16, the following data have been measured:
Calculate its area using Simpson's Rule:

$$h_0 = 0' \qquad h_4 = 101.34' \qquad h_8 = 0'$$

$$h_1 = 120.25' \qquad h_5 = 98.04' \qquad w = 30.00'$$

$$h_2 = 118.75' \qquad h_6 = 107.50'$$

$$h_3 = 97.63' \qquad h_7 = 102.12'$$

$$\text{Area } A = \frac{30.00'}{3}\left[(0 + 0) + 4(120.25' + 97.63' + 98.04' + 102.12') + 2(118.75' + 101.34' + 107.50')\right]$$

$$= 23{,}273.40 \text{ ft}^2$$

Please note, just as with the use of Eq. (2-14), as the number of strips into which the plot is divided increases, so does the accuracy of the results.

VOLUME CALCULATIONS

To the landscaper and builder volume calculations are commonplace. The work involves, for example, determining cut and fill quantities for earthwork excavations and estimating volumes of stockpiles of sand, gravel, top soil, coal, crushed stone, etc. An interesting use for this type of calculation is determining the volume of water in reservoirs and ponds. However, the most common use of volume calculations is perhaps estimating the cost of foundation excavation. The work usually involves a combination of field and office work, though some can be done on maps and drawings alone.

Many volume determinations are ascertained simply by multiplying the *area* of the end surface of the unit in question by its length (see Fig. 2-17). Please note that this is true only if the cross-sectional area is constant through the entire length. In situations where the cross-sectional area changes along the length of the unit, one of several variations of the same general procedure can be employed. The first alternative is to use the average of the areas of the two end sections; see Fig. 2-18 and Eq. (2-17):

$$\text{Volume } V = \frac{A_1 + A_2}{2} \times L \tag{2-17}$$

where A_1 = area of end 1
A_2 = area of end 2
L = length.

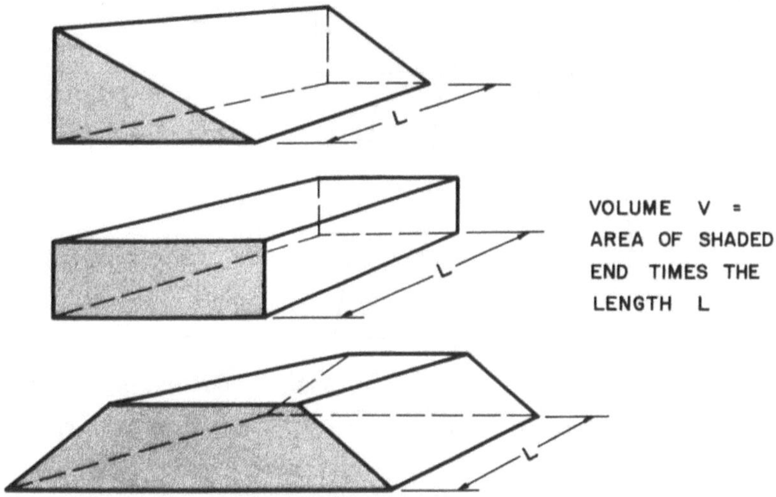

Fig. 2-17. Volume calculation given the area of the end sections.

Although this method is not exact when applied to many actual cases, it is still generally used because of the ease of application. A second alternative employs what is known as the *prismoidal formula*; Fig. 2-18 and Eq. (2-18):

$$\text{Volume } V = \frac{L}{6}(A_1 + 4M + A_2) \tag{2-18}$$

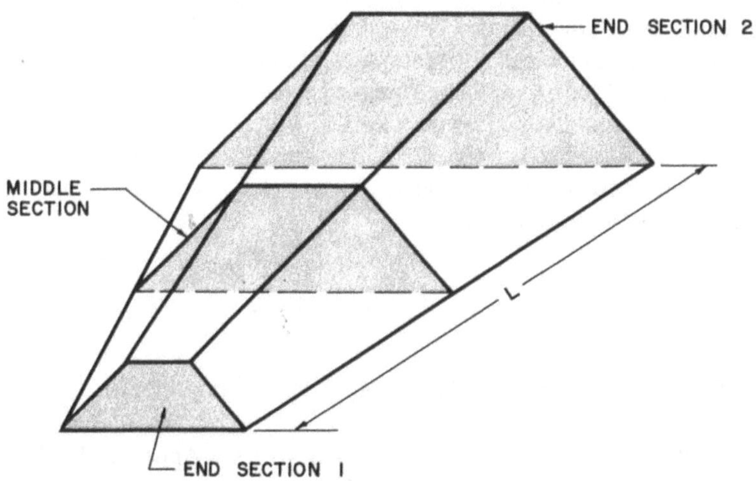

Fig. 2-18. Terminology for volume calculations when end sections are different sizes.

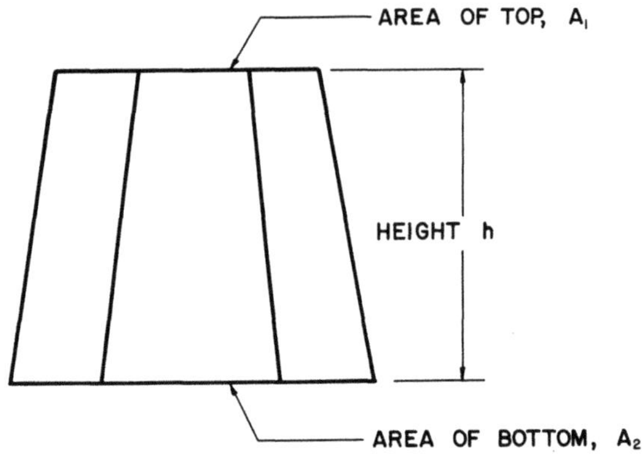

Fig. 2-19. Nomenclature for calculating the volume of a frustum of an *N*-sided pyramid.

where M is the area of the middle section. This method is more accurate but it requires measuring an additional cross section. Yet a third method uses a formula for calculating the volume of a frustum of a prism; see Fig. 2-19 and Eq. (2-19):

$$\text{Volume } V = \frac{h}{3}\left(A_1 + A_2 + \sqrt{A_1 \times A_2}\right) \tag{2-19}$$

This equation can be used for any *N*-sided frustum. It is doubtful, however, whether more than a few building or landscaping jobs requiring more than a five- or six-sided frustum will be encountered. It might be helpful here to present an example of how this procedure might be applied in an actual project. Suppose the local municipal golf course is adding nine holes to the existing facilities. Plans call for one of the tee areas to be elevated 3 ft above the general grade of the fairway. The top of the tee is to have rectangular dimensions of 20′ × 25′, while the base is to be 30′ × 35′ (see Fig. 2-20). How many cubic yards of fill are needed for this project?

Using Eq. (2-19) we get:

$$V = \frac{h}{3}\left(A_1 + A_2 + \sqrt{A_1 \times A_2}\right) \qquad h = 3\text{ ft}$$

$$= \frac{3}{3}\left(500 + 1050 + \sqrt{500 \times 1050}\right) \quad A_1 = 500\text{ ft}^2 \text{ (top of tee)}$$

$$= 2275\text{ ft}^3 \text{ (27 ft}^2 = 1\text{ yd}^3\text{; see Appendix D, Table 5)}$$

$$= 84.26\text{ yds}^3$$

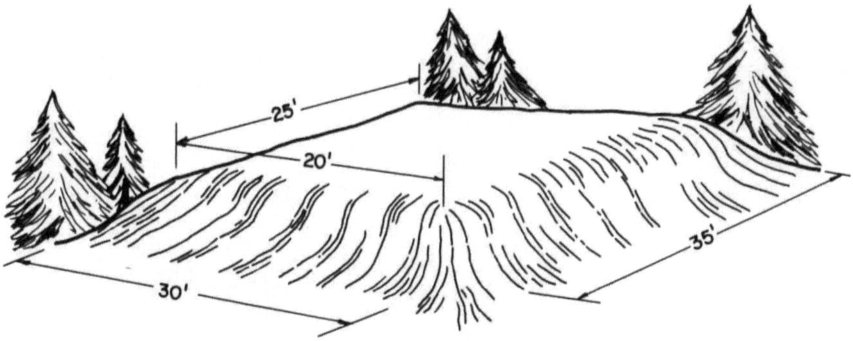

Fig. 2-20. Raised golf course tee (frustum of a 4-sided pyramid).

One of the more useful and frequently encountered volumes in the construction trades are those of a cone and frustum of a cone. Storage piles for example of topsoil, sand, gravel, etc. either dumped in the open or in storage bays frequently assume a rough or general conical shape; see Figs. 2-21 and 2-22 and Eqs. (2-20) and (2-21):

$$\text{Volume } V = \frac{\pi r^2 h}{3} \qquad \text{(cone)} \qquad (2\text{-}20)$$

$$\text{Volume } V = 1.0472h(R^2 + Rr + r^2) \qquad \text{(frustum of a cone)} \qquad (2\text{-}21)$$

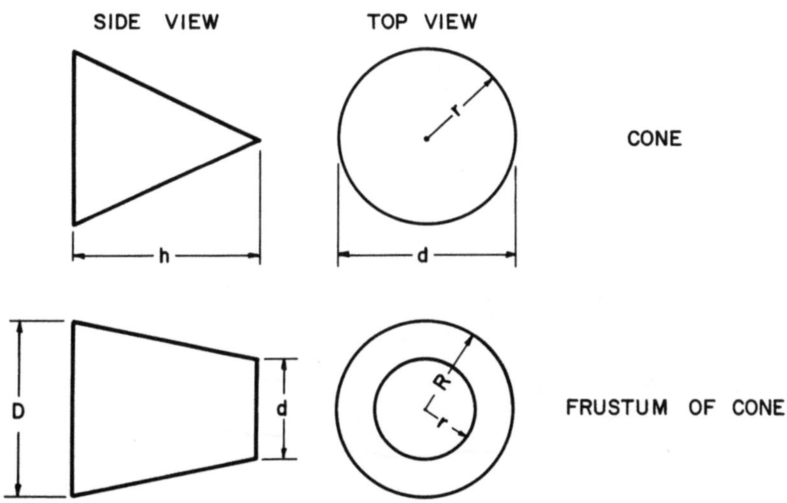

Fig. 2-21. Nomenclature used in determining the volumes of a cone and the frustum of a cone.

Fig. 2-22. Conical-shaped storage pile, the volume of which can be calculated by the equation for the volume of a cone.

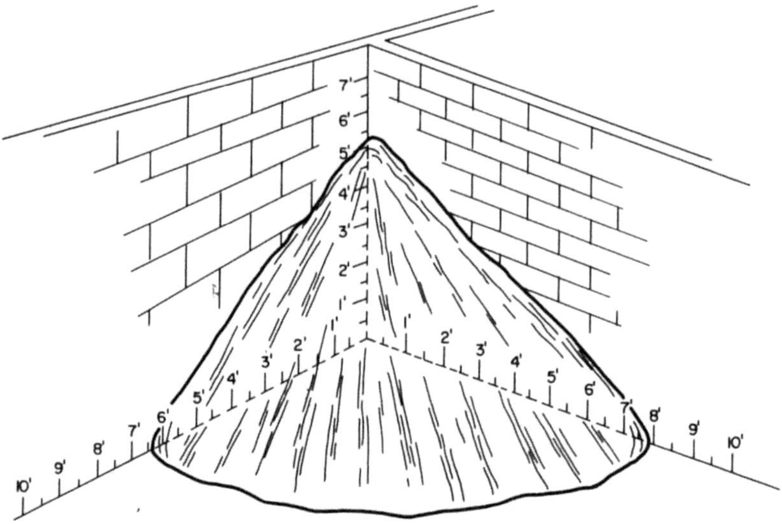

Fig. 2-23. A pile of gravel in a corner storage bay. Note that it is .25 of a complete cone. Note too the wall markings.

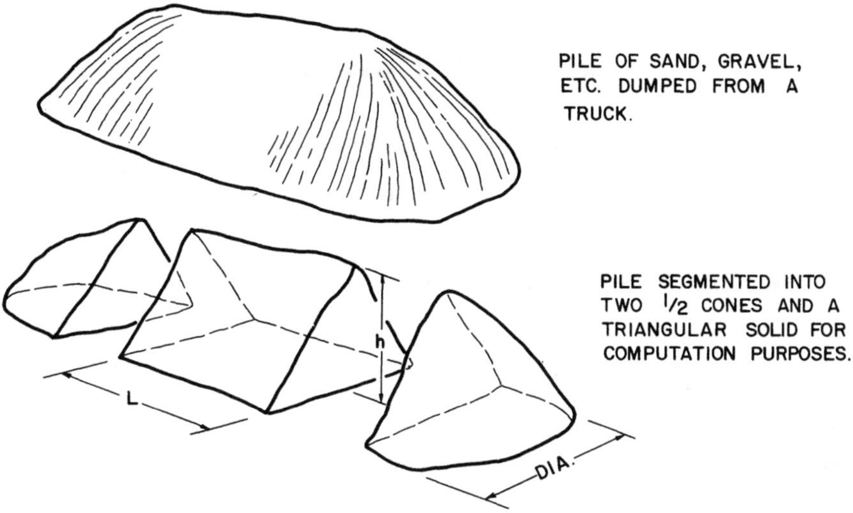

Fig. 2-24. Technique of calculating the volume of a dumped pile of sand, gravel, etc. by segmentation.

For materials piled in corner bays multiply volumes by 0.25 and for these materials piled against flat walls, multiply volumes by 0.50 (see Fig. 2-23). As is shown in Fig. 2-23, it might be convenient to mark the walls every half foot or so to easily determine approximate dimensions of the pile. If the base of the pile does not extend to the same point on the walls as in Fig. 2-23, simply average the two.

In situations where material is piled in the open (from a dump truck), it usually assumes a shape that is regular enough to permit a good estimate of volume (see Fig. 2-24). (If possible, it would be much easier and more accurate, however, to determine volume while the material is still in the truck where it is confined by flat regular sides.)

AREAS AND VOLUMES ON THE COMPUTER

The merits of using a microcomputer for calculations were highlighted at the end of the last chapter and should be obvious when applied to area and volume computations. If a computer is to be used, it is recommended that Appendix B be studied *before you start* the project to determine specifically what field data are required for the program in question. You will also need to make sure that all data, i.e., linear, angular, and area measurements are consistent in their units. Checking this information prior to beginning your project will give you some direction on how to approach the problem so that you can cut your time in the field as well as in the office.

3

Surveying Accuracy and Precision

ACCURACY VERSUS PRECISION

On several occasions in the previous two chapters mention was made of the notion of accuracy. Accuracy of measurements, calculations, estimates, etc. is of paramount importance in the building trades. Because it can often spell the difference between the financial success or failure of a project, it is necessary to know what it is and how to attain it in our work. There is also another concept that is frequently associated with and often confused with accuracy, i.e., precision. It, too, must be well understood.

To grasp concepts of accuracy and precision, two very basic assumptions have to be made. First, realize that for any physical measurement, whether linear or angular, a true value does, in fact, exist. Second, no measurement of that true value is ever exact. In fact, the only way to determine the true value of something, say the distance between points *A* and *B*, is to obtain an infinite number of observations or measurements. Because this is not possible, we must live with measurements or readings which closely approximate the true or absolute value. *Accuracy* can be defined then as the closeness of our measurements to a true value. On the other hand, *precision* is the "closeness together," or the "intensity of clustering," or the "packing" of our measurements or readings to *each other*, not necessarily to a true value.

As an illustration, suppose four marksmen are shooting at a target (see Fig. 3-1). The marksman in Fig. 3-1a is very precise because his shots are clustered very close together. But because he is not close to the true value, i.e., the bull's-eye, he is not accurate. Fig. 3-1b shows the hits of a marksman who is accurate but not precise. He is around the bull's-eye but his spread is great, i.e., he shows no precision. The shooter in Fig. 3-1c exhibits neither accuracy nor precision. Finally, Fig. 3-1d is the ideal situation where the marksman is both accurate and precise.

GAUGING ACCURACY IN HORIZONTAL, VERTICAL, AND ANGULAR MEASUREMENTS

In work in the field and in the office, accuracy as well as precision is the ideal goal. However, a question immediately arises as to how much accuracy and

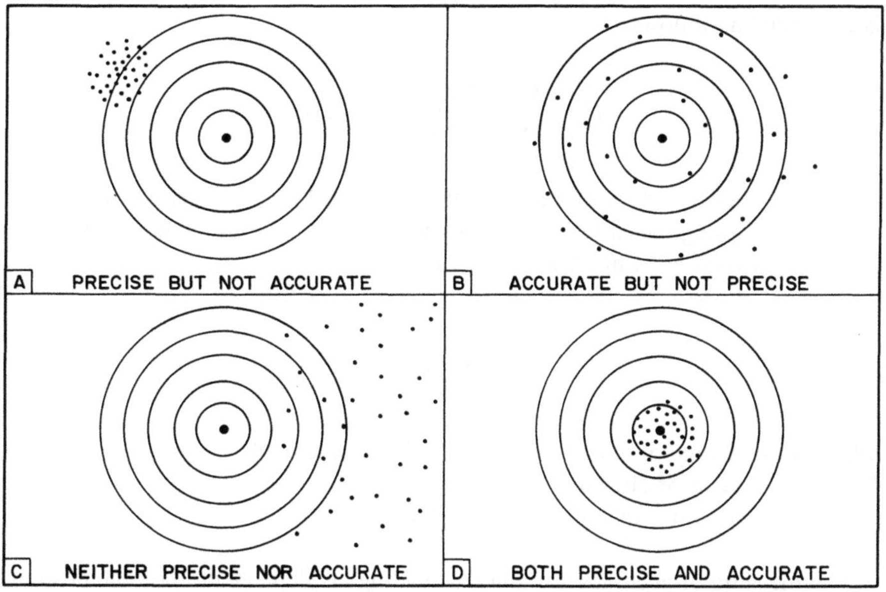

Fig. 3-1. Accuracy versus precision, represented by four marksmen shooting at a target. (Modified from Griffiths, J. C., *Scientific Method in the Analysis of Sediments* © 1967, McGraw-Hill by permission.)

how much precision is necessary for a particular project? Inordinate amounts of time and effort could be expended on attaining great accuracy and precision when it is not needed.

To determine how accurate the work of the builder and landscaper needs to be, and to ascertain the accuracy of work that is already completed, the concept of closure or error of closure needs to be understood. *Error of closure* is defined as the length of a line by which a close traverse fails to meet or close on its starting point, and in the case of an open traverse, the error of closure is the length of a line by which the traverse fails to reach its end point.

The degree of accuracy, then, for these horizontal measurements is commonly expressed as a ratio of the error of closure to the total distance measured:

$$H = \frac{C}{D} \tag{3-1}$$

where H = degree of accuracy for horizontal measurements
C = Error of Closure
D = Total distance measured.

For example, if an error of closure of 0.125 ft is made in laying out a 150 ft driveway from a garage to its intersection with the street, the degree of accuracy is computed as follows:

$$H = \frac{C}{D} = \frac{0.125}{150} = 1/1200$$

This means that when measuring a distance of 1200 ft, a 1 ft error will result.

For establishing measures of accuracy for vertical measurements, for what is commonly called leveling, see Eq. (3-2):

$$V = C\sqrt{\frac{5280}{f}} \tag{3-2}$$

where V = degree of accuracy for vertical measurements
C = enor of closure
f = total distance measured (in feet).

The measure of accuracy for angular readings is dependent upon the desired accuracy of the project in general. As an illustration, suppose you are laying out a large patio and linear measurements are being held to an accuracy of 1/1000. Therefore, any angle that is being measured should have a corresponding accuracy. So the allowable angular error will be such that the tangent of the angle will also be 1/1000 or 0.001. Using a 5-place table of trigonometric functions or a pocket calculator with trigonometric functions on it, note that the related angle is 3′. This angle then is the allowable angle of error given the standards set for the project.

It should be mentioned here that for most work done by landscapers and builders, such as layouts of buildings, driveways, decks, patios, utility lines, etc., that are within property lines, an accuracy of 1/5000 is usually sufficient. Greater accuracy, however, is required when working with property surveys, municipal surveys, etc. This work falls into the domain of the registered surveyor.

TECHNIQUES FOR INCREASING ACCURACY

Sources of Error

In order to attain the accuracy that is normally required in the building trades, it is not necessary to use the complex and sophisticated electrooptical surveying instruments on the market today. In fact, if care is taken, a common builder's transit that is in good condition can serve quite well. Discussed below are several techniques that should be part of the common, everyday procedures followed in our work. They are presented here so that maximum accuracy can be obtained from instruments that a typical landscaper or builder, who does both his own field work and office work, is apt to have at his disposal.

It might be useful to list a few areas where inaccuracies, i.e., errors frequently originate. This list is certainly not exhaustive, nor is it in any particular order of importance, but it serves to point out that if work is not carried out with deliberate care, errors can creep in from many places. Some sources of error are:

1. Instruments in poor condition, i.e., dirty, out of adjustment, damaged, etc.
2. Instruments improperly used
3. Not following proper techniques and/or procedure, especially in the field.
4. General carelessness, usually the result of hurrying or fatigue
5. Entering erroneous notations into field book, calculator, or computer
6. Misreading instruments, e.g., tapes, map scales, rods, level bubbles, verniers
7. Errors from natural conditions, i.e., wind, cold, heat
8. Errors in calculations, e.g., using wrong equations, decimals misplaced, carrying nonsignificant digits
9. Inexperienced instrument operator, photographer, pilot, draftsperson
10. General blunders, e.g., adding when subtraction is needed, turning left instead of right

Most textbooks on surveying, aerial photography, and general field methods, will list many more items.

How to Use Verniers

It might be said that one of the most important and simplest devices for obtaining greater accuracy is the *vernier*, invented by Pierre Vernier in the 1600s. In general, a vernier is a device for accurately determining readings that are smaller than the smallest divisions on the scale of an instrument. Please be aware that verniers are found on many types of instruments, not only those used by surveyors and photointerpreters.

A vernier consists of a small auxiliary scale located immediately adjacent to the main scale of the instrument. The subdivisions on the vernier scale are 0.9 as long as the smallest subdivisions on the instrument's main scale (see Fig. 3-2). Stating this another way, 9 of the smallest subdivisions on the main scale are equal to 10 divisions on the vernier scale (see Fig. 3-2).

To read verniers, first suppose that the 0 on the vernier scale falls somewhere between two of the smallest divisions on the main scale (see Fig. 3-3a). This indicates that the value being determined in our example is between 17 and 18.

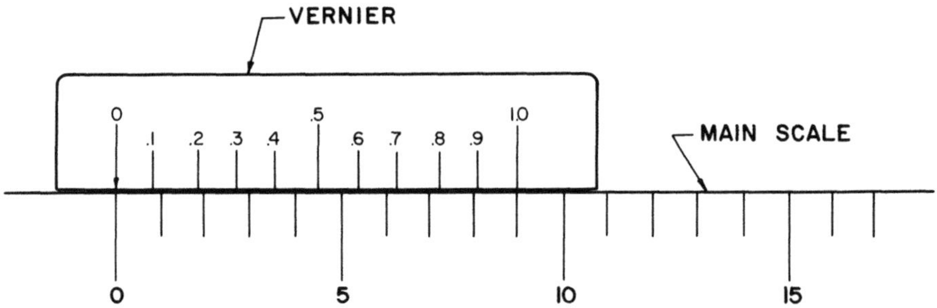

Fig. 3-2. Vernier scale and main scale. Note their subdivisions.

To ascertain exactly where the correct value lies between those two values, find a line on the vernier scale that is collinear (lines up) with a line on the main scale. In our illustration it is .5, so the final exact reading is 17.5. Study Fig. 3-3b and Fig. 3-4 carefully and compare your readings with those indicated. We now can determine with more certainty the last digit of our readings, which without the vernier, would just be a guess.

On instruments that have circular scales such as transits, drafting machines, some planimeters, etc., it is necessary to read the scales in two directions. With

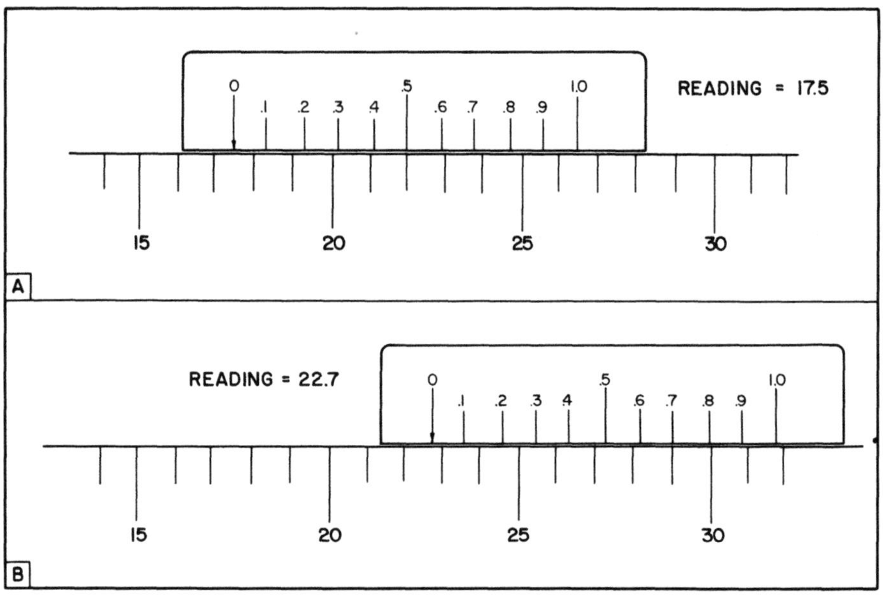

Fig. 3-3. Examples of reading scales with verniers. Scale A reads 17.5, scale B reads 22.7.

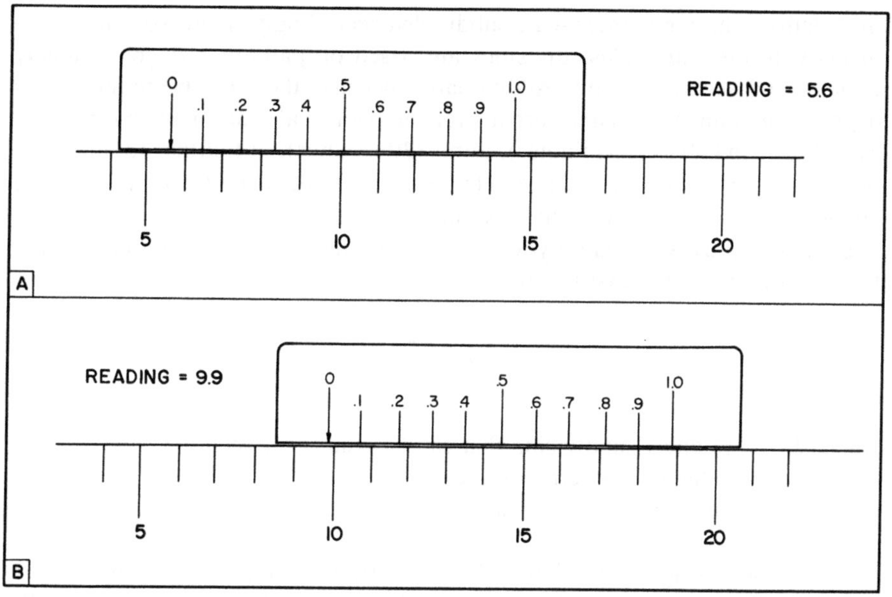

Fig. 3-4. Examples of reading scales with verniers. Scale A reads 5.6, scale B reads 9.9.

instruments such as these, a double vernier is used (see Fig. 3-5). A double vernier is simply two verniers sharing the same 0 line to accommodate readings in either direction. While referring to Fig. 3-5, note that if the main scale is rotated to the left (counterclockwise), use the left-side vernier and ignore the right side. Use the other vernier when rotating the scale to the right.

In our illustrations to this point we have used scales that were subdivided

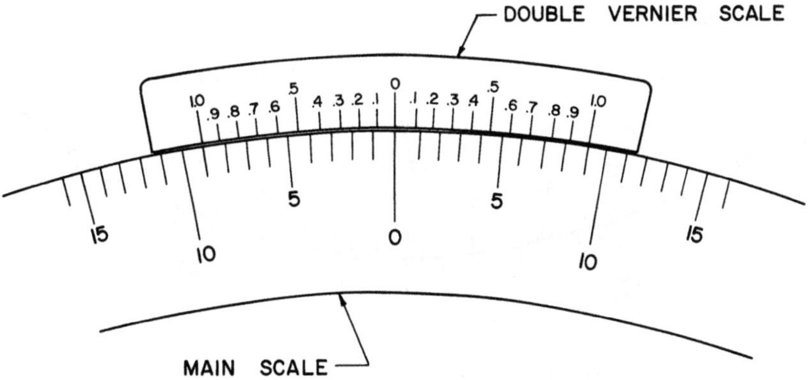

Fig. 3-5. Double vernier scale used on instruments with circular scales.

into tenths, with the vernier scale subdivided accordingly. However, on instruments with circular scales, divisions are based on parts of a degree, namely minutes and seconds of arc. Additionally, because there are 60 minutes in a degree, common subdivision formats for the main scales are made by dividing each degree in halves (30 minutes), thirds (20 minutes), or fourths (15 minutes). So the smallest subdivision that can be read on a scale using the accompanying vernier is going to vary with the instrument.

Equation (3-3) will enable you to determine just what accuracy can be expected from a specific instrument:

$$D_s = \frac{S}{n} \tag{3-3}$$

where D_s = smallest subdivision that can be read, i.e., accuracy
S = value of the vernier scale
n = number of vernier subdivisions.

For example, using Figure 3-6 and Eq. (3-3), note that the vernier scale is graduated to 15 minutes and the number of divisions is 45. Therefore, the smallest readable division is:

$$D_s = \frac{S}{n}$$

$$= \frac{15'}{45} = 0.3' \text{ or } 20''$$

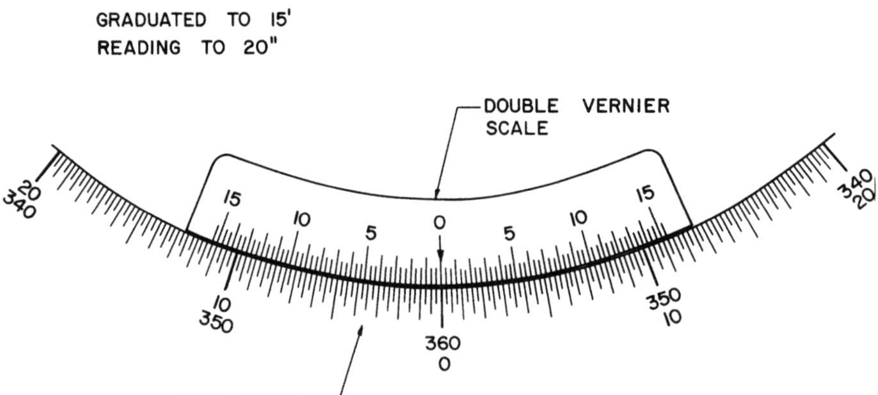

Fig. 3-6. Double vernier scale and main scale; graduated to 15′ reading to 20″.

It is important to note that to avoid error and confusion, care should be taken to determine the limits of an instrument *before* it is used.

Accuracy in Taping

The builder and landscaper, of course, do not need to carry the accuracy that a professional surveyor does in municipal, highway, property, or geodetic surveys. Nevertheless one should strive for as much accuracy as possible given the equipment at hand. Next to being able to read verniers properly, taping is probably the most important, useful, and commonplace measuring procedure employed in the building trades. It is important, then, to explore those areas where accuracy and precision can be increased without expending large amounts of extra effort and time. In view of this we will not dwell extensively on field procedures, but will examine a few techniques that will enhance accuracy.

First, it is of great importance to understand that distances shown on maps and charts are horizontal projections of the ground distances onto a horizontal plane (see Fig. 3-7). The ground distance between points *A*, *B*, and *C* is shown on a map, plan, or chart (the horizontal plane) as the distance between *A′*, *B′*, and *C′*. Therefore, when taping over irregular ground, the slopes are broken into a series of horizontal segments as shown in Fig. 3-8. When segments are measured that are less than the length of the tape, the procedure is known to professional surveyors as *breaking tape*. Note carefully that at the downhill end of each segment, the exact position on the ground is read on the tape where a plumb line suspended directly over the desired point intersects the tape. Do not try to read the tape without the plumb line. Also, implied in Fig. 3-8 is the fact

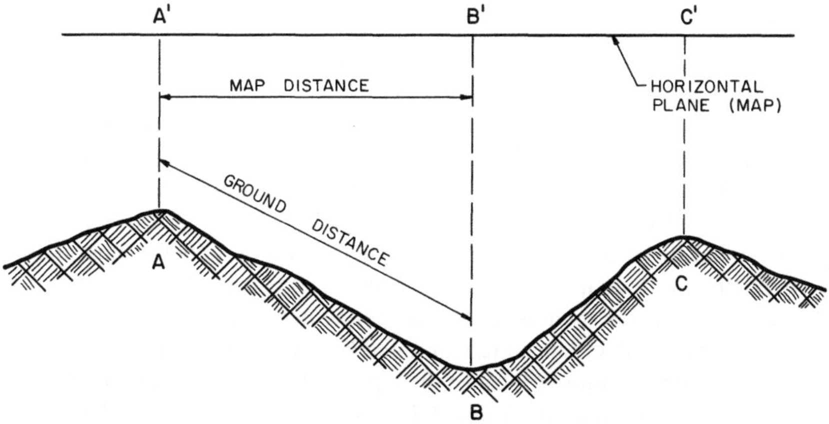

Fig. 3-7. Relationship between distances on the ground versus map distances.

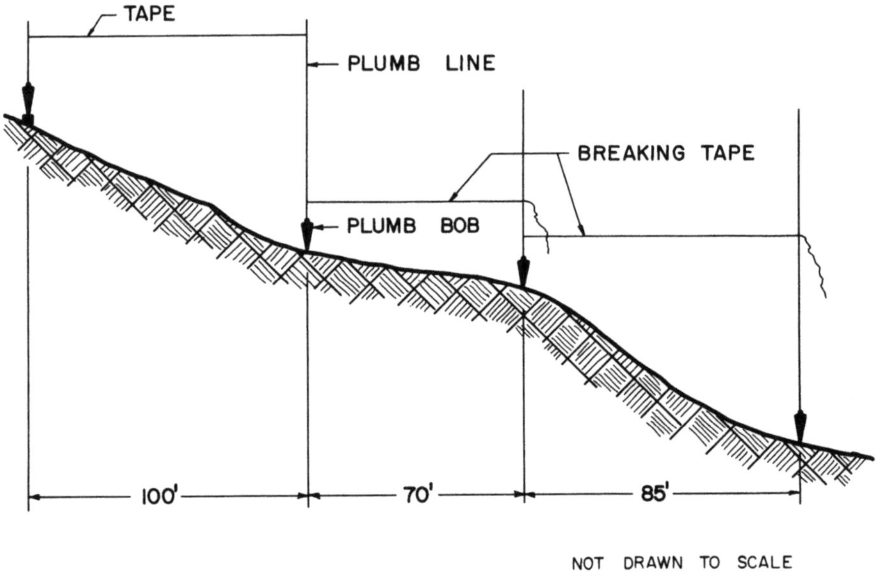

Fig. 3-8. Taping on a slope. Note the procedure of breaking tape.

that it is much easier to measure downhill than uphill, i.e., place the 0 end of the tape at the uphill end of the segment.

Keep in mind, too, that when taping over irregular ground, the tape is often unsupported along its midsection and therefore sags. This is unavoidable. To keep error due to sag to a minimum, the tape should be tightly pulled by the same amount each time a measurement is made. Usually a 15 lb tension (either determined with a spring balance or estimated) is sufficient. The spring balance allows for the exact same tension to be applied every time.

Yet another simple but important procedure to help with accuracy is the alignment of the tape. When the distance between two points must be measured in segments, or if the distance is longer than the tape, it is important that the intervening points be *on line*, i.e., in a straight line (see Fig. 3-9). Notice that in measuring the distance between A and Z, the intervening points H, J, and K are not on line. Thus, measured distance $AH + HJ + JK + KZ$ will be greater than the actual distance AZ. For accuracy of 1/3000 or 1/5000, tapes should not be off line more than 1%, i.e., 1 ft in 100 ft, 0.5 ft in 50 ft, etc.

To attain accuracies such as these, another source of error can be variations in air temperature. Steel tapes are calibrated by the manufacturer at 68°F, and any change in temperature will cause the tape to change length slightly because of expansion and contraction. Table 3-1 lists the corrections for a 100 ft steel tape at various temperatures. These corrections are based on a coefficient of

Fig. 3-9. Taping between points *A* and *Z*. Note the increased distance by not keeping the tape along line *AZ*.

expansion of 0.0000645 per degree F. To determine a correction value, multiply the appropriate value from Table 3-1 by the number of hundreds of feet in the measured length. For example, if you are taping 456 ft and the temperature is 90°F, the temperature correction is calculated as follows:

$$(4.56)(0.015) = 0.068 \text{ ft} = \text{temperature correction}$$

Because the temperature is above the standard 68°F, the correction is added to the tape reading.

The last item that should be mentioned here that will help increase taping accuracy is the fact that on occasion the tape itself may be in error. In view of this, it should be checked against a standard now and then. Many registered, professional surveyors have tapes that have been tested by the National Bureau of Standards. One of these tapes would serve as a good standard.

Table 3-1.

°F	Corrections per 100 ft	°F	Corrections per 100 ft	°F	Corrections per 100 ft
−13		+33		+80	
−9	−0.050	+37	−0.020	+84	+0.010
−6		+41		+87	
−2	−0.045	+44	−0.015	+91	+0.015
+2		+49		+95	
+6	−0.040	+53	−0.010	+99	+0.020
+10		+56		+103	
+14	−0.035	+60	−0.005	+107	+0.025
+18		+64		+111	
+22	−0.030	+68	0.000	+115	+0.030
+25		+72		+118	
+29	−0.025	+76	+0.005	+122	+0.035

Table data from Kissam, P., *Surveying Practice*, © 1971, reproduced by permission, McGraw-Hill.

Reading a Rod with Accuracy

Level rods are constructed illustrating a wide variety of numbering systems and styles. However, the one that seems to be the most popular is the *Philadelphia rod*. The Philadelphia rod is graduated in feet, tenths, and hundredths by a series of black stripes and red and black numbers painted on the face of the rod (see Fig. 3-10). This system of markings is simple and very easy to read, with little chance of making errors. Nevertheless, several practice examples are illustrated in Fig. 3-11. The markings and numbers on the rod are sufficiently large so that they can be read directly from the sighting instrument, except when shooting over unusually large distances, about 400 feet or so.

In the event very accurate work needs to be done, or when intervening vegetation or other obstructions make reading difficult, a rod target is sometimes used to assist the instrument operator (see Fig. 3-12). Note that some targets have verniers on them which permit accurate readings to thousandths of a foot. Note too, that it is important that the rod be held plumb while readings are being taken, otherwise accuracy attained by correct and careful reading will be lost. This is particularly true if readings are being made near the top of the rod (see Fig. 3-13). Some rods have bull's-eye level bubbles attached to them when extra accuracy is required.

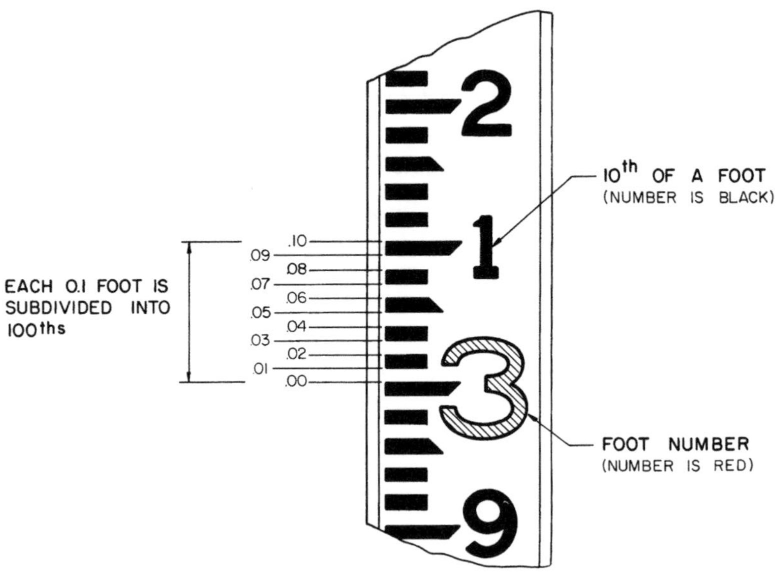

Fig. 3-10. Markings on the face of a Philadelphia rod.

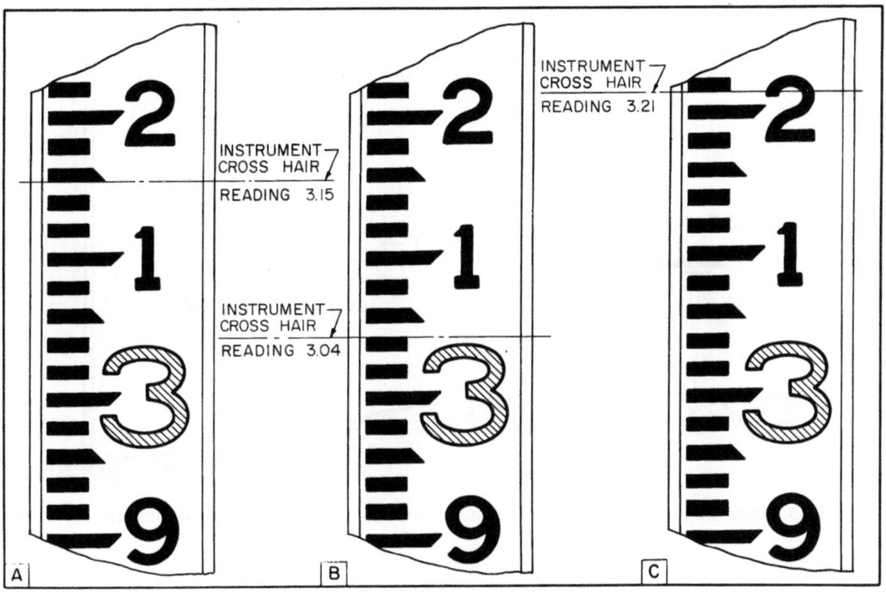

Fig. 3-11. Examples of reading a Philadelphia rod.

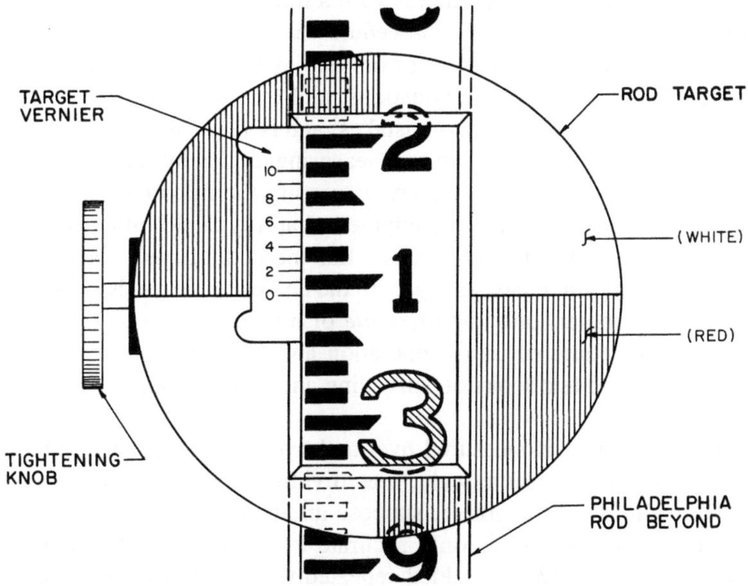

Fig. 3-12. Philadelphia rod with a target and vernier. The reading is 3.085.

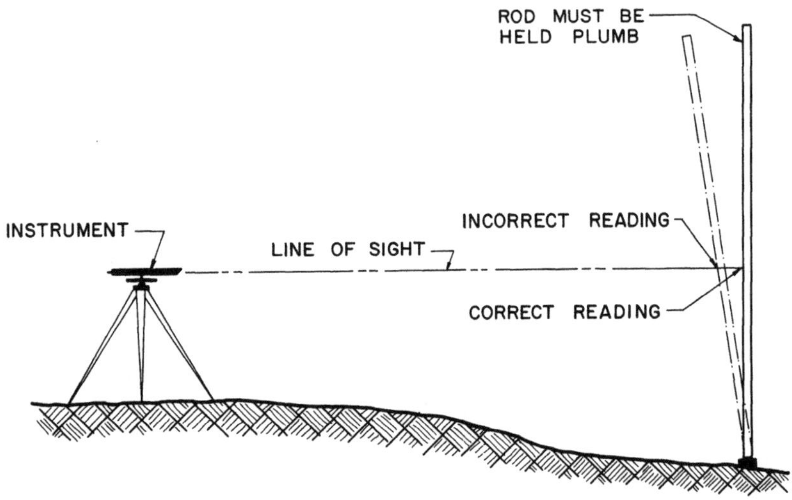

Fig. 3-13. Obtaining correct readings on a plumb rod. Note the error is greater near the top of the rod.

Angles by Repetition for Increased Accuracy

The accuracy of angle measurements with a transit can be greatly increased by a method called *repetition*. The procedure reduces errors due to the fact that verniers cannot be read precisely, errors due to nonadjustment of the transit's horizontal axis, errors on the graduations, and others. It is assumed, however, that the instrument user makes no mistakes, and is very careful in the procedure, otherwise there is no justification for measuring angles by repetition.

Here again, as a reminder, we are extracting the maximum accuracy from relatively common, inexpensive equipment that a typical builder or landscaper is likely to possess. For example, using this method a transit that has an accuracy of ± 1 minute can measure angles to the nearest 10 seconds if care is taken. Note that at this point, only the *principle* of repetition is to be discussed. The step-by-step procedure of transit operation and the recording of notes can be found in most good textbooks on surveying and in some transit operation manuals.

Basically, an angle is measured over and over again, accumulating the successive readings on the same scale, *then* dividing the results by the number of repetitive readings. For example, suppose an angle of $21° 13' 17''$ needs to be measured, but your transit is only accurate to 1 minute. Column A of Table 3-2 shows angle $21° 13' 17''$ added repeatedly 2, 3, and 4 times. Column B shows the same angles measured with our $1'$ transit. As our transit readings are divided by their corresponding repetition, note in Column C how they converge

Table 3-2.

Reading Number	Multiple Values of Angle 21°13′17″ (A)	Cumulative Transit Readings with 1′ Transit (B)	Values Calculated from Transit Readings (C)
1	21°13′17″	21°13′	21°13′00″
2	42°26′34″	42°27′	21°13′30″
3	63°39′51″	63°40′	21°13′20″
4	84°53′08″	84°53′	21°13′15″

to within 02″ of the original "true value" of 21° 13′ 17″. Thus using a 1 minute transit we have been able to extend its accuracy and increase its value as a tool. It should be mentioned that accuracy accrued by 4 repetitions is usually more than is needed for the building trades. Nevertheless the procedure should be understood as it can be employed with very little extra effort and time.

Significant Figures and Rounding

While the previous portion of this chapter deals with increasing accuracy and precision, it is appropriate to discuss at this point how to correctly carry and apply this acquired accuracy to calculations, drawings, and ultimately to and from the field. We want to be as accurate as possible giving the maximum amount of information, but care must be taken not to give any misinformation. It is tempting to use figures with many digits to the right of the decimal point; they look impressive, but in fact they may be meaningless or even misleading. In view of this, it is necessary to understand what is commonly referred to as *significant figures*. In any measurement some of the numbers are meaningful, and with some of the numbers there is uncertainty, especially if they have been estimated. Note that the number of significant digits used to express a measurement primarily depends on the measuring instrument.

As a very simple example, suppose you are going to measure a person's height with two different tapes. One tape is graduated in feet only, the other is graduated in feet and tenths of feet (see Fig. 3-14). Tape A shows 5.6 feet. The reading is certainly between 5 and 6 and is estimated to be about six-tenths of the way from 5 to 6. Tape B, however, reads 5.65 feet. On it, the line is estimated to be half way between 5.6 and 5.7. The .6 and .7 on this tape are known with certainty. Thus, the first measurement has two significant figures and the second, three, because of the difference in the two tapes. So with all measurements, whether in the field or in the office, the number of significant digits will be determined by the graduations on the instrument.

While the concept of significant figures is important it is also important to understand how to handle them in calculations. For example, if we add the two measurements from Fig. 3-14, a meaningful sum would not be 11.25 ft but 11.2 ft. The 5 in the sum is *not* significant, i.e., it has no meaning. This is true because the reading of 5.6 feet from Tape A is not accurate enough to specify anything about a third digit. Because the second digit is already estimated, it would be impossible to estimate to an even finer degree. A simple rule applicable to calculations involving addition and subtraction states:

> Digits that are *not* significant in any of the numbers being added or subtracted are *not* significant in the resulting sum or difference.

A different rule illustrates how to handle significant figures involved in multiplication and division. It states:

> The number of significant digits in a product or quotient cannot be greater than the number in the *least* accurate term entering into the calculation.

For example, in determining the area of a rectangle, the length was measured to be 145.25 feet and the width 55.625 feet. According to the multiplication/division rule, the area is 8079.53 sq. feet not 8079.5312 sq. feet. The least accurate measurement has only two digits to the right of the decimal point, hence the product can have no more than that.

The procedure of *rounding* numbers should be mentioned here as a method

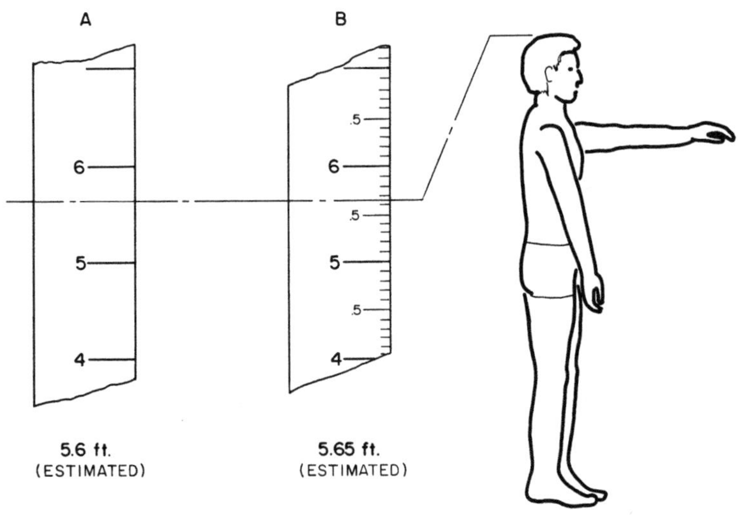

Fig. 3-14. The number of significant figures is determined by the graduations on the measuring instrument.

Table 3-3.

A		B
176.7297	becomes	176.73
8.1667	becomes	8.17
12.3456	becomes	12.35
123.6090	becomes	123.61
123.4999	becomes	123.50

Table 3-4.

A		B
12.345	becomes	12.34
123.455	becomes	123.46
123.545	becomes	123.54
123.505	becomes	123.50

for ridding our calculations, sums, products, and quotients of meaningless digits. Suppose the entries in Column A of Table 3-3 must be rounded off to hundredths, i.e, two digits to the right of the decimal point. Note the results in Column B.

If the last number to be dropped is exactly 5, it is common practice to round to the *nearest even* number. Column A of Table 3-4 is to be rounded to the nearest hundredth.

This procedure has logic to it in that in the long run over the course of many calculations, there tends to be an equal number of round-ups and round-downs. As a final note, rounding should be performed on the final answer, *not* on the intermediate step.

4

Some Fundamentals of Photogrammetry

INTRODUCTION

Photogrammetry is the science of making measurements on photographs. Because *aerial photographs* are taken from a high vantage point, they can record images of very wide areas and are thus quite valuable to the builder or landscaper. Moreover, if the measurements are made carefully enough, the data can complement those derived by conventional surveying techniques. Like surveying, then, aerial photographs can be used to determine distances, angles, areas, volumes, elevations, slopes, sizes and shapes of objects, and for planimetric and topographic mapping. In fact in some respects, aerial photographs can be even *more* useful than maps. For example, in reading maps the interpreter is limited by the lines, colors, and symbols chosen by the cartographer or draftsperson. An aerial photograph, on the other hand, shows the actual ground surface in its various colors *or* in shades of gray, along with individual trees, bushes, buildings, and very subtle breaks in slope. Aside from the limits of resolution of the photograph itself, then, a skilled interpreter may be able to gather much more information from an aerial photograph than from a map. In this chapter we will not explore the techniques of interpretation, but will concentrate on the geometry and mechanics of the photographs that will assist us in making accurate and precise measurements.

There are some limits and drawbacks to using aerial photography. For example, unlike maps, the photograph does not come with a ready-made, ready-to-use scale. One has to be generated. In some projects, aerial photography may not yield the required accuracy, or may not be economically feasible. Nevertheless, it can serve well in many aspects of the building trades if used properly and under the correct circumstances.

TYPES OF AERIAL PHOTOGRAPHY

Aerial photography falls into two categories: vertical and oblique. Vertical aerial photographs are more common and the easier of the two to work with in terms of making measurements. An aerial photograph is said to be *vertical* if

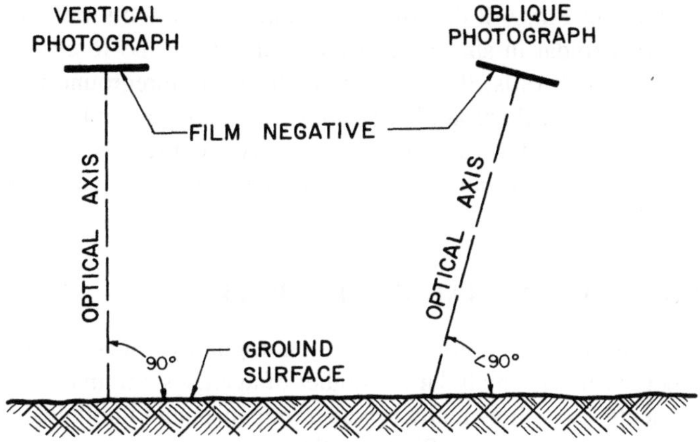

Fig. 4-1. Types of aerial photographs.

its optical axis is plumb. The *optical axis* of a photograph is an imaginary line extending at right angles through the center point of the photograph to the ground (datum) (see Fig. 4-1). An *oblique* aerial photograph is one in which the optical axis is inclined at some angle with reference to the datum (see Fig. 4-1). It should be noted that there are two varieties of oblique photography, *high obliques* and *low obliques*, depending on the inclination of the optical axis. The optical axis of a high oblique aerial photograph is tilted to such an extent that the horizon is visible on the photograph, whereas in a low oblique, the horizon is *not* visible (see Figs. 4-1 and 4-2).

Although oblique aerial photographs show the ground as one would normally view it from the window of an aircraft in flight, they are more complex in their

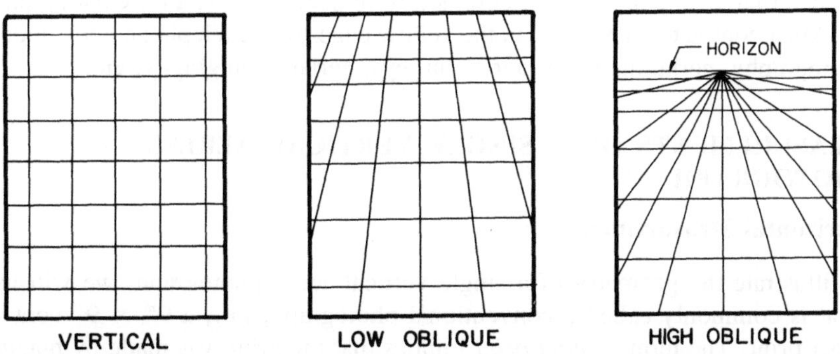

Fig. 4-2. Square grid sections as they appear on three types of aerial photographs. (Redrawn from Wolf, P. R., *Elements of Photogrammetry* © 1983 McGraw-Hill, by permission.)

geometry. Notice in Fig. 4-2 that the grid squares viewed on the oblique photographs are trapezoidal in shape and vary in size from the foreground to the background. In other words, the *scale* varies from the foreground to the background. Views on vertical aerial photographs, on the other hand, are quite similar to map views, and as a result have essentially a constant scale. Therefore, making measurements and transferring data to and from maps can be more easily done.

AVAILABILITY OF AERIAL PHOTOGRAPHS

As discussed here, the availability of aerial photography refers to existing photographs. There is an extremely large number of agencies within the federal and state governments, as well as many private companies and individuals, that are engaged in aerial photography. Photographic coverage already exists for virtually all of Canada and the U.S. In fact, some areas have been photographed several times at various scales, so in many areas, the user can pick and choose what best suits the project in question.

The National Cartographic Information Center (NCIC), which is a division of the U.S. Geological Survey, maintains an Aerial Photographic Summary Records System (APSRS).* This system has on record the most complete and up-to-data tabulation of aerial photography covering the U.S. It can identify those agencies within the federal and state governments and in the private sector that hold the negatives of a particular area. APSRS can also supply data about scale, type of photography, date of photography, camera, etc. The author strongly recommends that before beginning any project in which aerial photography might be useful, NCIC should be queried to ascertain what is available for a particular area.

Because there may be some private companies or state highway departments that do not send information to NCIC, or are tardy in doing so, it is recommended that local photographic contractors be queried as well for existing photos. Most contractors are listed in the yellow pages of the telephone book under: photography, aerial mapping, aerial surveys, aerial photography, etc.

MEASUREMENTS ON A SINGLE VERTICAL AERIAL PHOTOGRAPH

Horizontal Measurements

To illustrate the geometry of a single vertical aerial photograph, we will use what is commonly called a conventional photograph, i.e., a 9″ × 9″ contact paper print. The term *contact print* implies that the print was made by placing

---

*Inquiries and requests regarding APSRS should be sent to the National Cartographic Information Center, U.S. Geological Survey, 507 National Center, Reston, Virginia 22092.

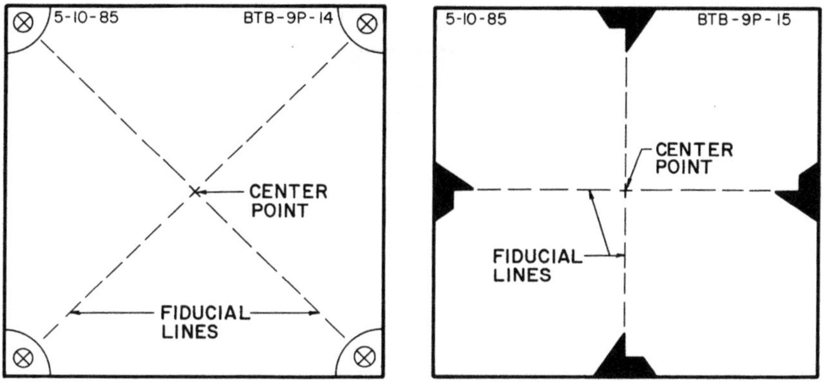

Fig. 4-3. Examples of corner and side fiducial marks. Their size is exaggerated for illustration purposes.

the negative in direct contact with the photographic paper then exposing it. In the transfer from the negative to the print there is no enlargement and hence no loss of detail. This is the most common variety of print available from most private and public aerial photographic sources, especially agencies of the federal government. The photograph will exhibit *fiducial marks* along its edges or corners that serve to locate the center point of the photograph by joining lines from opposite sides of the photograph (see Fig. 4-3). The photographs will also show project code numbers (or letters), film roll numbers, and photograph numbers along one edge (see Fig. 4-3).

The terminology and geometry of a single vertical aerial photograph are shown in Fig. 4-4. Several items of importance in the figure should be noted. First, note that angles α and α_1 are equal. Second, note that because of the equal angles, triangle *ABC* is *similar* to triangle *DCE*. Therefore, any measurement made on *ABC* is going to be proportional to a similar measurement made on *DCE*. Thus, the following relationship can be established.

$$\frac{f}{H} = \frac{i}{O} \tag{4-1}$$

where f = camera focal length (feet)
H = aircraft flying height (feet above the ground)
i = distance between two points on the photograph image (in feet)
O = the ground distance between the two points (in feet).

You need not use feet as the units. Whatever units are chosen, they must be the same on both sides of the equation.

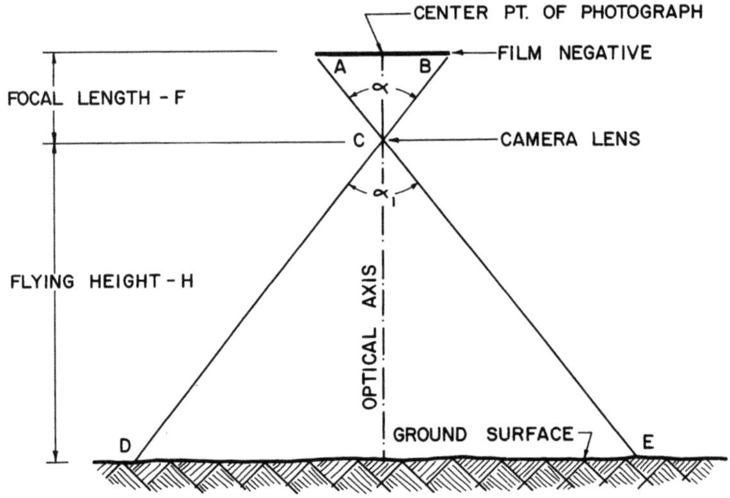

Fig. 4-4. Geometry of a single aerial photograph.

Figure 4-5 is a portion of a conventional photograph taken from a flying height of 10,000 feet with a camera having a focal length of 6 inches. How far is it (ground distance in feet) between points *A* and *B*? The photo distance is 3″. Appendix D, Table 1, might be helpful here. Using Eq. (4-1) and keeping units the same on both sides of the equation we get:

$$\frac{f}{H} = \frac{i}{O}\ \frac{0.5'}{10{,}000'} = \frac{0.25'}{0}\ 0 = 5{,}000.00'$$

From Eq. (4-1) and using the nomenclature in Fig. 4-4, we can express the difference in size between the *image* and the *ground* as a ratio, or still better, as a ratio scale (see Eq. (4-2) and Appendix D, Tables 1 and 4):

$$\text{Photo scale} = \frac{\text{Focal length}}{\text{Flying height}} = \frac{f}{H} = \frac{AB}{DE} \tag{4-2}$$

This photo scale can be expressed in written fashion as being directly proportional to the camera focal length and inversely proportional to the flying height above the ground. For example, the photograph in Fig. 4-5 has a ratio scale

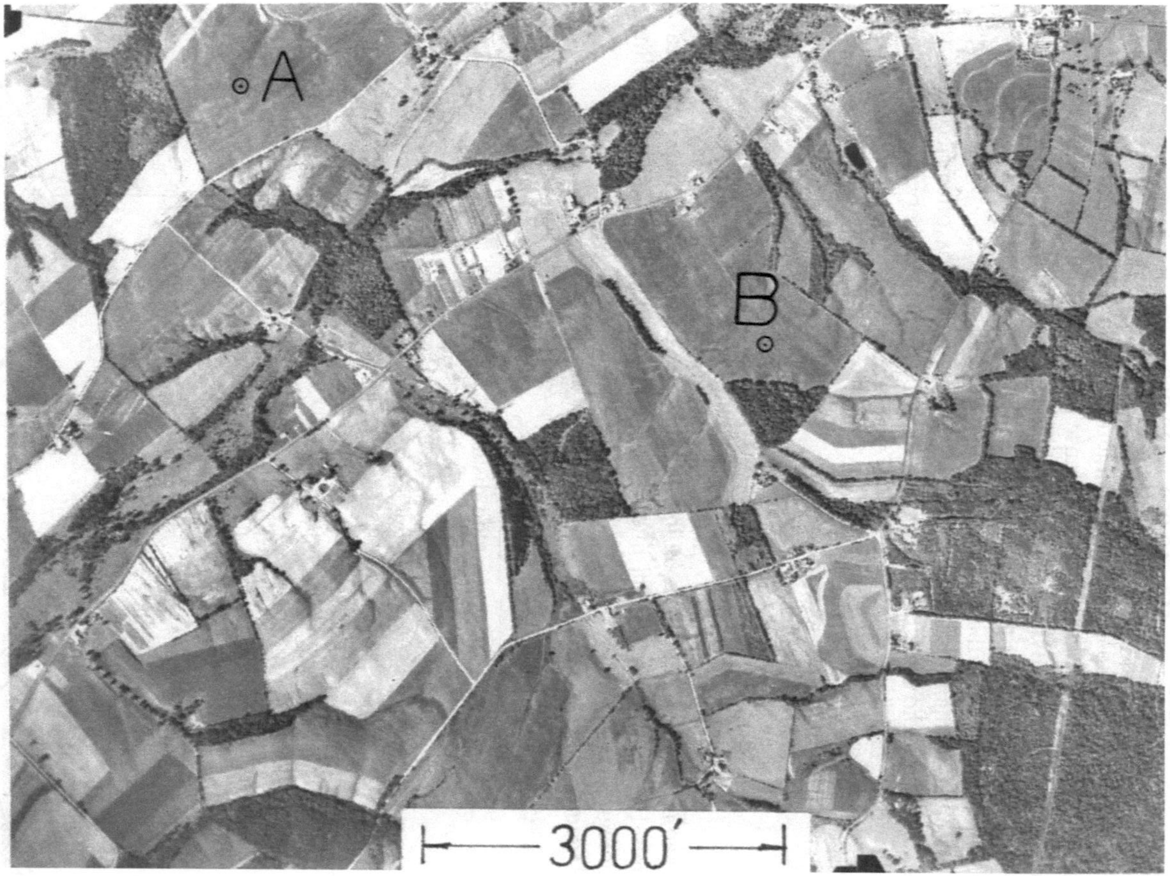

Fig. 4-5. A conventional aerial photograph; exposed with a 6″ focal length camera from a flying height (above the ground) of 10,000 ft.

that is calculated by:

$$\text{Scale} = \frac{f}{H} = \frac{0.5'}{10{,}000} = \frac{1}{20{,}000} \quad \text{or} \quad 1{:}20{,}000$$

In the above example notice that flat terrain was used. Because in most areas the terrain is not perfectly flat, and because an aircraft in flight is rarely (if ever) perfectly stable, the flying height above the ground is constantly changing (see Fig. 4-6). When using Eq. (4-2) to ascertain scale, please be aware that what is being calculated is the *average* or *nominal* photo scale.

In order to maintain a high standard of accuracy, it is often necessary to determine a photo scale that is *exact*. If the ground distance between two points on a photograph is already known, the exact scale relationship for *that particular photograph* is the ratio between the (known) ground distance and the (measured) distance on the photograph:

$$\text{Exact scale} = \frac{\text{Photo distance}}{\text{Ground distance}} \tag{4-3}$$

As an illustration, suppose a football field is viewed on a photograph and measures 1.125 inches from goal line to goal line. What is the exact scale of the photograph?

From Eq. (4-3):

$$\text{Exact scale} = \frac{\text{Photo distance}}{\text{Ground distance}} = \frac{1.125''}{3600.00''} = \frac{1}{3{,}200}$$

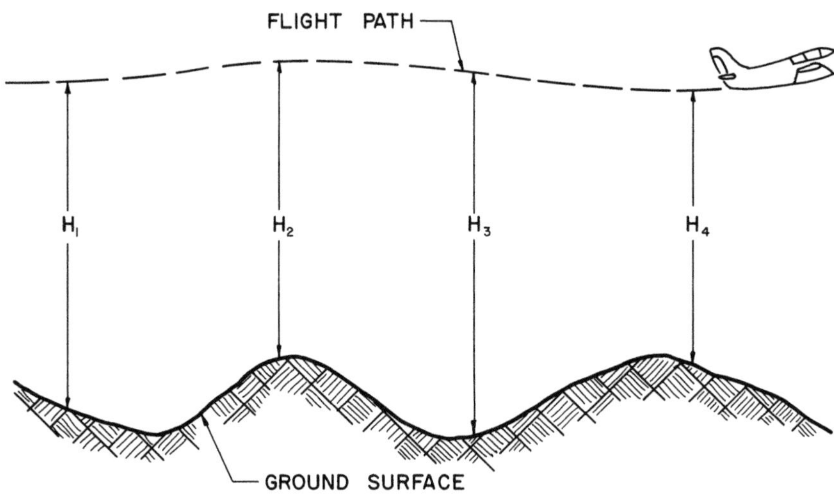

Fig. 4-6. Flying height (above the ground) changes with each photograph.

Table 4-1.

Feature Seen on Photo	Actual Size on Ground
Railroad tracks (Fig. 4-7)	4′8″ (rail to rail)
Traffic lane (Fig. 4-8)	12′ (1 lane of 4 lane interstate highway, for example)
Tennis court (Fig. 4-9)	78′ (end line to end line)
Standard railroad car width (Fig. 4-10)	10′ (standard car width)
Baseball field	90′ (between bases)
Utility poles	200′ (be *very careful* with these)
Football field (Fig. 4-11)	300′ (goal line to goal line)
Football field	360′ (end line to end line)
Section road (Fig. 4-12)	5280′ (corner to corner)

In this case, we were able to calculate the exact scale without measuring the ground distance because it is common knowledge that a football field (in the U.S.) is 100 yd long (300 ft or 3600 inches). Table 4-1 lists some additional features that can easily be seen and measured on an aerial photograph for which, by common knowledge, the size (ground distance) is already known.

Often, however, one of the items in Table 4-1 is not visible and yet the exact scale needs to be determined. It can still be calculated if a good map covering

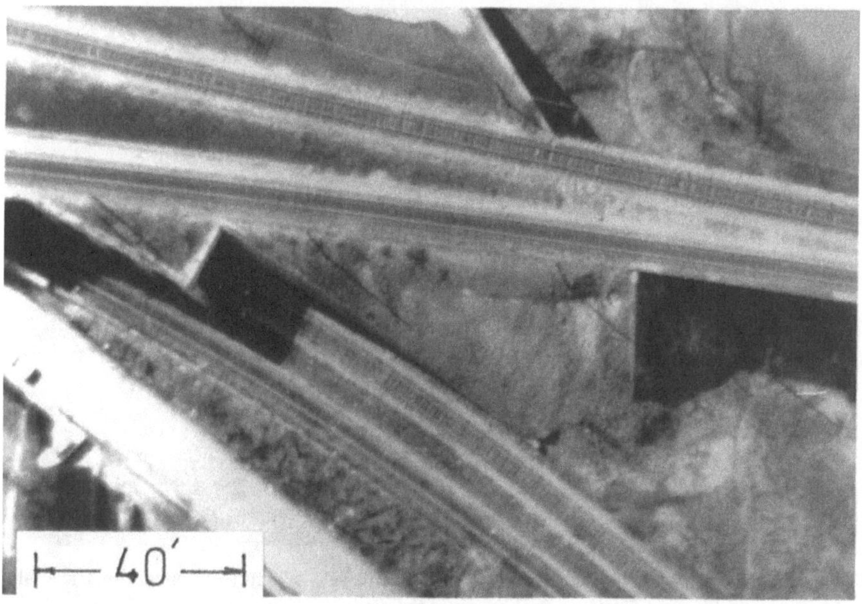

Fig. 4-7. The spacing between rails on standard gauge tracks in Canada and in the United States is 4 ft 8 in.

Fig. 4-8. A single traffic lane is 12 ft. Both lanes (not including berms and medial strips) are 24 ft.

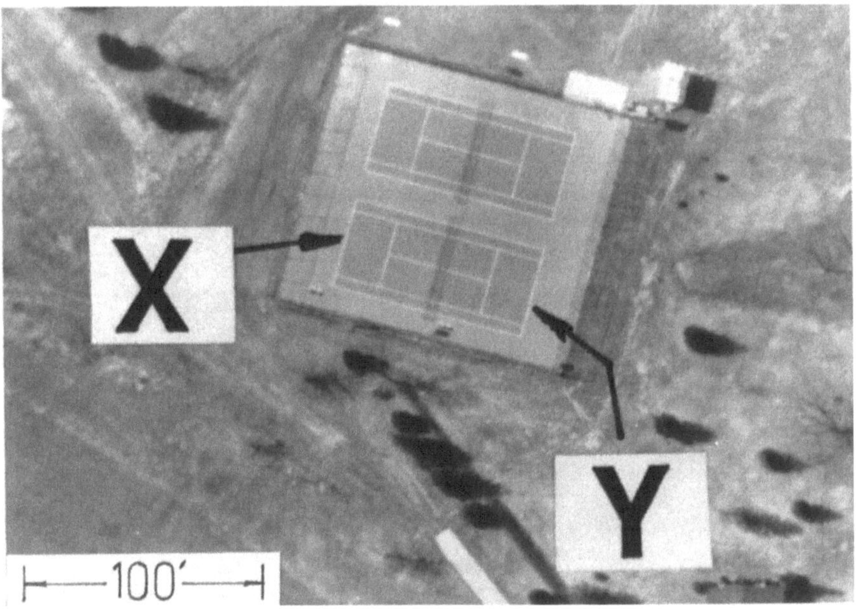

Fig. 4-9. Tennis court length from end line to end line (*X* to *Y*) is 78 ft. (Reprinted from Ciciarelli, J. A., *Practical Physical Geology, Problems and Solutions*, © 1986, Gordon and Breach Science Publishers by permission.)

Fig. 4-10. The standard width of a railroad car in Canada and in the United States is 10 ft. This includes boxcars, hopper cars (shown above), flat cars, gondolas, etc.

Fig. 4-11. A football field, goal line to goal line (*X* to *Y*) in the United States is 100 yds. (reprinted from Ciciarelli, J. A., *Practical Physical Geology, Problems and Solutions*, © 1986, Gordon and Breach Science Publishers, by permission.)

Fig. 4-12. The distance between *X* and *Y* (section corners) is 5,280 ft.

the same area as the photograph is available. First measure the distance between any two points that can be easily identified *both* on the photograph and on the map; then use Eq. (4-4).

$$\text{Photo scale} = \frac{\text{Photo distance}}{\text{Map distance}} \times \text{Map scale} \tag{4-4}$$

As an example of this procedure, suppose the distance between two street utility access ports (manhole covers) seen on photographs are 3.125 inches apart. On the street department's map (scale 1 : 1200) the same two manholes are 5.375 inches apart. What is the photo scale? By using Eq. (4-4) we can get the scale:

$$\text{Photo Scale} = \frac{3.125}{5.375} \times \frac{1}{1200} = \frac{1}{2064}$$

In summary, by using any of the Eqs. (4-1) through (4-4) it is possible to make horizontal measurements on a single vertical aerial photograph. Obviously, however, not every landscaping or building project lends itself to the use of aerial photographs. But some will, and horizontal dimensions will be the measurement most likely made.

Vertical Measurements

It is possible, too, within certain limits, to determine vertical distances, i.e., elevation differences on a single photograph. First, however, it is necessary to understand a phenomenon inherent in all vertical photographs called *relief displacement* or sometimes *radial displacement*. The author prefers the latter as it is more descriptive.

Unlike maps where terrain points are positioned as though they are viewed from directly above (see Fig. 3-7 of Chapter 3), terrain points on an aerial photograph are viewed from a *central point*, i.e., the camera lens. For example on a map, a vertical object such as a flagpole will appear as a single point regardless of where it appears on the map. This characteristic is referred to as *orthogonal projection*. On a vertical aerial photograph, a vertical object such as a flagpole will appear as a short line with the top radially displaced outward, away from the photograph's center point (see Figs. 4-13 and 4-14). The greater the distance from the center point, the greater will be the displacement, and vice versa. (*Note*: The only place on the photograph where a vertical object appears as a point is at the center point, i.e., directly below the camera.)

By virtue of this radial displacement, it is possible to measure heights of

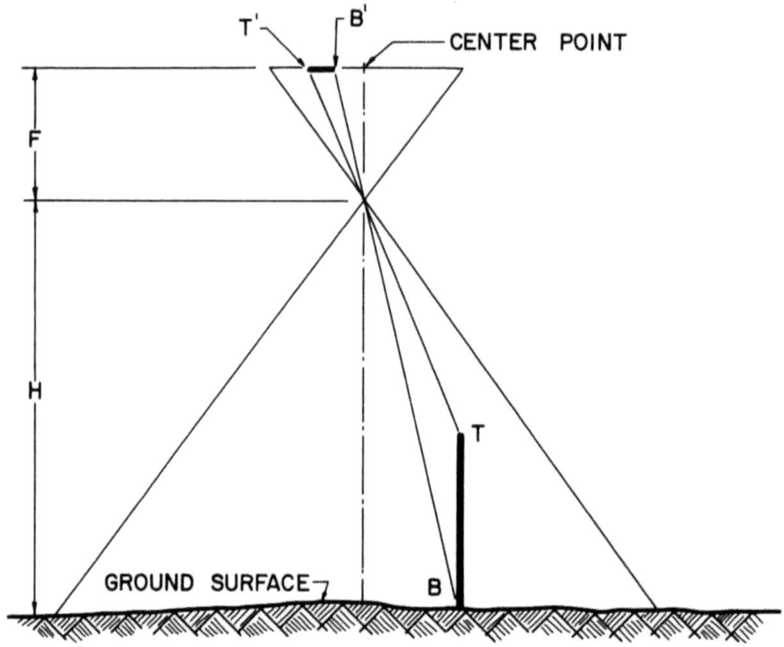

Fig. 4-13. Radial displacement of the top of a object away from the center point of the photograph.

Fig. 4-14. A portion of an aerial photograph illustrating radial displacement. The center point is located to the right and slightly toward the top of the figure. Notice that the top of the vertical water tower appears tilted away from its base.

objects; see Eq. (4-5):

$$h = \frac{dH}{r} \tag{4-5}$$

where h = object height (in feet)
d = relief displacement, i.e., distance between top of object and the bottom
r = distance from center point to lower point of the object (*Note*: d and r must be in the same units)
H = flying height (in feet).

If the flying height is not known, it can be calculated using Eq. (4-6):

$$H = \frac{f}{s} \tag{4-6}$$

where H = flying height (above the ground)
f = focal length
s = scale of the photograph.

It is important to note that this procedure which uses radial displacement is only applicable to vertical objects where the upper point is directly above the lower point, e.g., trees, smoke stacks, cliffs, buildings. Again, note the special case when the object is located at the center point. The value of r will be zero, and it will not be possible to measure d, as both the top point and bottom points will be coincident. For this case, this procedure cannot be used.

To illustrate this, suppose a flagpole has a radial displacement of 0.125″, the distance from the center point to the bottom of the pole is 2.50″, and the flying height above the ground is 1100′. Using Eq. (4-3) we calculate:

$$h = \frac{(0.125'')(1100')}{2.50''} = 55 \text{ feet}$$

A second method of measuring heights on a single photo uses shadow lengths (see Fig. 4-15). The height of an object can be calculated by using Eq. (4-7):

$$h = \frac{Hl(\tan \alpha)}{f} \tag{4-7}$$

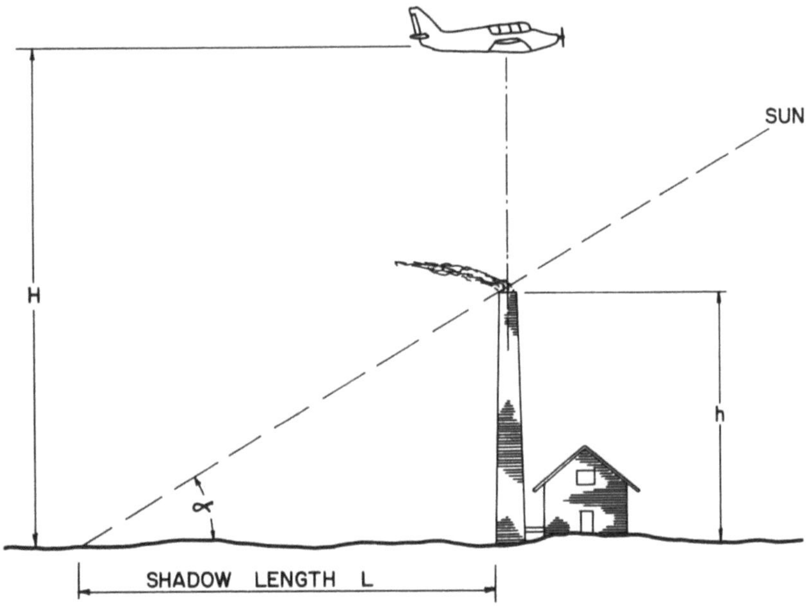

Fig. 4-15. Determining the height of a vertical object from shadow lengths.

where H = flying height
h = height of the object
l = length of shadow on photo sun angle
f = focal length.

For this method to be accurate, the shadow must fall on level ground.

GEOMETRY AND VIEWING OF STEREOSCOPIC PAIRS

A dictionarylike definition of *parallax* states that it is the apparent displacement of an object as seen from two different points. A simple demonstration will nicely illustrate the concept. Hold a pencil vertical and very still at arm's length, look at it with your left eye closed and at the same time notice the background, e.g., the far wall. Now close your right eye and look at the pencil with the other. Notice how it apparently shifts position against the background. This apparent motion is parallax. Alternately blinking your eyes rapidly will more dramatically illustrate this effect. The same effect can be observed, with one eye closed, moving your head slowly from side to side. The parallax is even greater if the pencil is held closer to your eyes.

If an observer were watching the ground through the viewfinder of an aerial camera as the aircraft is moving forward, objects on the ground would exhibit the same apparent shift. If photographs were taken at regular intervals of time, the parallax observed would, of course, be recorded on the photographs. This parallax, which is caused by the movement of the aircraft, is called *stereoscopic parallax* (see Fig. 4-16). Observe how the object on the ground shifts positions on the two adjacent photographs. The two photos are really like our eyes looking at the pencil at arm's length. To insure continuity of coverage, and to insure that each point on the ground is photographed from at least two adjacent stations, a strictly controlled, regular pattern is flown by the aircraft doing the photography (see Fig. 4-17).

If we select two adjacent photographs (called a *stereoscopic pair*) from the flight line in Fig. 4-17, and a *stereoscope* (Fig. 4-18), we would be able to view the ground in three dimensions. The analogy again is viewing the pencil held at arm's length. This time, however, we are using both eyes, hence we can see the pencil from *two* different stations and can see it in three dimensions against the background. The three-dimensional model of the ground that can be seen on the stereoscopic pair will reveal hills, valley, slopes, cliffs, subtle changes in slope, and tree heights.

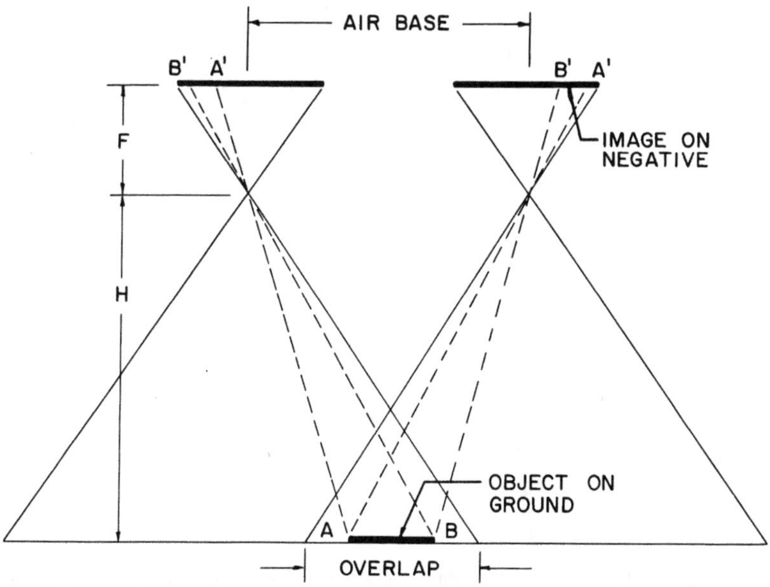

Fig. 4-16. Stereoscopic parallax. Notice how the position of the object on the ground changes on the two photographs. (Redrawn from Ciciarelli, J. A., *Practical Physical Geology, Problems and Solutions*, © 1986, Gordon and Breach Science Publishers, by permission.)

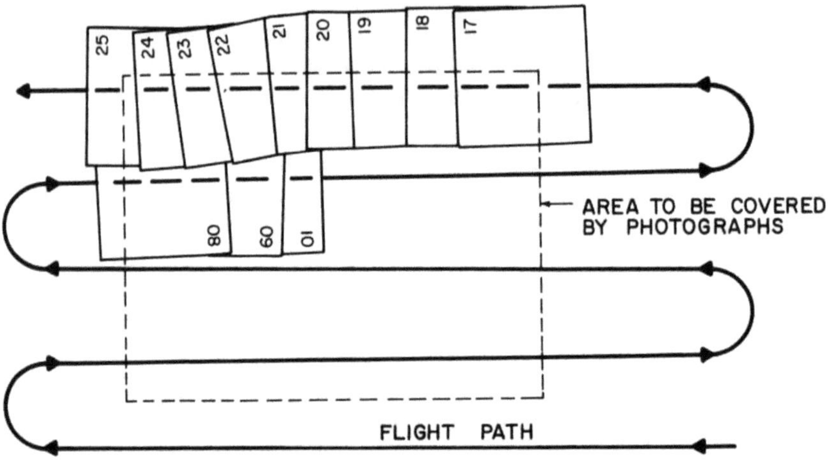

Fig. 4-17. A regular flight path followed by an aircraft engaged in aerial photography. (Redrawn from Ciciarelli, J. A., *Practical Physical Geology, Problems and Solutions*, © 1986, Gordon and Breach Science Publishers, by permission.)

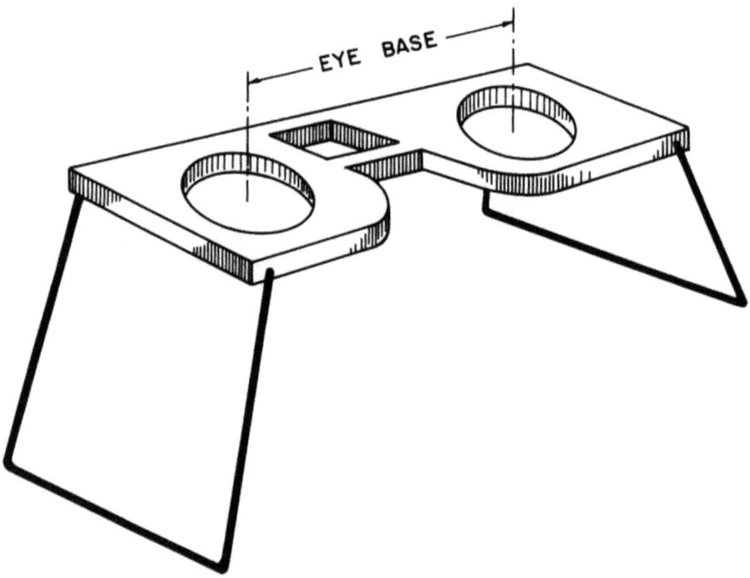

Fig. 4-18. A pocket stereoscope. The distance between the lenses can be adjusted to accommodate the user's eye base (interpupillary distance), i.e., the distance from the center of one eye to the center of the other.

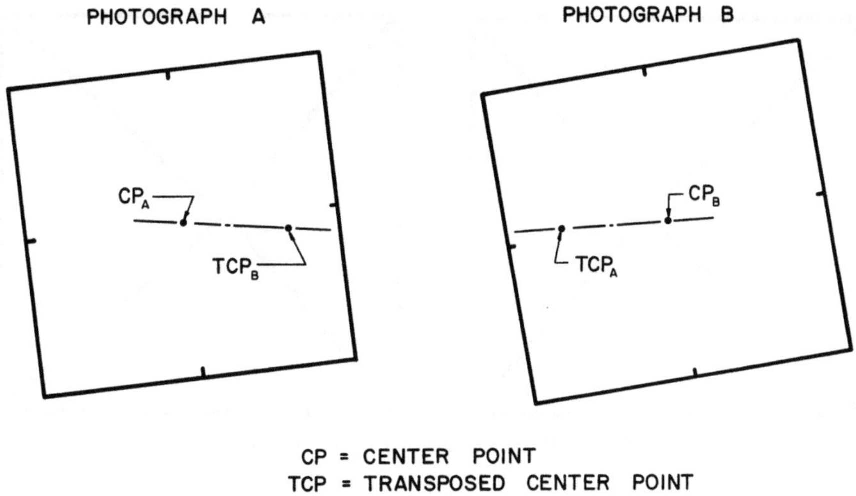

Fig. 4-19. Location of center points and transposed center points.

With that in mind, it should be apparent that stereoscopic viewing can be very valuable to the landscaper, builder, and engineer. However, it is important that the photographs comprising the stereoscopic pair be properly oriented and properly spaced while viewing them. This is important in order to avoid eye strain and to produce the maximum stereoscopic effect. The guidelines listed below should be followed while referring to Fig. 4-19.

1. Locate the center point (CP) on both photographs of the stereo pair. The *center point* is located at the intersection of the two lines connecting fiducial marks at the edges or corners of the photograph. *Note*: Use a soft No. 2 pencil.
2. Inspect the center points closely identifying a feature on the ground marked by the center point. It may be a tree, corner of a field, a bend in a road, a building, etc. Notice that the point on photograph A can be seen on Photograph B, and vice versa. These center point locations identified on the opposite photographs are called *transposed center points* (TCP).
3. Mark with a small dot the transposed center points.
4. Connect CP_A and TCP_B with a line. Do the same with CP_B and TCP_A.
5. Orient and position the photographs so that CP_A, TCP_B, TCP_A, and CP_B fall along a straight line. (This is the flight line of the aircraft.)
6. Maintaining the straight line from Step 5, overlap the photographs so that the same object seen on two photographs is *directly below the lenses* of a pocket *stereoscope*. *Note*: The pocket stereoscope would be adjusted to

Fig. 4-20. Stereogram of a 4-sided pyramid.

match your *eye base*, i.e., the distance from the center of one eye to the center of the other eye when they are focused at a distant object.

7. Keeping the top edge of the stereoscope parallel with line CP_A, TCP_B, TCP_A, and CP_B, the terrain should come into three-dimensional view. If you experience some difficulty, adjust the separation of the photographs slightly.

Figure 4-20 is a stereogram of a 4-sided pyramid. Using a pocket stereoscope, try to view it in three dimensions. Try it also on Fig. 4-21 and 4-22.

VERTICAL MEASUREMENTS BY THE PARALLAX METHOD

It is valuable in itself to be able to observe the stereoscopic model, but what needs to be done now is to make accurate vertical measurements from the stereoscopic pair. The key to making these measurements is parallax, which is exhibited on the photographs as radial displacement.

We will again use the example of a flagpole viewed on two adjacent photographs. The top of the pole, of course, will be radially displaced outward from the center point of each photograph with respect to the bottom of the pole (see Fig. 4-23). The higher the pole and/or the greater the distance from the center point, the greater the displacement, and vice versa. Because of this relationship between the height of an object, parallax, and flying height, it is possible to state Eq. (4-8). Please note that the deriviation of this equation can be found in most good textbooks on photogrammetry and is beyond the scope of this book.

$$h = \frac{H'}{ab - \Delta P} \Delta P \qquad (4\text{-}8)$$

Fig. 4-21. A stereoscopic pair. *Note:* Use a lens stereoscope.

Fig. 4-22. A stereoscopic pair. *Note:* Use a lens stereoscope.

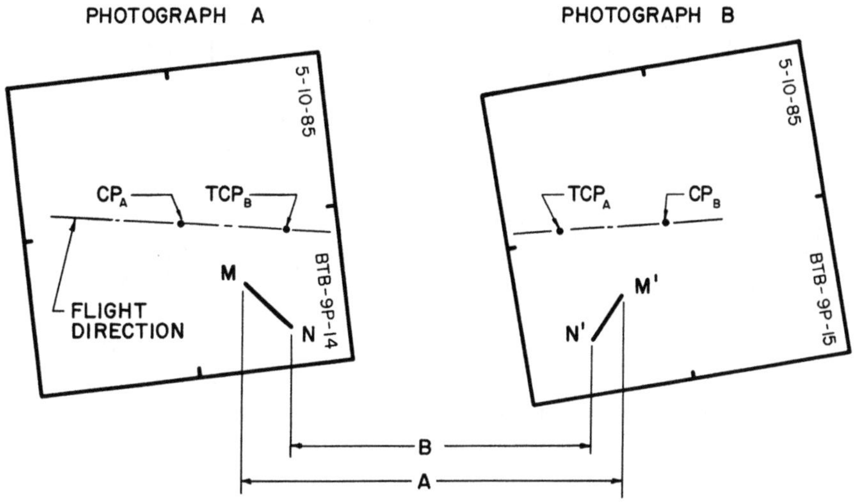

Fig. 4-23. Measuring parallax on a stereoscopic pair. *Note:* The spacing of the photographs is determined by the type of stereoscope used.

where h = height of object (in feet)
H = flying height (in feet)
ab = adjusted photobase (in mm or inches)
ΔP = Parallax difference (in mm or inches).

Parallax difference (ΔP) is A − B in Fig. 4-23 and can be very accurately measured with a very simple photogrammetric instrument called a *stereometer* or a *parallax bar* (see Fig. 4-24). Because the details of the actual procedure of using stereometers will vary slightly with the model, it is not discussed here. The adjusted photobase *ab* in Eq. 4-8 is the distance $CP_A + TCP_B + CP_B + TCP_A/2$ in Fig. 4-19.

Equation (4-8) is a relationship that will produce very reliable measurements when using a parallax bar. It should be mentioned here that a parallax bar is a tool that every landscaper and builder who uses aerial photography should have at his disposal. They are relatively inexpensive, easy to use, and give accurate results. Depending upon the scale of the photography, it is possible to measure elevation difference within a few feet.

Once it becomes possible to measure heights of objects and elevation differences, a number of possibilities open up to the landscaper and builder. For example, it is now possible to calculate the angle of a slope, an often used piece of information, see Eq. (4-9):

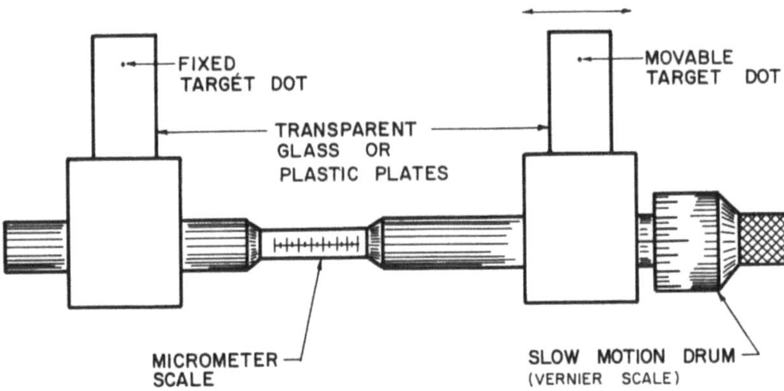

Fig. 4-24. Schematic diagram of a stereometer or parallax bar.

$$\text{tangent of the slope angle} = \frac{H \Delta P}{a(ab + \Delta P)} \qquad (4\text{-}9)$$

where d is the ground equivalent distance, i.e., photo distance/photo scale.

Of course, once the slope angle is determined, it can be converted to % grade, gradient in ft/mile, or any other mode of expression. Thus, if slope angles and horizontal distances can be measured, volumes can also be ascertained by applying data obtained by these simple photogrammetric procedures to equations presented in Chapter 2.

In situations where accuracy requirements are not very great, topographic maps can be made using a stereoscope and parallax bar. Using Eqs. (4-8) and (4-9), many spot elevations situated in key locations can be determined. This method, however, is time-consuming and very laborious, especially for large areas.

5
Aerial Photography as a Surveying Tool

In the last chapter we learned that it is possible to make accurate horizontal and vertical measurements on aerial photographs by following a few simple but rigorous photogrammetric procedures. This chapter attempts now to show how the aerial photographs themselves can be used as maps and how details viewed on the photographs can be accurately transferred to existing base maps. Recall that a map is merely a graphic representation of the earth's surface, whereas on an aerial photograph the actual ground surface can be seen with all its details. These details, such as buildings, trees, and shrubbery whose location may not appear on the map, can be transferred to the map at their correct position.

Please note that in this chapter we are working with *planimetric maps*, i.e., those that show only distance and direction. *Topographic maps*, which show elevations as well, and which require the compilation of abundant vertical control, are much more difficult to produce. They usually require expensive plotting machines which are out of the realm of the landscaper, architect, and builder. Moreover, topographic mapping from aerial photography is only cost-effective when large areas need to be mapped. In general, with small plots on which those in the building trades usually work, topographic mapping with a transit or level and rod is more practical. These techniques will be discussed in a later chapter.

MAP SCALE AND NORTH DIRECTION

Before any meaningful work can be carried out, a scale has to be generated, and a north reference has to be established. While techniques for determining scale were covered in the last chapter, the four simple steps below are used for ascertaining an approximate north reference on an aerial photograph. The only item needed is a map covering the same area as the photograph.

1. Select two points that are easily visible on the map as well as the photograph.

2. On the map, connect the two points by a straight line. Using the map's north arrow determine the azimuth of the line.
3. Locate the same two points on the aerial photograph and also connect them with a lightly drawn line. *Note:* Use a No. 2 pencil. A harder pencil may damage the photo emulsion. The azimuth of the line on the photograph in Step 2 *roughly* equals the azimuth of the line on the photograph.
4. With a protractor, true north can be determined from the azimuth of the photograph line.

If magnetic north is desired on the photograph, simply transfer the magnitude and direction of the declination data from the map or other source to the photograph. As a word of caution, do not attempt to use the north line determined by this method as a reference line for very accurate work. Because of radial displacement due to relief, this is only approximate.

TRANSFERRING PHOTOGRAPHIC DATA TO PLANIMETRIC MAPS

Simple Direct Sketching

The simplest and easiest way to transfer information from a photograph to an existing map is by simple sketching. Details from the photograph are traced onto an overlay of semitransparent tracing paper or mylar. However, in order to maintain a good level of accuracy, the position of a number of ground objects that can be seen on the photograph must first be accurately plotted on the overlay at a scale equal to the scale of the photograph. These are used as guides while sketching. It is important that the horizontal position of these *control points* be known, i.e., their mutual spatial relationship or their horizontal separation must be known. These needed data must be ascertained by scaling from a good map, direct field survey, etc. The control points may be a road intersection, corner of a building, bend in a stream, a large tree, or whatever is convenient and somewhat permanent.

Once the control points have been labeled and carefully plotted to scale on the semitransparent sheet, the overlay is then placed atop the photograph. Its position is adjusted so that the points on the photograph coincide as closely as possible with those plotted on the overlay, and the assembly secured. Please note that it will probably be impossible to get *all* of the control points to coincide with their respective images on the photograph. If this occurs, readjust the overlay so that the control points nearest the area you are tracing or nearest the area that is most critical to you coincide most accurately. This lack of coincidence of the photo and overlay points is in part due to the fact that the photograph may not be truly vertical, and it will be particularly noticeable in areas of high relief.

Mapping with Projection Instruments

The technique described in the previous section produces a map whose scale is equal to the scale of the photograph. Often, however, this may not be satisfactory. By using simple projection devices it is possible to produce maps at any scale and/or, if desired, fill in photographic details on an existing map of any scale. The projection devices described below are well within the wherewithal of most landscapers and should be used routinely in projects requiring aerial photographs.

The projectors are classified into two categories. The first type is based on the camera-lucida principle that allows the user to view the photograph and the base sheet simultaneously (see Fig. 5-1). As described in the previous section, ground control points should be plotted on an overlay. These control points on the overlay are again used as guides during sketching. After as many of the overlay points and photo points as possible are brought together by shifting the overlay around, further refinement can be accomplished by adjusting the legs of the projector. The up and down movement caused by the leg adjustments creates slight scale changes that may, in turn, bring to coincidence additional points.

If very large scale changes are needed, a second type of projector called an *opaque projector* or a *reflecting projector* can be used. The photograph is placed

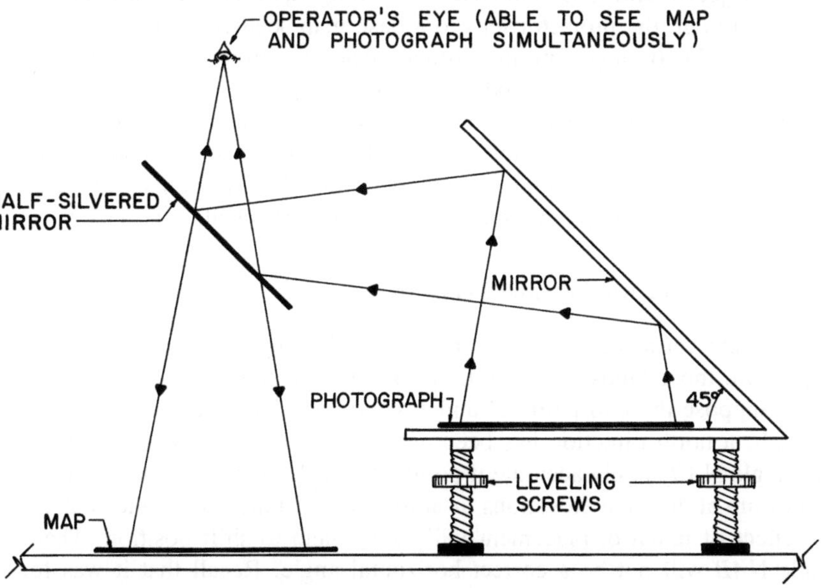

Fig. 5-1. Optical principles of a projector based on the camera-lucida principle. (Redrawn from Wolf, P. R., *Elements of Photogrammetry*, © 1983, McGraw-Hill, by permission.)

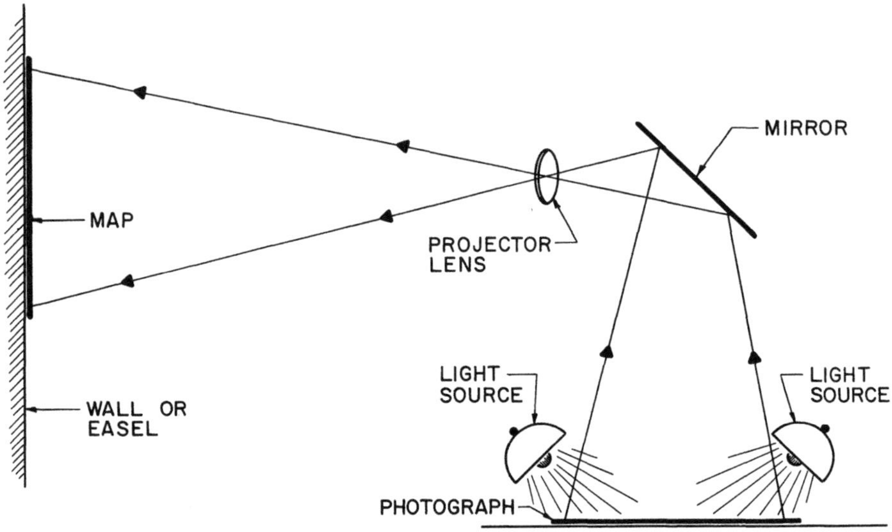

Fig. 5-2. Optical principle of a reflecting projector.

in the projector and is illuminated by a strong light source. Its image is then carried through the projector lens to a base map secured to an easel or a wall (see Fig. 5-2). By adjusting the distance from the projector to the easel, large scale changes can be accommodated. Again, keep in mind that an overlay containing ground control points should be compiled to maintain accuracy. Here, as with the procedures described above, the accuracy of the final map (or existing map revision) is to a large extent determined by the density of the control points, i.e., the more points, the more accurate the results, and vice versa.

The Principle of Radial-Line Triangulation

On a single vertical aerial photograph the only angles that are true horizontal angles are those whose vertices lie at the photograph's center point (see Fig. 5-3). Suppose the two points *A* and *B* have been identified on the aerial photograph, a north direction has been determined. It is important to understand that angle N*AB* is only accurate if points *A* and *B* are at the same elevation. For if they are at different elevations *or* at different distances from the center point, the effect of radial displacement will cause them to shift position. Therefore, angle N*AB* will not be a correct horizontal angle. Recall that it was for this reason that the north-line determination described in a previous section was only approximate, and should not be used where good accuracy is required. Angle *A*CP*B* on the other hand, is an accurate horizontal angle.

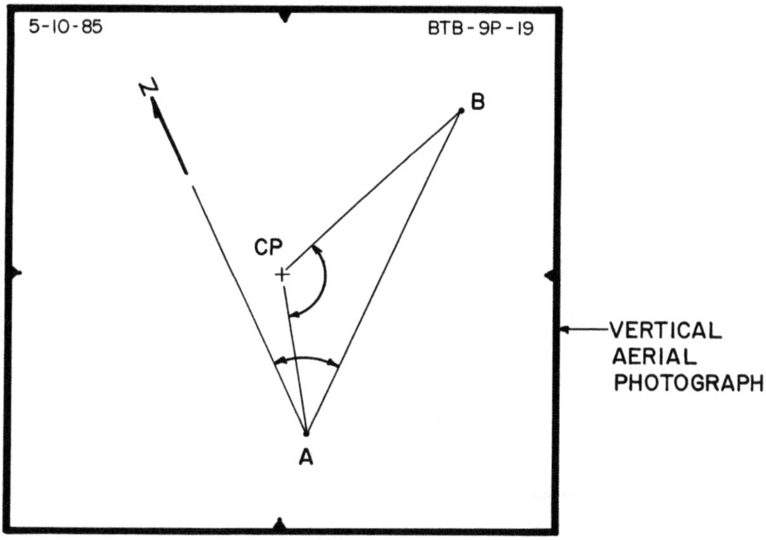

Fig. 5-3. On this vertical photograph, angle *ACPB* is a true horizontal angle.

This principle can be said to be scale independent. It is not affected by variations in flying height, size of atmospheric distortion, etc. Hence it is quite useful for accurate and direct positioning of points on the photograph and subsequent transfer of points to maps.

Two-Photograph Radial Triangulation

Radial-line triangulation can be quite useful to the landscaper and builder particularly when a limited number of features need to be accurately plotted. The procedure is quite simple and accurate and requires no special equipment. It is not recommended, however, if more than one stereoscopic pair is involved.

Follow closely Fig. 5-4 and the following steps. Keep in mind that the highest possible accuracy and care is required.

1. Find the center points and transposed center points of each photograph of the stereoscopic pair. If needed, refer back to the instructions in the previous chapter.
2. Align the photographs atop the map so that flight line (the line connecting CP_A-TCP_B-TCP_A-CP_B) is *absolutely* collinear. Be aware that the separation of the photographs determines the scale of the final map, the greater the separation the greater the scale.
3. Once collinearity of the flight line has been established and the desired separation is attained, secure the photographs. They must *not* be moved until the work is completed.

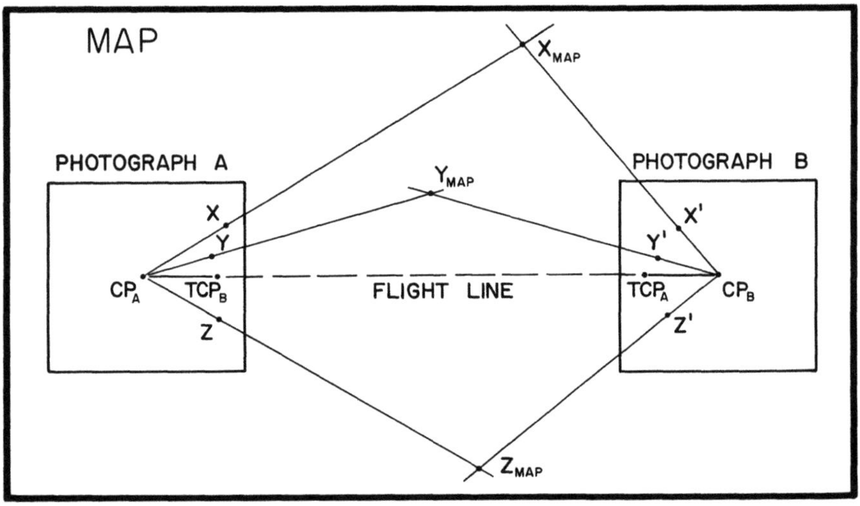

Fig. 5-4. The principle of two photograph triangulation. (From Lattman, L. H., personal communication, by permission.)

4. To determine the map position of any point(s) that appear(s) on both photographs, draw a line from the center point through the point on the photograph onto the map. The lines from both photographs intersect at the correct map position.

Paper Strip Triangulation

The paper strip method of triangulation is yet another simple procedure that can be used to transfer points accurately from a vertical aerial photograph to a base map. It requires only one photograph, a small strip of paper, and the final base map onto which the information is to be plotted. This procedure can be very valuable in the building trades because many, if not most, of the project areas are small enough to be visible on a single photograph. Follow the procedure below while studying Fig. 5-5. Suppose we want to transfer points *A* and *B* at their correct planimetric spacing to a map:

1. Mark the same four readily identifiable and random points on the map *and* photograph. Although the points can be any object anywhere on the photograph and map, ideally they should be near the corners.
2. Secure a paper strip across the photograph as shown in Fig. 5-5.
3. On the photograph, draw lines from known point 1 to known points 2, 3, and 4. *Also* draw lines from point 1 to points *A* and *B*. (These points, of

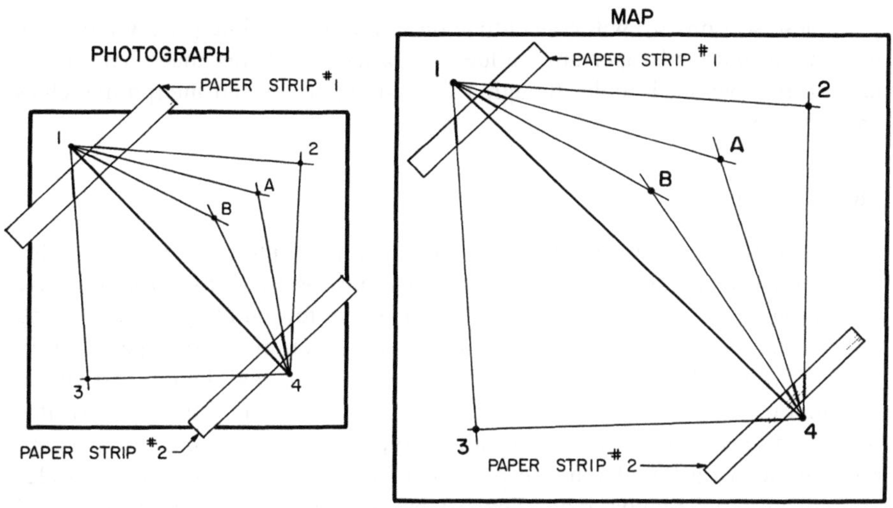

Fig. 5-5. Illustration of the paper strip method of triangulation. (From Lattman, L. H., personal communication, by permission.)

course, do not appear on the map.) Make sure that all the lines cross the paper strip, and if there are a large number of them, they should be labeled to avoid confusion.

4. On the map draw lines from point 1 to points, 2, 3, and 4.
5. Transfer the paper strip to the map and slide it around until lines 1-2, 1-3, and 1-4 coincide. Then secure the paper strip to the map.
6. Now draw lines on the map across the paper strip to points *A* and *B*.
7. Repeat the procedure from point 4. The lines for points *A* and *B* cross the map at their correct planimetric position.

OTHER USES OF AERIAL PHOTOGRAPHS

Before embarking on a discussion of some of the techniques of aerial photographic interpretation, a few comments should be made regarding the process of interpretation. *Interpretation* is defined as the process of logically assessing the meaning or sense of objects recognized on photographs. Aerial photographs exhibit an overall view of an entire area which allows the user to see features, patterns, and trends which cannot be seen on the ground, and they permit views from an entirely different perspective. Because of this they can, within limits, be an effective tool in planning, identifying existing or potential hazards, decision-making, etc. It is, however, sufficiently noteworthy to once again state that work on aerial photography is no substitute for field work.

In these sections we will concentrate on a few areas of interpretation that are germane to the landscaper and builder. No attempt is being made here to exhaustively cover each topic, though a few salient points and interpretive clues are presented.

Slope Stability

It is estimated that damage from all types of slope failures in Canada and the U.S. annually exceeds a billion dollars or more. Some areas, of course, are more prone than others, but it is safe to assume that few building or landscaping projects can be conducted without some consideration given to slope stability. Aerial photographs can be useful in site investigations both before and after a potential problem has been identified. Although it is beyond the scope of this book to discuss the mechanics of various types of earth movements, it is appropriate to list a few diagnostic patterns of landslides or potential landslides viewed on aerial photographs. The list below is modified from Shuster and Krizek (1978); note the figures indicated in the list:

1. Land masses undercut by streams, road cuts, railroad cuts, or building excavations (see Fig. 5-6)
2. Steep slopes having large masses of loose soil and rock
3. Sharp line-of-break at the scarp (up slope end) or the presence of tensions, cracks, or both (see Figs. 5-6 and 5-7)
4. Hummocky surface of the sliding mass downhill from the scarp
5. Unnatural topography, such as spoon-shaped trough in the terrain (see Fig. 5-8)
6. Seepage zones (see Fig. 5-9)
7. Elongated undrained depressions
8. Closely spaced drainage channels
9. Accumulation of debris in drainage channels or valleys
10. Appearance of light tones where vegetation has not been reestablished (see Figs. 5-8 and 5-9)
11. Abrupt change in photograph tones from lighter to darker, the darker tones indicating higher moisture content (see Fig. 5-9)
12. Distinctive and abrupt changes in vegetation indicative of changes in moisture (see Figure 5-10)
13. Inclined trees, headstones in cemeteries, utility poles, fence posts, and walls due to creep

As can be gathered from the above list, sometimes the slide itself may not be visible, but must be inferred from clues seen on the photographs.

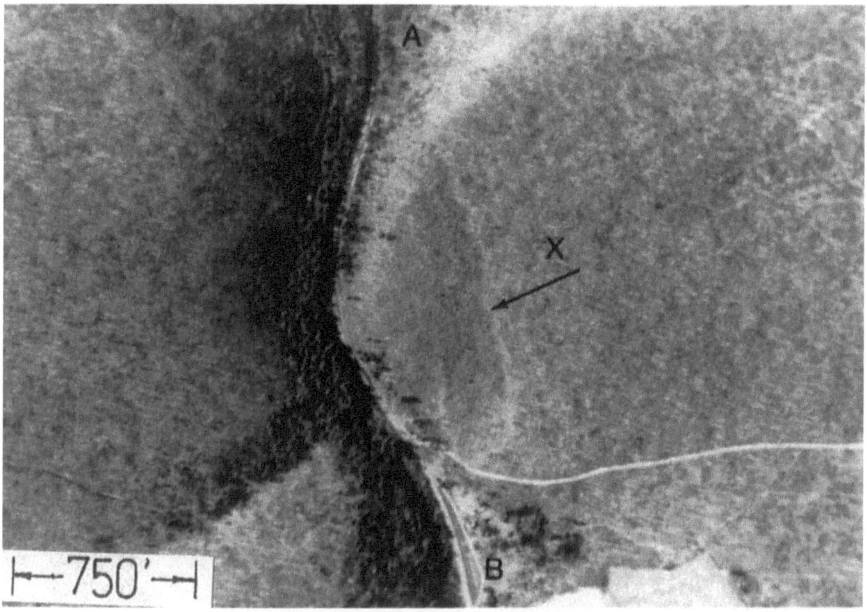

Fig. 5-6. The stream and road cut (between *A* and *B*) have under cut a landmass which has moved slightly creating a small scarp at *X*.

Fig. 5-7. Tension cracks at the upslope end of a small landslide block. Note their location at *X*.

Fig. 5-8. Spoon-shaped depression in the terrain, and difference in vegetation indicating a small

Fig. 5-9. Note the abrupt change in soil tone along the points of the arrows. This indicates a line of springs (a seepage zone), the largest of which is at *X*.

Fig. 5-10. Abrupt change in vegetation at the edge of a slide.

Fractures in Bedrock

Fractures in bedrock can have an important effect on construction activity. This is true because fractures of all types in bedrock tend to break rock material into smaller pieces which renders it structurally weaker. Fracturing also creates avenues of percolation for surface and groundwater. Hence the fractures are frequently the controlling factors in slope movement subsidence, seepage and percolation of water, etc., so their location must be identified and their significance evaluated if they are within, or close to, a project area.

Fractures can be classified into two main categories. First, a *fault* is defined as a rock break along which there has been some lateral movement or displacement. The second type of fracture is termed a *joint*; this is a break in the rock along which has been no lateral movement, i.e., merely a crack or parting in the rock. Joints are more numerous than faults and can be especially troublesome when large numbers are concentrated in a small area. In any construction project where the bedrock needs to be sound, and where large numbers of joints and/or a fault(s) are encountered, a competent geologist or engineer should be consulted. Joints and faults rarely affect directly small landscaping projects, as these projects are usually restricted to the soil and to the unconsolidated weathered mantle which sit atop bedrock.

Fractures are almost always expressed as linear features on aerial photographs. It should be noted, however, that the fractures that readily show up on vertical photographs are the ones that dip at a high angle or are vertical. Large numbers of faults and joints are inclined at shallow angles, and hence are not exhibited as linear features. Of course, for these field work is essential. It should also be pointed out here that fractures range in size from very large-scale features many miles in length to small-scale ones measuring only a few hundred yards in length.

The linear features which indicate some sort of fracturing may be expressed in a number of ways and are listed below. Be careful, however, when looking for faults and joints that you are not confused by linear features that are the result of human activity, e.g., fence lines, buried pipelines, property boundaries, plowing furrows, abandoned roads or trails. The fracture identification list below is modified from Ray (1960); note the figures referenced in the list:

1. Linear tonal changes (either light to dark or vice versa) (see Figs. 5-11 and 5-12)
2. Alignment of vegetation
3. Straight segments of streams (see Fig. 5-13)
4. Alignment of ponds, swamps, springs, sink holes, etc.

Fig. 5-11. Linear tonal change in the soil indicating the surface expression of a fracture in the bedrock.

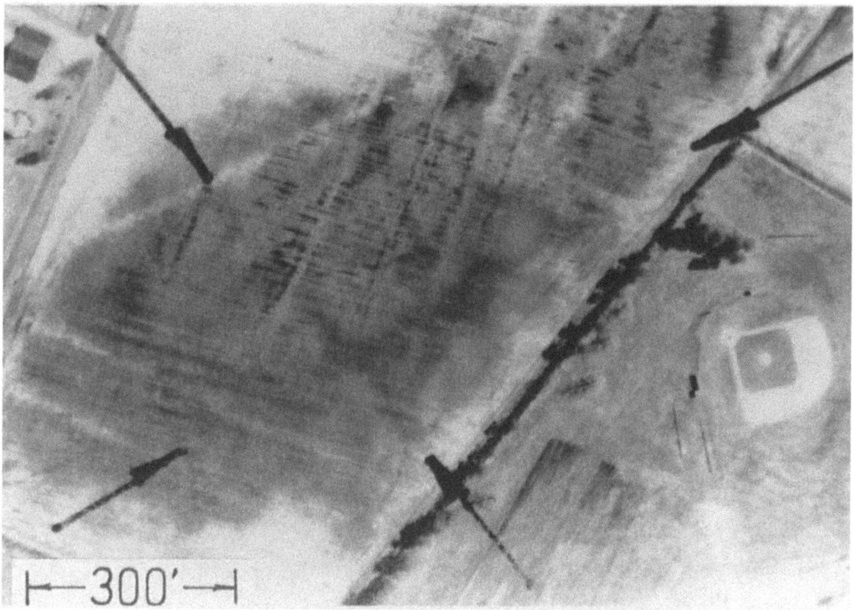

Fig. 5-12. Intersecting linear features in the soil indicating intersecting fractures in the bedrock below.

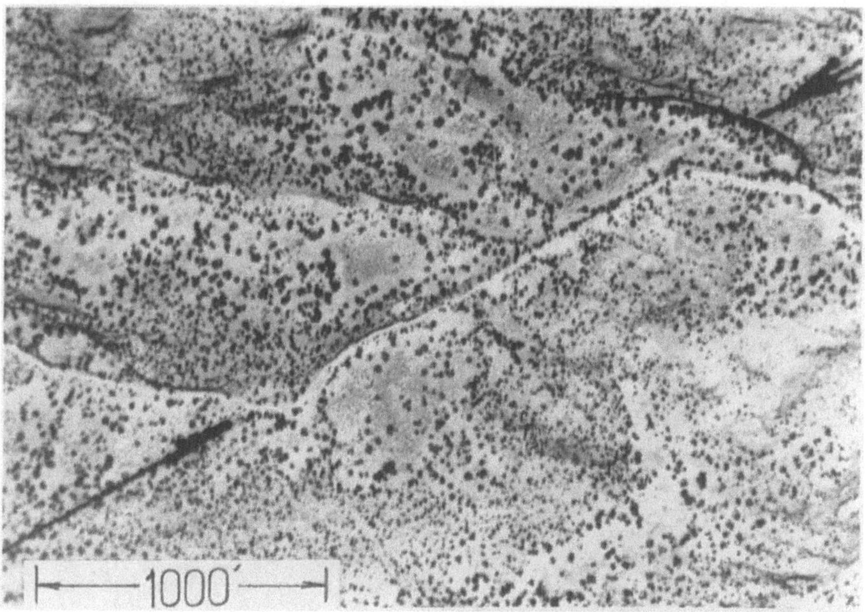

Fig. 5-13. The straight stream segment seen between the arrows is likely due to fracture control.

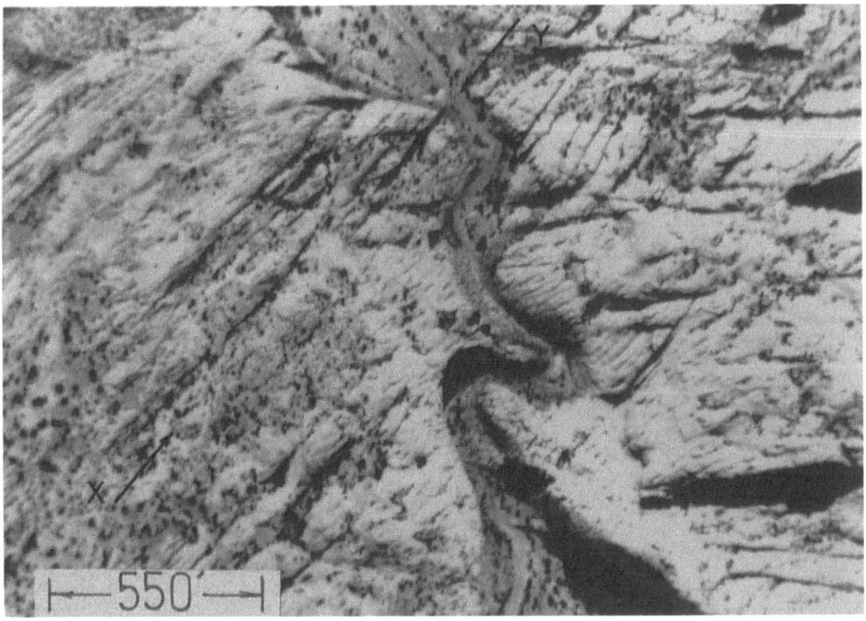

Fig. 5-14. Joints in the exposed bedrock are easily visible oriented parallel to the arrows *XY*.

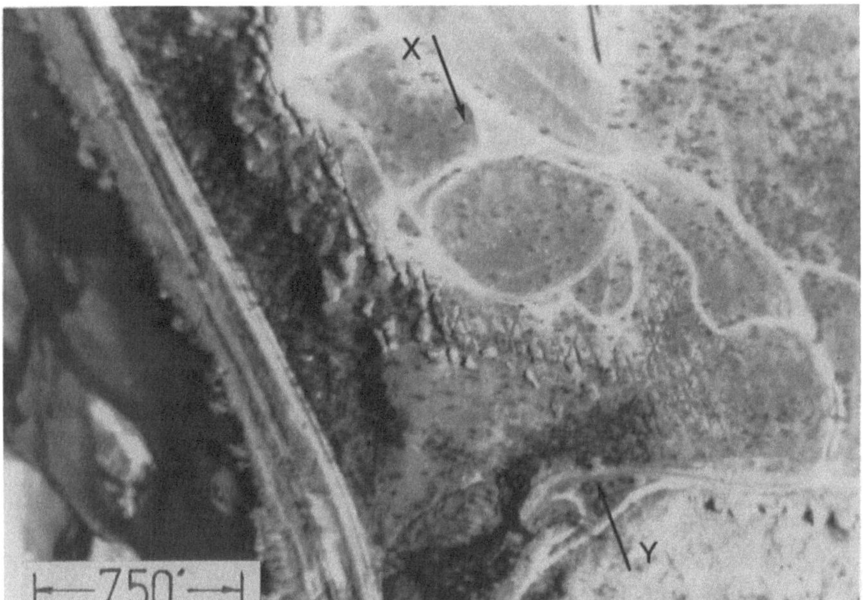

Fig. 5-15. Joints in the bedrock are visible roughly oriented parallel to the arrows *XY*.

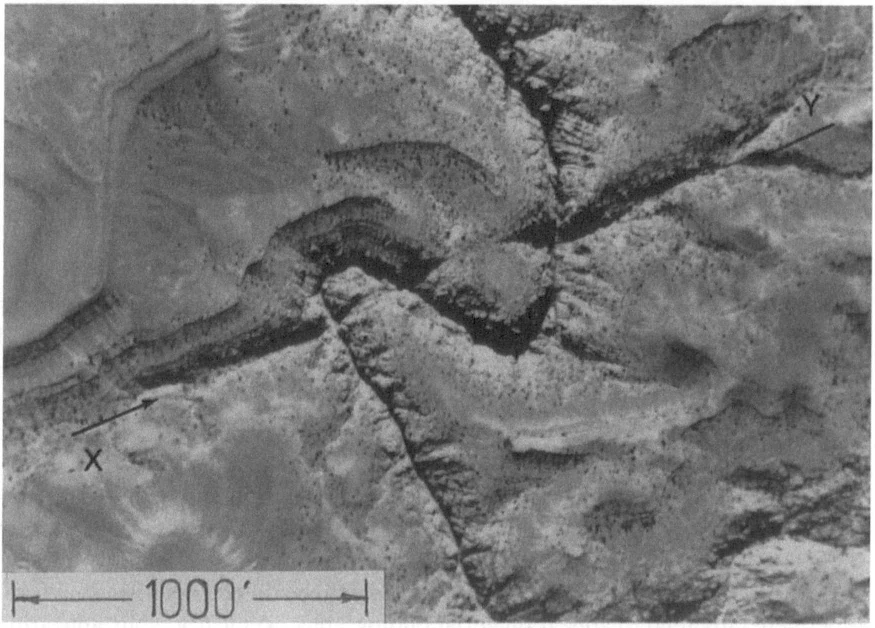

Fig. 5-16. Note the reversal of the stream course as it crosses fault *XY*.

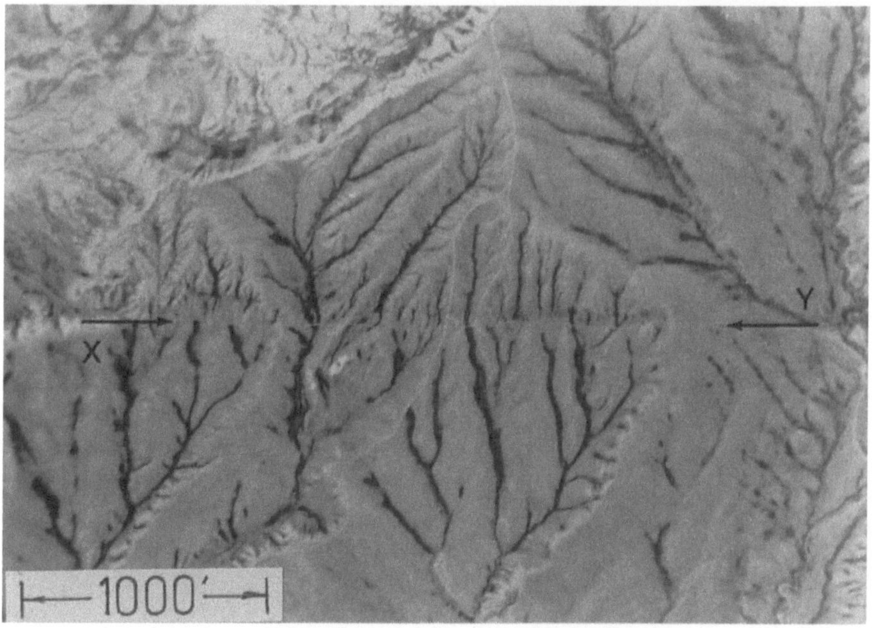

Fig. 5-17. A slight break in slope is visible along line *XY*.

5. Occasionally joints themselves are actually visible (see Figs. 5-14 and 5-15)
6. Local reversal of a stream course (see Fig. 5-16)
7. Slight break in slope (or waterfall in streams) (see Fig. 5-17)
8. Offset stream or gully channels

Soils

As with the case of all photo interpretation work, soil studies on photographs does *not* eliminate the need for field work. While we can make some general assessments regarding soils seen on aerial photographs, it is very difficult to accurately identify soil materials because they are so numerous and variable even within a very small area. If possible, and where available, builders should first closely consult soil surveys published by the U.S. Department of Agriculture for the county in which the work is being done. Although county soil surveys are primarily for agricultural purposes, the soil descriptions contained in them can be valuable and applicable to many disciplines.

We will examine here a few of the ways in which soil characteristics and soil types generally tend to appear on aerial photographs. Typically, soils appearing

Fig. 5-18. Dark soil tones indicate a high moisture content. Note drainage channels draining water from the wet area.

MATERIALS	CROSS SECTIONS	PROFILES
PLASTIC OR COHESIVE MATERIALS (CLAY, SILT WITH HIGH ORGANIC CONTENT) SHALE CLOSE TO SURFACE POSSIBLE.	 GULLIES ARE SHALLOW WITH SMOOTH-ROUNDED SHOULDERS & BOTTOMS.	 GULLIES ARE LONG (APPROX. I MILE) AND GRADIENTS ARE UNIFORM AND VERY GENTLE.
SILTS, SANDY CLAYS, POSSIBLY VOLCANIC ASH OR POSSIBLY POORLY CEMENTED SANDSTONES MODERATE-COHESIVE MATERIAL.	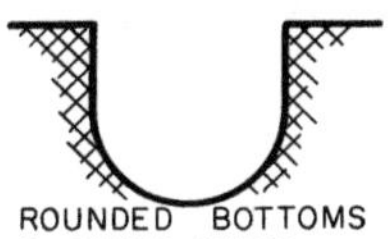 ROUNDED BOTTOMS & STEEP SIDES.	 GULLIES SHOW COMPOUND GRADIENTS AND WELL DEFINED SLOPE CHANGE AT HEAD, SEE A.
NON-COHESIVE GRANULAR MATERIAL, SIZE OF MATERIAL SHOWS GRADATION FROM COARSE (GRAVELS) TO FINE (SANDS).	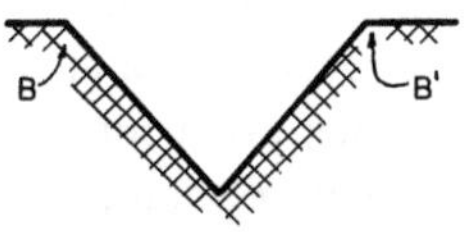 V-SHAPED GULLIES, WELL DEFINED EDGES, SEE B-B'.	 GULLIES ARE SHORT, STEEP AND EXHIBIT A UNIFORM GRADIENT.

Fig. 5-19. Gully forms typically associated with various types of surficial deposits. (Modified from Strandberg, C. H., *Aerial Discovery Manuel*, © 1967, John Wiley & Sons, by permission, and from Lattman, L. H. personal communication, by permission.)

in dark photographic tones are wetter and have a higher water retention capability then those of lighter tones (see Fig. 5-18). From this, the drier, well-drained soils are likely to be composed of a high percentage of sands and/or gravels, whereas the more moisture-laden soils are apt to be composed of clays. Frost (1946) indicates that soils with a mottled appearance in his study area in Indiana are almost always composed of gravel.

Frequently, soil types exhibit a set of landforms that can be seen and identified on a photograph. There if these landforms can be recognized, a corresponding soil type can be inferred. The landforms that are most sensitive to variations in soil material are the headwater channels of streams, sometimes referred to as gullies. Fig. 5-19 shows various gully forms that tend to be associated with various types of surficial deposits.

Please keep in mind that because so many variables enter into soil formation, the general guidelines mentioned above cannot be transposed to all places and to all situations. They are meant to give the interpreter and overall indication of what an unidentified soil might be.

6

Using Your Own Aerial Photography

To this point we have been using existing, commercially available aerial photographs. This is photography done by government and private companies for routine mapping, crop and lumber inventories, road construction, and planning. However, in some situations available existing photography is unsatisfactory. The most common problem is that photographs are taken from too high an altitude to be useful for small project areas. So if photography is to be used, the only alternatives open to the landscaper and builder are to enlist an aerial survey company to take the needed photographs, or to take their own photographs. In this chapter, we will examine both alternatives, and discuss factors that need to be considered in each undertaking.

PHOTOGRAPHS FROM COMMERCIAL AERIAL SURVEYS

Scale and Coverage

There are many considerations that need to be reviewed when ordering aerial photography from a commercial aerial survey company. Though many of these considerations are the concern of the professional photogrammetrist, the user of the photographs *must* be very clear in communicating to the photocontractor the purpose of the photography, the scale required, ground coverage and whether stereoscopic coverage is needed.

If the photographs are going to be used to transfer details to an existing base map, the photo scale needs to be chosen carefully. The scale of the photographs should be determined by the intended scale of the final map. The scale change capability of available sketching projectors will control the allowable variation between the photoscale and the final map scale. Note that in order to maintain good accuracy when using projection-type sketching instruments, enlargements should not exceed five or six times. Thus, an ideal photoscale should be no more than about one-fifth of the intended map scale. For example, a final map scale of 1″ = 50′ (1:600) necessitates a photoscale no smaller than about 1″

= 250′ (1:3000). Given the equipment at his disposal the contractor would then determine the flying height for the job.

On the other hand, if detailed measurements are to be made directly on the photographs, then request a scale which permits the smallest objects of interest to be easily seen and measured. It is difficult to accurately measure objects whose image is smaller than 1 mm unless some magnification and specialized measuring tools are used.

The area to be covered by the photographs should be accurately indicated to the photocontractor so that the flight path and the number of photographs needed can accurately be determined. As an illustration of how to determine the required number of photographs be aware that in usual photographic coverage for stereoscopic viewing there is a 60% forward overlap along the flight line and a 30% side lap with respect to the adjacent flight line. In view of this, the total *new* area covered by each photograph is smaller than the area covered by a single photograph standing alone. The new area per each photograph can be determined using Eq. (6-1):

$$A_n = A_p(100 - O_f)(100 - O_s) \tag{6-1}$$

where: A_n = new area on each photograph
A_p = total area on each photograph
O_f = forward overlap
O_s = side lap

With this in mind, suppose a large airport construction and landscaping project is to be undertaken, and it covers an area of 2.94 sq. miles. If each photograph covers an area of 0.75 sq. mile (given the chosen scale), how many photographs will be needed? From Eq. (6-1):

$$A_n = (0.75)(0.40)(0.70) = 0.21 \text{ sq. mile}$$

If an area of 2.94 sq. miles needs to be covered, the total number of photographs required for the project is:

$$14 \text{ photographs} = \frac{2.94 \text{ sq. miles}}{0.21 \text{ sq. mile/photograph}}$$

The 14 photographs calculated here are needed for complete stereoscopic viewing of the entire project area. If stereoscopic viewing is not required, then half the number will suffice. It is not advisable, however, to give up stereoscopic coverage because of the small additional cost.

Ground Control

Once the scale and coverage have been decided upon, the next step for the user is to establish ground control. *Ground control* consists of any points whose ground positions are accurately known with respect to each other and/or to a reference datum, *and* can be easily seen on an aerial photograph. The points may be corners of buildings, fence corners, utility poles, manhole covers, road intersections, sidewalls, fire hydrants, etc. Where few existing features are available, artificial points or targets can be placed on the ground at strategic locations (see Figs. 6-1 and 6-2). For most low-altitude photography that would be useful to the landscaper and builder, inexpensive materials can be used to construct targets. Plywood, canvas, stones, and paint (on paved areas) all work well. Old automobile tires painted white and placed atop the control points are particularly good as they are durable, very inexpensive, and reusable.

Because landscapers and builders very rarely get involved in constructing topographic maps from aerial photography, vertical control (the elevation of the points) is seldom necessary. *Horizontal control*, the correct horizontal spatial relationship of the points on the ground and on the photographs, is usually all that is needed. The reference for the ground control points originates from ex-

Fig. 6-1. An example of an artificial photogrammetric target painted on asphalt pavement. Aside from surveying and photogrammetric considerations, targets should be placed in favorable locations for viewing from the air. Note the glasses for scale.

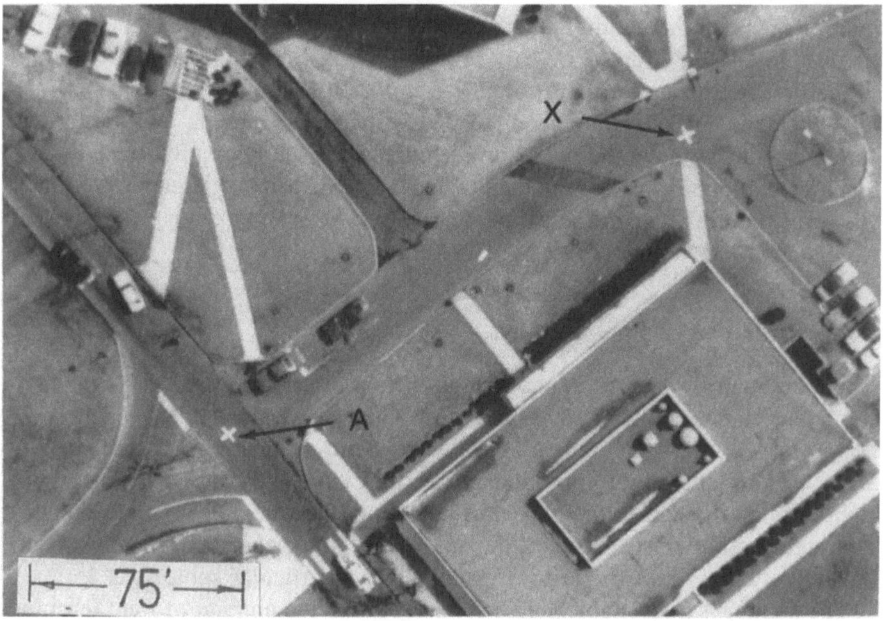

Fig. 6-2. Examples of artificial targets viewed from the air. They are similar to the one shown in Fig. 6-1.

isting reliable control stations. It is from these primary control stations that other subordinate or secondary points are located by conventional surveying techniques (see Chapter 9). It is important that the ground control points be accurately and carefully determined, for their accuracy or lack of it will be transferred to the photographs and subsequently to maps and plots.

For simple problems such as scale determination only two widely spaced points per photograph are needed. However, for sketching planimetric maps, five or six points per photograph should be adequate. If possible, the points should be close to the same elevation so as to minimize the effect of radial displacement and also evenly and widely spread throughout the project area.

TAKING YOUR OWN PHOTOGRAPHS

In some instances it may not be necessary or practical to hire a photocontractor for needed aerial photography. It can be done by anyone with a good hand-held camera and access to an aircraft. Preferably the camera should have the capability of adjustable exposure settings to accommodate variations in light conditions. Most 35 mm cameras satisfy this requirement. The only aircraft requirement is that the photographer should have a good unobstructed view of the ground; low-wing aircraft, i.e., those whose wings are situated on the bottom of the fuselage are not usually suitable or practical.

While aerial photography acquired with a hand-held camera in a light plane certainly will not have the precision of professional photography, nevertheless much valuable information and many measurements can be gleaned from them. However, before doing the actual photography, a number of principles and some techniques need to be discussed.

Some Principles of Photography

There are many variables in photography. Some can be controlled by the photographer, some cannot. Perhaps the most important controllable variable is the exposure setting of the camera. Because the film's sensitivity to light is constant, and because light conditions vary from subject to subject, the flow of light through the lens to the film must be controlled accurately. This is done by altering the shutter speed and adjusting the lens aperture. These adjustments collectively determine the exposure setting for a particular subject.

The shutter is a mechanical device that controls the amount of light reaching the film by the amount of time that it remains open. Most shutters can be set to open for 1 second, 1/2 second, 1/4 second, 1/8 second, 1/15 second, 1/30 second, 1/60 second, 1/125 second, 1/250 second, 1/500 second, and 1/1000 second.

Operating in conjunction with the shutter is the aperture of the lens. The lens aperture varies like the pupil of an eye, enlarging to allow more light to enter and contracting to restrict the flow of light. The size of the aperture is measured on a scale of numbers called *f-stop* numbers. The larger the *f*-stop number, the smaller the lens opening and vice-versa (see Fig. 6-3).

It is possible then to allow the correct amount of light to pass through the lens to the film by using various combinations of shutter speed and *f*-stops. The proper combination (there can be several correct combinations for a particular scene) can be determined by means of a light meter. A *light meter* is simply a small electronic device that measures the amount of light being reflected from the subject. The meter is calibrated to convert the reflected light to suggested *f*-stops and shutter speeds. Most modern 35 mm cameras have light meters built into the camera body, while some meters are hand-held, self-contained units and can be used independently of the camera.

In general photography, the camera must also be focused on the subject to get a clear, crisp image. However, in aerial photography the distance from the camera to the ground is great enough that the camera can be prefocused to infinity and left at that setting during the flight.

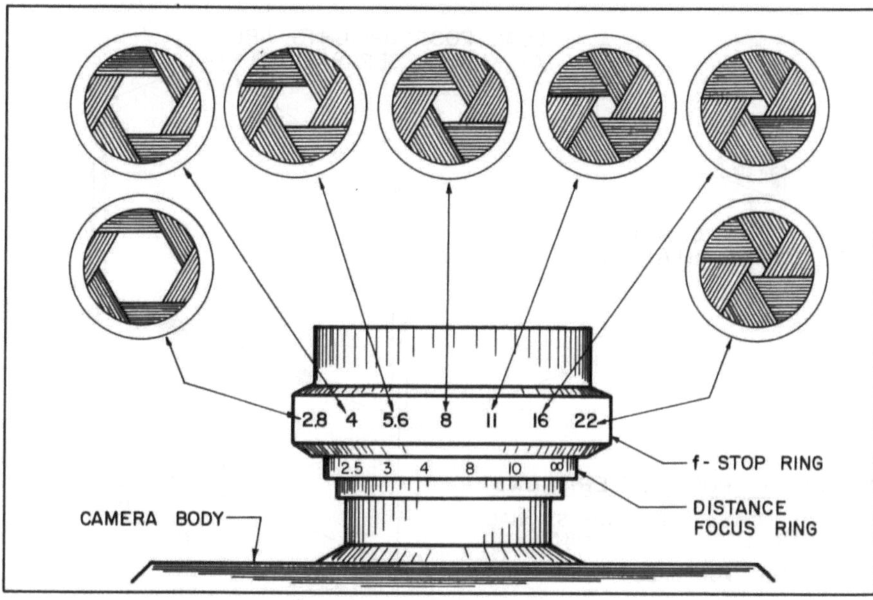

Fig. 6-3. The relationship between *f*-stops and lens aperture. (Redrawn from *Photography for Visual Communicators*, © Lynne Bentley-Kemp, by permission.)

Camera/Lens Calibration

You are undoubtedly aware that one of the elements needed in making measurements on an aerial photograph where there is no ground control is the focal length of the camera. If in your work you find it useful and necessary to take your own photographs, you will need to determine the focal length of your own lens and camera.

A simple procedure for doing this is to take a photograph of an object of known size and use the relationship represented by Eq. (4-1) from Chapter 4 to calculate focal length. The known object can be anything that can be easily and accurately measured, e.g., the distance between two utility poles, the width of a house, a front porch, window, etc. Substituting in Eq. (4-1), the size of the object (O) in our example here is the distance between, say, the upright of a goal post on a football field; and the flying height (H) is the distance from the camera to a line connecting the two uprights (see Fig. 6-4). The size of the image (i) can be measured directly on the negative or on a contract print. Note that if an enlargement is used, the size of the object must be divided by the amount of the enlargement. Using Eq. (4-1) and solving for focal length (f)

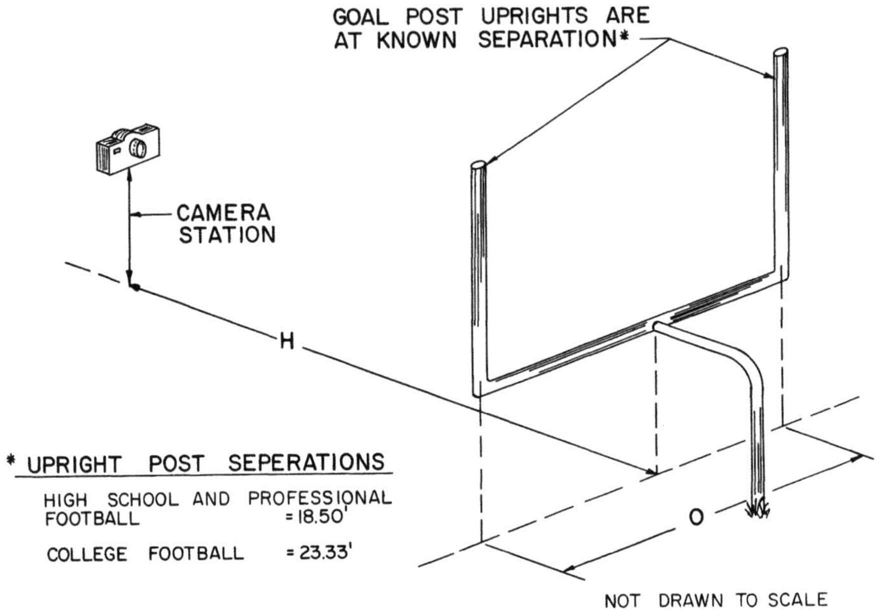

Fig. 6-4. Determination of the camera focal length using the arrangement shown above.

we get:

$$f = \frac{Hi}{O} \tag{6-2}$$

Be aware that what has been determined here is the calibrated focal length for a particular lens and camera. If a different lens is employed, then the procedure must be repeated for the new one. A word of caution here about using variable-focal-length "zoom" lenses. Because these lenses have an infinite variation in focal length they should be used for photogrammetric purposes only if the focal length is preset to a calibrated value and held there for the entire photomission.

The so-called wide-angle lenses where the field of view is very wide should not be used because of associated image distortion. Other than these two caveats, most "normal" 55 mm or mild telephoto (135 mm) lenses can be used.

AERIAL PHOTOGRAPHS WITH A HAND-HELD CAMERA

Photography from a small aircraft, though simple and useful, does require a few techniques and procedures that are not necessary with ground-based photography. The following are suggestions that should help in providing high-quality photographs. Depending on conditions, aircraft, and camera equipment, variations of these suggestions may prove useful.

First, all flights should be made under ideal weather conditions when there is no cloud cover, visibility is greater than 12 miles, and when atmospheric turbulence is at a minimum. Minimum turbulence is necessary to ensure a stable camera platform.

Second, the photographs should be taken through the window of the aircraft perpendicular to the flight direction. In order to eliminate glare, it is advisable to hold the camera lens as close to the window as possible without touching it. Contact with the window should be avoided in order to minimize camera motion due to vibration of the aircraft. Ideally, a soft rubber lens hood should be placed on the lens; this will help against glare, and if the window is accidentally bumped the camera will not be jarred significantly.

Third, all of the photographs should be taken with the camera pointed away from the sun. Moreover, to emphasize very slight breaks in slope that may be useful to the landscaper, the flights should be made when the sun is not too high in the sky, i.e., around mid-morning or mid-afternoon.

Fourth, occasionally replication of a particular photograph is required either to employ a different film/filter combination or to retake a poor quality photograph. To replicate an aerial photograph exactly, the aircraft must reoccupy the same position where the first photo was taken. This can be accomplished in the following manner. Subsequent flights must be made at the same time of day so that the angle of the sun above the horizon is approximately the same. With the pilot keeping the aircraft at exactly the same altitude as the previous flight, all that is necessary is to place the shadow of the aircraft (easily seen from the air) over a designated landmark at each pass over the target (see Fig. 6-5).

Film and Filter Selection

Yet another variable that can be controlled by the photographer is the selection of an appropriate film and filter. Photographic films, although sensitive to nearly all visible light, are particularly sensitive to certain portions of the spectrum. Infrared film, for example, is more receptive to radiation in the red end of the visible spectrum than it is to radiation in the middle and blue portions. Hence, objects reflecting in the red portion of the spectrum will be emphasized with respect to objects of other colors.

Photographic filters, on the other hand, readily transmit light in one part of the spectrum and suppress or exclude light from other parts. The filters themselves are simple pieces of colored glass that are attached to the front of the camera lens. The degree to which a filter will transmit radiation in a particular part of the spectrum is usually expressed as a percentage of the total transmis-

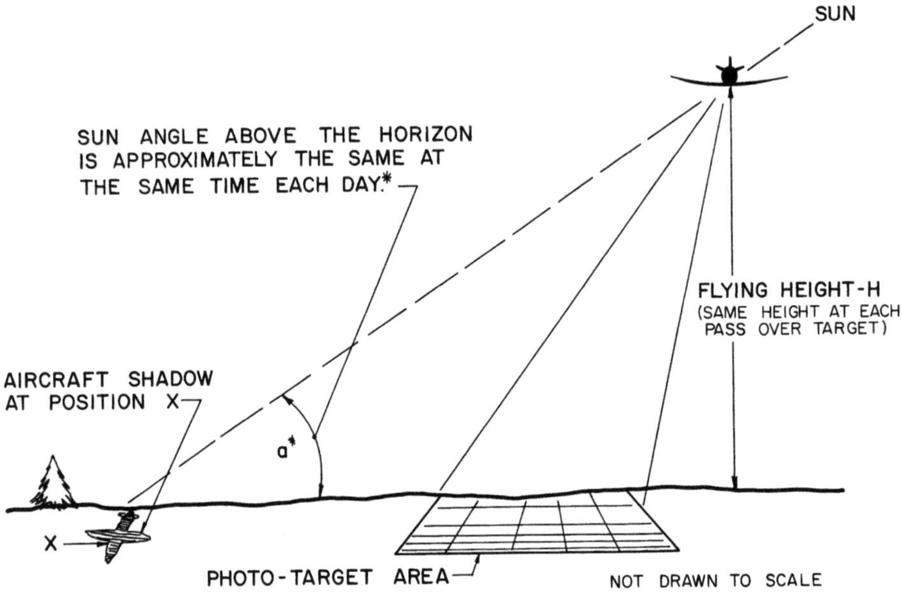

Fig. 6-5. The position of the aircraft in the air can be replicated by keeping *H*, *a*, and *X* the same at each pass over the target. This angle is the same at the same time each day only within a several week span of time.

sion. So by combining the spectral sensitivity of the film and the percent transmissibility of the filter, narrow portions of the spectrum can be emphasized to enhance particular subjects. For example, if a project area has several reddish soils that need to be distinguished from one another, a film-filter combination should be chosen that accentuates that end of the spectrum. Fig. 6-6 shows filter transmittance curves for several filters. Note how selected portions of the spectrum can be emphasized.

From the author's experience, few if any generalizations can be made concerning optimal film-filter combinations employed in field work. Each project area has its own set of circumstances and factors which may require the use of film-filter combinations particularly adapted to the problem in question. General recommendations are that the spectral expression of the subject should be first studied. Film-filters that enhance the portion of the spectrum in which the important features of the subject should be used. One researcher (Fisher 1962) concluded that color photography almost always contains information that is not recorded on conventional panchromatic photography. So this option should also be considered. His conclusion is based on studying geologic subjects.

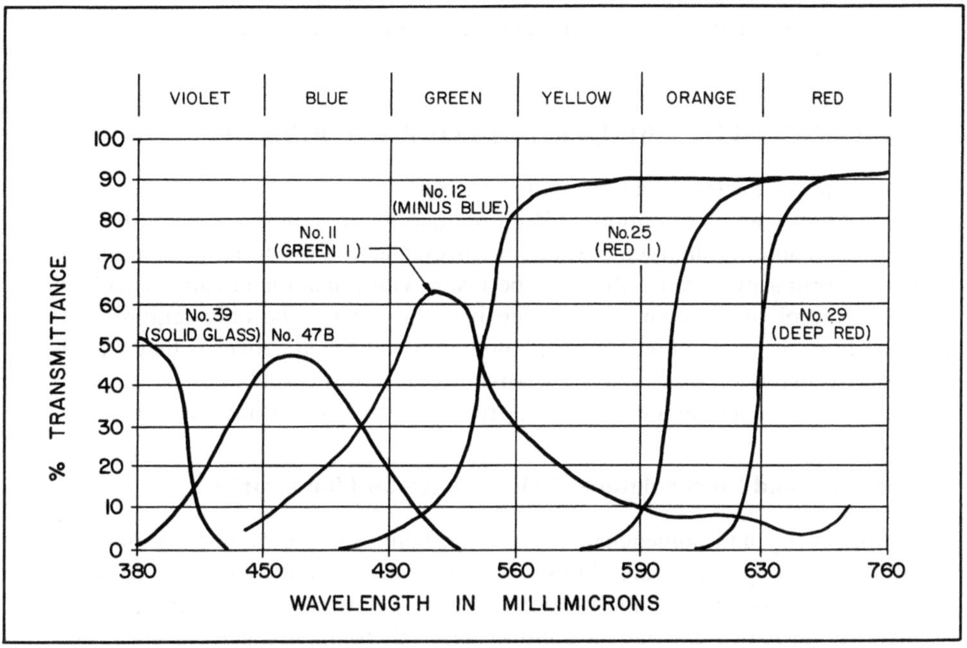

Fig. 6-6. Light transmission of various optical filters. (Redrawn from *Kodak Filters for Scientific and Technical Uses*, © 1985, Eastman Kodak Co., by permission.)

Exposure Settings

In work done by the present author it was discovered that correct exposure readings can be established and set *prior* to taking off, thereby eliminating some work while airborne. The procedure used employs a light meter reading taken away from the sun and in a grass-covered area before taking off. It was presumed that the grass would have approximately the same reflectance as the overall terrain observed from the aircraft in flight. Exposure readings were not taken while airborne, as atmospheric scattering tended to give incorrect readings.

In a test procedure, a number of representative test areas were chosen and photographed. For each area, three photographs were taken in succession, one at the exposure value obtained from the light meter, and at two other exposures, one *f*-stop above and one below this value. The shutter speed was not changed. No appreciable difference was seen in the resulting prints. It can be recommended from this that most aerial photographs taken with a hand-held camera can rely on exposure readings obtained from grassy areas immediately prior to take-off. Note, however, airborne exposures still should be taken periodically as a check.

WORKING WITH OBLIQUE AERIAL PHOTOGRAPHY

Recall from Chapter 4 that there are two types of oblique aerial photographs, high obliques and low obliques. High oblique photography, on which the horizon is visible, is the type that can be of greater use to the builder and landscaper, primarily because they can be taken with a hand-held camera without strict control requirements and expensive equipment. The only requirement, aside from the principles of good photography described above, is that the horizon be visible on each exposure. Also, because of complexities in scale determination on oblique photographs, ground control targets are recommended.

Geometry and Nomenclature of Oblique Aerial Photography

The geometry and nomenclature described here, and the procedures outlined later for plotting points, are almost verbatim from Lattman and Ray (1965) and Rich (1947) with some modifications. Fig. 6-7 illustrates some general terms that are needed to understand oblique aerial photography. It is strongly suggested that you follow Figs. 6-8, 6-9, and 6-10 closely while studying the geometry described here.

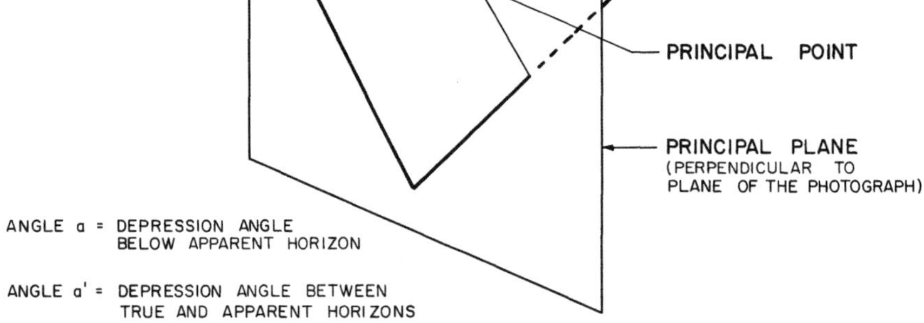

Fig. 6-7. Geometry and nomenclature for oblique aerial photographs. (Adapted from *Aerial Photography in Field Geology* by Lattman, L. H. and Ray, R. G., © 1965 by Holt, Rinehart, & Winston, Inc., reprinted by permission of the publisher.)

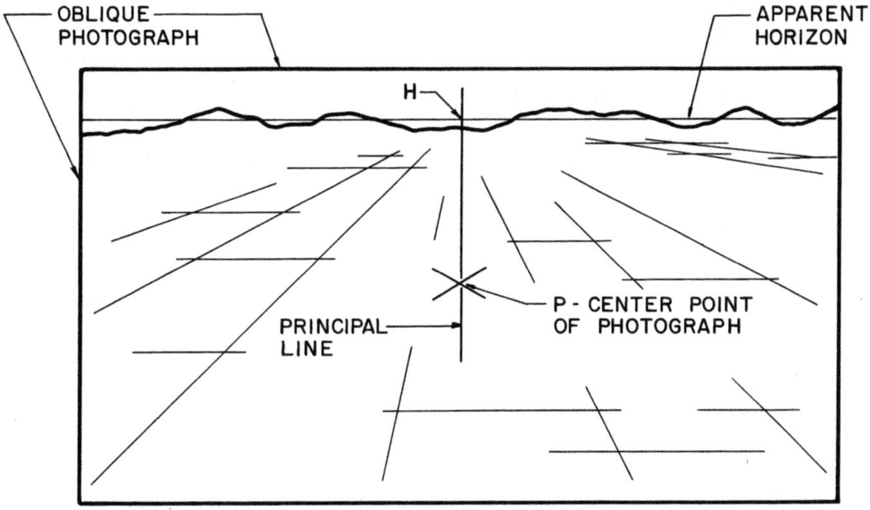

Fig. 6-8. Schematic diagram of an oblique aerial photograph. The view is perpendicular (face on) to the photograph and parallel to the principal plane. (Adapted from *Aerial Photography in Field Geology* by Lattman, L. H. and Ray, R. G., © 1965 by Holt, Rinehart, & Winston, Inc., reprinted by permission of the publisher.)

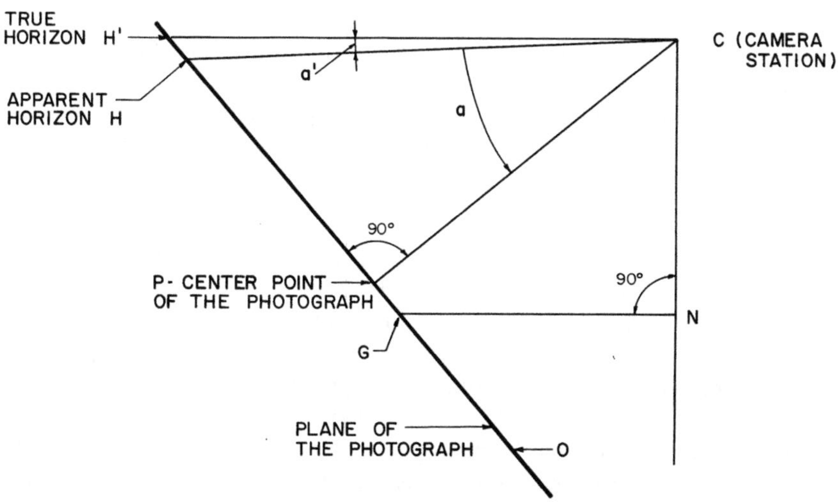

Fig. 6-9. Geometry of the principal plane of an oblique aerial photograph. This view is perpendicular (face on) to the principal plane and parallel to the plane of the photograph. (Modified from Rich, 1947.)

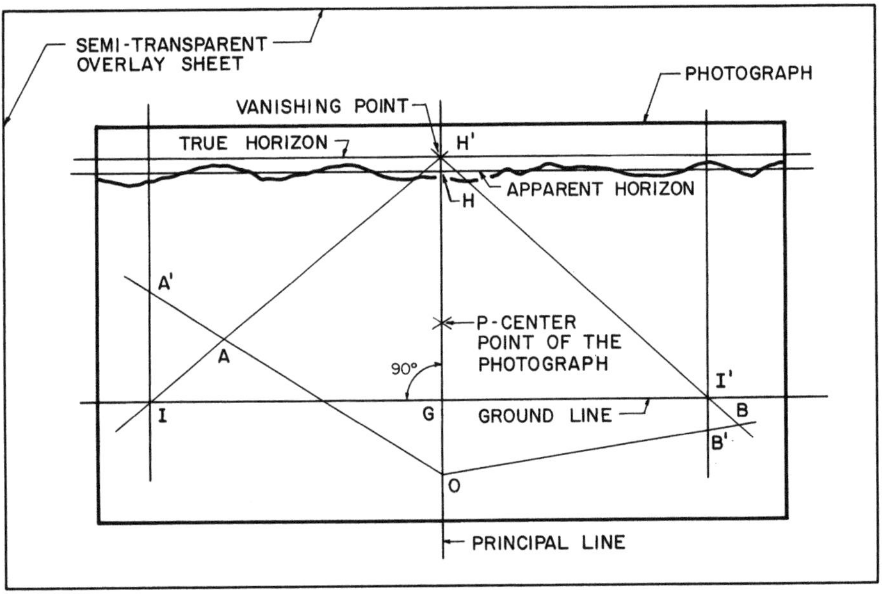

Fig. 6-10. Graphic construction method for determining the correct planimetric location of photopoints on an overlay. Points A and B are points on the photograph, while their correct planimetric positions plotted on the overlay are A' and B', respectively. (Adapted from *Aerial Photography in Field Geology* by Lattman, L. H. and Ray R. G., © 1965 by Holt, Rinehart, & Winston, Inc., reprinted by permission of the publisher.)

The first objective is to define a few terms rigorously and to construct the geometry of the principal plane. The *principal plane* is an imaginary plane perpendicular to the plane of the photograph, perpendicular to the horizon, and contains the optical axis. It is necessary to define the principal plane in order to ascertain the correct map positions of selected points visible on the photograph.

a. Locate and label the principal point (center point) of the photograph at the intersection of diagonal lines drawn from the corners of the print. Label this point P. (*Note:* The corners of the print must be the *actual* corners of the negative, i.e., the print should not be masked (see Fig. 6-8).
b. Draw the apparent horizon line on the photograph. This is the apparent junction of the earth and sky (see Fig. 6-8).
c. From P draw a line on the photograph perpendicular to the apparent horizon. Label the intersection of this line with the apparent horizon as H (see Fig. 6-8).
d. To determine the camera depression angle below the horizon, use Eq. (6-3):

$$\tan a = \frac{PH(\text{inches})}{f(\text{inches})} \tag{6-3}$$

where *PH* is the length of the line from step c, and *f* is the focal length. (*Note:* If an enlargement is used, multiply *f* by the amount of the enlargement.)

e. On a separate sheet of paper, locate any point and label it *C* (camera station) (see Fig. 6-9).
f. From *C*, lay off line *CP*, the length of which is equal to the focal length of the camera lens. The line should be slanted downward at an angle equal to angle *a*. (*Note:* Again, if an enlargement is used, *CP* is drawn to a length multiplied by the amount of the enlargement. See Fig. 6-9).
g. From *P* draw a line perpendicular to *CP*, and from *P* lay off *on this line* the distance *PH* from Fig. 6-8. Angle *PCH* is angle *a*, the depression angle of the camera calculated above.
h. To determine the true horizon H', first determine angle a' from Eq. (6-4):

$$a' \text{ (minutes of arc)} = 0.98\sqrt{\text{flying height}} \tag{6-4}$$

where flying height is in feet.

i. Plot a' with *C* as the vertex of angle HCH'. (*Note:* H' is above *H*).
j. From *c* drop a line perpendicular to CH', and along this line lay off *CN*, representing the flying height above the terrain at a scale desired for the final map. If the flying height is not known, *CN* may be laid off at any length. This length, however, fixes the scale of the plotting to be done. The photoscale must then be determined from ground data.
k. From *N*, lay off a line perpendicular to *CN* until it intersects the extension of $H'P$ along $H'PG$ so as to define point *O*, the orthocenter. (The orthocenter is the observer's position in the plane of an oblique photograph.) Line $H'PGO$ represents the plane of the photograph as viewed parallel to it.

Photoscale and Map Positions

With terms now defined, it is possible to comment on photographic scale as well as plotting points on a map in their correct planimetric position. With high oblique aerial photographs, objects on the ground vary in distance from the camera. Minimum distances are in the foreground, with maximum distances near the apparent horizon. This, of course, causes a considerable scale change, i.e., the closer a point is to the horizon, the smaller its scale, and vice-versa. It is possible to calculate the scale at any one point given focal length, flying height, and photomeasurements. However, because of the variation in distance from the foreground to the horizon, these calculations are complex and fall in the domain of the photogrammetrist. It is recommended that ground control be used to calculate the scale at different locations on the photograph. Please note that in flat terrain, the scale is constant along any line parallel to the horizon.

One method of plotting the correct planimetric positions of photopoints on a map requires that the following steps be followed. Again, these procedures are almost verbatim from Lattman and Ray (1965), and they will enable us to accurately plot photopoints planimetrically correct using an overlay of semitransparent material.

1. Preparation of Overlay
   a. Draw a vertical line representing the principal line near the center of the overlay sheet.
   b. Near the top of this line, draw a line perpendicular to the first line. This second line is the apparent horizon. Label the intersection of these two lines *H*.
   c. From work done in defining the principal plane (see Fig. 6-9) lay off on the vertical line *HH'*, *HP*, *H*, *G*, and *HO*. (Recall this notion from Figs. 6-8 and 6-9.)
   d. Draw a line through *G* parallel to the apparent horizon. This is referred to as the ground line.
   e. Place the overlay on the photograph so that *P* on the overlay coincides with the center point of the photograph. Keeping these points in coincidence, rotate the overlay until the horizon line through *H* coincides with the apparent horizon on the photograph.
   f. Tape down the photograph and overlay sheet securely so that coincidence established in step e is not lost.
2. Plotting map points (closely follow Fig. 6-10). (Assume points *A* and *B* on the photograph are to be correctly located on a map.)
   a. Draw a line on the overlay from *A* to *H'*. Extend this line in the opposite direction to the ground line. Label this intersection *I*. (Because this line intersects the horizon at *H'*, the vanishing point, it is parallel to *OH'*. The map distance between these parallel lines can only be determined where the photograph scale is known, i.e., along the ground line. Thus, *GI* is the map distance between these parallel lines.)
   b. Draw a line through *I* parallel to *OH'*.
   c. Draw a line from *O* through *A* and extend it some distance beyond. (*Note:* Angle *AOH'* is a true map angle.)
   d. Extend *OA* until it intersects the line drawn in step b. Label this intersection *A'* which is the true or correct map position on the overlay of photopoint *A*.

The same procedure is followed in ascertaining the correct map position of photopoint *B*. Many other points can be correctly transferred to maps or used to construct maps, but it is necessary to keep track of all of the lines drawn on the overlay, particularly if a large number of points are involved.

7

Some Fundamentals of Surveying

While this chapter treats and explains a variety of field tasks commonly performed by surveyors, builders, landscapers, etc., it is not intended to cover all aspects of the science of surveying. Only the most fundamental and most commonly used procedures, e.g., distance measurements, angle measurements, and leveling will be explained. Most projects will, of course, require a combination of these techniques and procedures.

DISTANCE MEASUREMENTS

Because there are several types of distance measuring equipment and techniques readily available, the suitability of each is determined by: (1) the accuracy required for the particular job at hand, (2) the equipment available, and (3) the size and physical nature of the area being measured.

Four methods will be mentioned here. Please keep in mind that whatever method is chosen, it must be carried out with as much accuracy and precision as the technique allows.

Pacing

For first approximations it is possible to ''pace off'' a particular distance. It should be possible to pace over smooth terrain to an accuracy of 1% or 2%. Before pacing can be used, one has to ascertain by length of a typical pace, i.e., paces need to be calibrated. This should be done over a two- or three-hundred foot measured course. The procedure is simple: just pace off the measured course counting the number of paces as you go. Then divide the number of paces by the total distance. As a check, and for increased accuracy, pace the measured course in both directions.

Be sure that your paces or steps are not unnaturally long, short, or in some other way awkward. They should be as close to your natural gait as possible. Note that pacing over hilly and/or uneven ground will greatly *decrease* accuracy.

Measuring Wheel

A second, more accurate, and very convenient method for measuring distances involves a *measuring wheel*. It consists of a wheel attached to a handle that is rolled along between two points. The wheel is constructed so that each revolution requires a linear movement of 10 ft. An odometer is attached to the wheel to count revolutions as it is rolling over the course being measured. Over even terrain it is realistic to expect accuracies of about 0.5% (1/200) using this method.

Stadia

The stadia method for measuring horizontal distances is also very simple and convenient. In addition, accuracies of 1/500 to 1/1000 can be attained. Only a transit and a stadia rod are required for this procedure. Inscribed in the optics of the telescopes of most transits, in addition to the centering horizontal and vertical crosshairs, are two stadia hairs. They are inscribed horizontally above and below the central hair (see Fig. 7-1). The stadia hairs are spaced so that on level terrain they intercept 1 foot (s) on an appropriately marked rod held 100 feet (D) away (see Fig. 7-2). Therefore, to find the distance in feet between the instrument and rod, multiply the intercepted rod distance by 100 (stadia constant K). For example, if 2.75 feet are intercepted on the rod by the stadia hairs, then the rod distance is 275 feet. While referring to Fig. 7-2, note Eq. (7-1):

$$D = Ks + c \qquad (7\text{-}1)$$

where: D = horizontal distance between rod and instrument
K = stadia constant (usually 100)
s = stadia interval (read from rod between stadia hairs)
C = correction factor (supplied by the manufacturer. $C = 0$ for all internally focusing instruments).

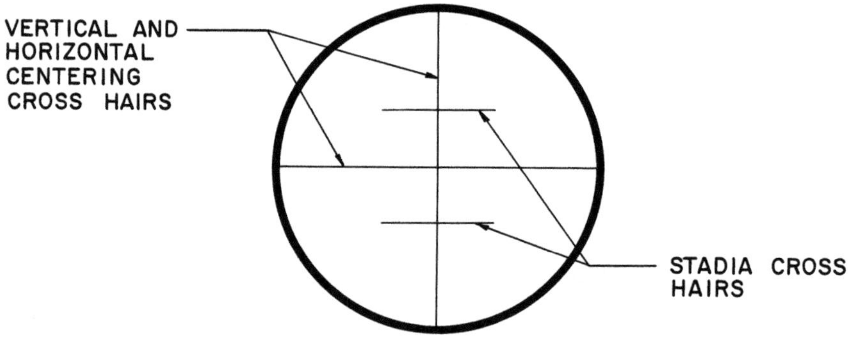

Fig. 7-1. View through a transit telescope equipped with stadia hairs.

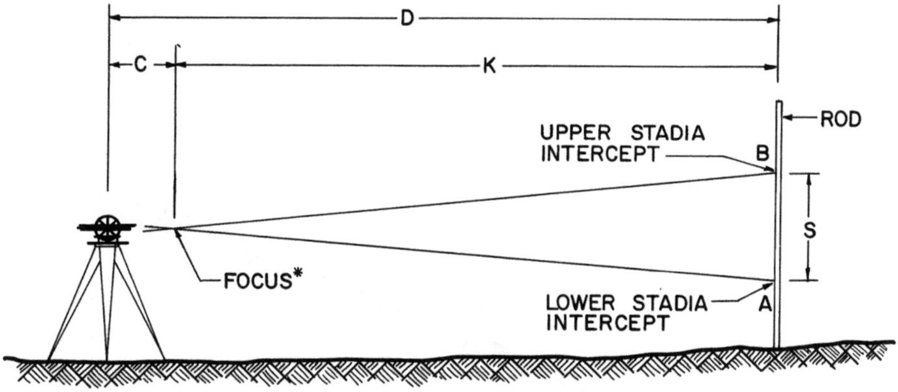

Fig. 7-2. Principles of using stadia for distance measurements.

As can be seen in Fig. 7-2, the instrument and rod are at the same elevation. Usually, however, they are at different elevations in which case the sight is said to be inclined. In actual field work in the building and landscaping trades, any sight having a vertical angle of more than 3° can be considered inclined. So in order to obtain the true horizontal distance instead of the slope distance between the points, several arithmetic operations must be performed on the stadia observations. Note that horizontal distances and differences in elevations are seldom calculated in the field using formulas. Stadia reduction tables (see Appendix A, Table 3) and stadia slide rules and calculators are commonly used as they provide easier and quicker solutions. The formulas are, however, included and explained here in order to understand the rationale of the tables and calculators.

Any inclined sight involves, first determining the slope distance, and second converting the slope distance to a horizontal distance, and if needed, a difference in elevation (see Fig. 7-3). Because standard procedure calls for always holding the rod vertically rather than perpendicularly to the line of sight, and because angle BAF is *very* small, angles $DB'F$ and $DA'F$ can be considered as right angles. Therefore, $S' = S \cos a$. Using that relationship and Eq. (7-1), the slope distance is $H' = Ks \cos a + c$. The slope distance and the difference in elevation are calculated by Eqs. (7-2) and (7-3):

$$H = (Ks \cos a + c) \cos a \tag{7-2}$$

$$V = (Ks \cos a + c) \sin a \tag{7-3}$$

As an illustrative example, suppose stadia readings were made and the values are: lower stadia hair = 4.56, upper stadia hair 7.79, and angle a is + 5°30′.

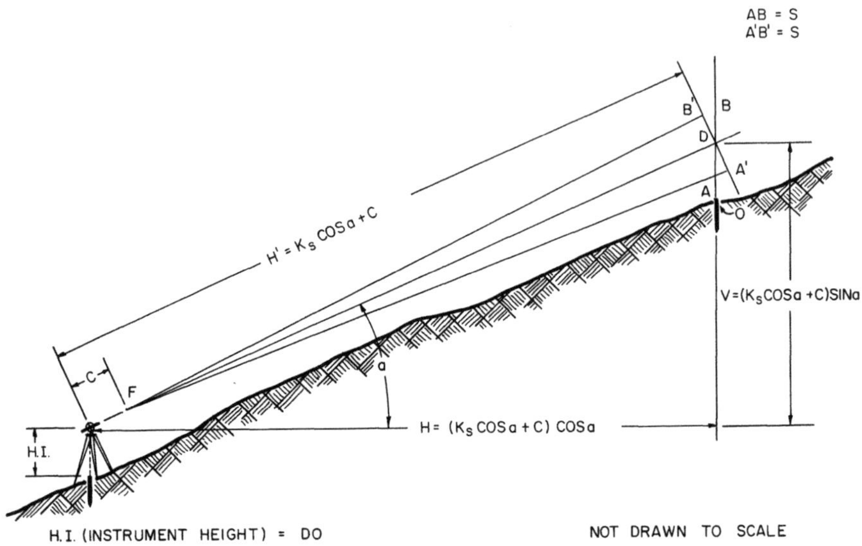

Fig. 7-3. Principles of an inclined stadia sight.

(The instrument's $K = 100$, and because the instrument is internally focusing, $C = 0$.) What is the horizontal distance between the instrument setup and rod? What is the difference in elevation between the two points? Using Eq. (7-2):

$$H = (Ks \cos a + C) \cos a$$
$$= (100 \times 3.23 \times .99540 + 0)\ .99540$$
$$= 320.035^*$$

For the elevation difference, using Eq. (7-3):

$$V = (Ks \cos a + C) \sin a$$
$$= (100 \times 3.23 \times .99540 + 0).09585$$
$$= 30.817^*$$

Using the stadia reduction tables in Appendix A to determine H, multiply the stadia interval s by the correction factor for the appropriate angle. In our example, for 5°30′, the correction is 99.08. Thus, $H = 99.08 \times 3.33 = 320.028'$.

*For actual field practice, and according to the rules of significant digits, three places to the right of the decimal would not be carried. It is shown here for the purposes of comparing actual calculations and the use of the stadia reduction tables.

For calculating the difference in elevation V, the correction is 9.54; thus $V = 9.54 \times 3.23 = 30.814'$.

Taping

Aside from expensive and sophisticated electrooptical equipment not readily available to the typical builder, taping is by far the most common and most accurate distance measuring procedure. Accuracies of 1/1,000 to as great as 1/25,000 are possible if care is taken and a little common sense is used.

Equipment needed aside from the tape itself (usually a 100-ft steel tape) is minimal. Two plumb bobs, taping pins, and range poles are standard. The plumb bobs are used, of course, to project points on the ground up to the tape and/or vice versa (see Chapter 3). The taping pins are steel rods about 18″ long that temporarily mark the ends of the tape if the distance being taped is greater than 100 feet. Range poles are usually 8′ long and are used to mark the beginning and ending points of the course, and are used to sight on to keep the tape on line (see Fig. 3-9, Chapter 3). The rear tapeperson located at the starting point gives hand signals to the head tapeperson to keep him on line with the range pole at the end of the course.

The actual method and procedure of taping, the vocal and hand signals, duties of the rear and head tapepersons are discussed in most surveying textbooks, and are not included here. However, a few special field techniques and uses of the tape should be mentioned. These techniques are quite helpful and are discussed because of their frequent use in the landscaping and building trades.

Perhaps one of the most used procedures is establishing an offset line parallel to a line of sight. Suppose the distance between points A and B needs to be determined, but because of intervening trees, shrubs, and other obstacles it is impossible to measure directly (see Fig. 7-4). First, construct two perpendiculars to line AB. Two methods are shown. For the first method, set points X and Y along line AB, or its extension, so that $XB = BY$. Using the tape, swing arcs on the ground with radius XO equal to radius YO. Line OB is perpendicular to AB. Another way of ascertaining a perpendicular to AB is by using the 3:4:5 triangle method (see Chapter 2). By using multiples of 3, 4, and 5, e.g., 15, 20, and 25, set point M 15 ft from A. Then swing arcs of 25 ft and 20 ft from points M and A.

Suppose in an example similar to the one above, an additional requirement is needed, i.e., that the perpendicular to AB must pass through specified point P (see Fig. 7-5). With the tape slightly longer than the perpendicular distance from P to line AB, swing an arc from P marking points C and D. Point E which is where the perpendicular meets AB is equal to one-half the distance CD.

It is also possible using only a tape to measure an inaccessible distance, say across a pond, building, vegetation, etc. There are several ways to accomplish this, depending on the obstacle itself, surrounding terrain, and other nearby

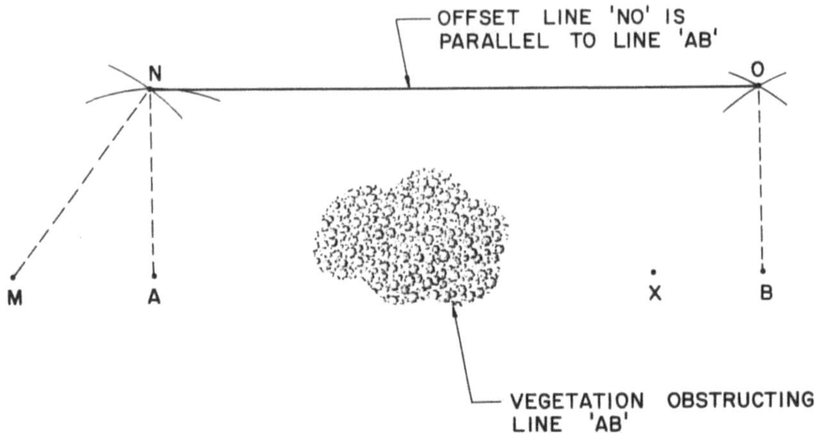

Fig. 7-4. Two methods of constructing an offset line parallel to an obstructed line of sight.

obstacles. They all, however, employ the principle of similar triangles. In Fig. 7-6a we need to measure the distance across a small pond from point *A* to point *B*. Set point *X* at some convenient location. Measure a record *AX* and *BX*. Lay off *CX* equal to $\frac{1}{2}BX$ and *DX* equal to $\frac{1}{2}AX$. The length of *CD* then will equal $\frac{1}{2}AB$. Please note that *any* proportional relationship can be established. Also note that great care should be taken when measuring *CD*. If an error is made in measuring, it will be increased by the proportion used when applied to *AB*. However, in order to avoid this magnification of error, and if it is practical, equal triangles can be used (see Fig. 7-6b). In this case, lay off $GO = OF$ and $HO = OE$. The taped distance *EF* will equal *GH*.

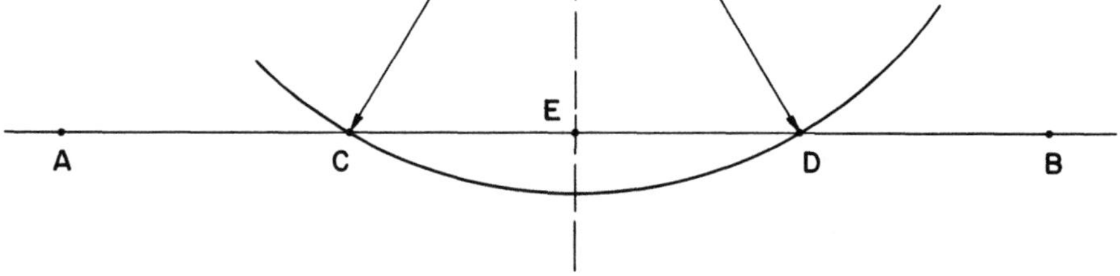

Fig. 7-5. Constructing a perpendicular *AB* through point *P*.

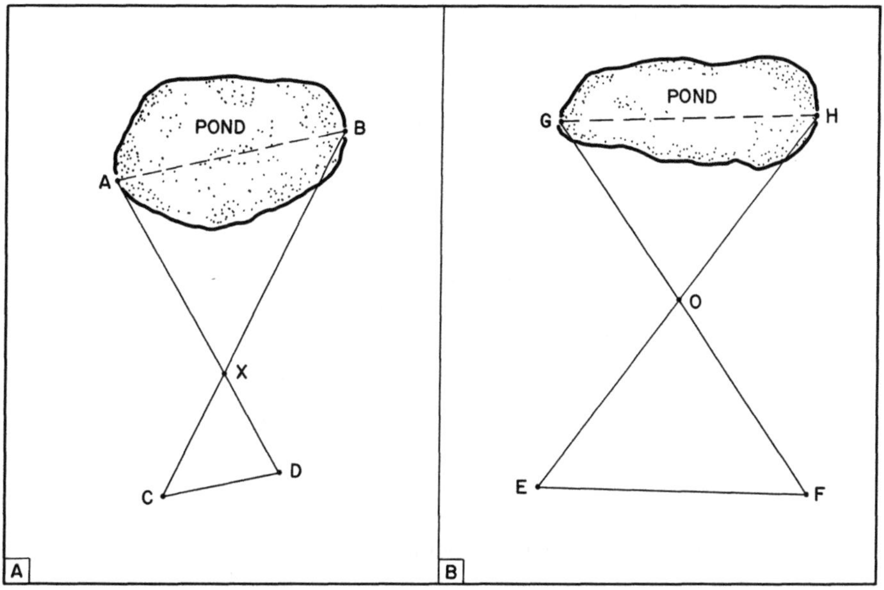

Fig. 7-6. Two methods to measure inaccessible distance using similar triangles.

To illustrate the versatility of this method, similar triangles can be laid out in an arrangement different from the above examples, but with the same results. Suppose in Fig. 7-7 the length of line *AB* needs to be known, but again an obstacle prevents its direct measurement. The following procedure, again using similar triangles, will allow us to obtain the unknown segment *MN*. Set point *P* where convenient, i.e., off to one side. Distance *PM* is measured, point M' is set midway between *P* and *M*. Point N' is located in the same manner. By the principle of similar triangles, distance $M'N' = \frac{1}{2}MN$.

Yet another application of taping is to determine the value of an angle that has already been laid out. The procedure uses only trigonometry and a tape. Suppose angle a (*BAC*) is already established on a project, but its value is unknown. To determine the angle, lay off 100 feet (see Fig. 7-8). Along line *AB*, also layoff 100 feet along line *AC*. Mark points *B* and *C* and tape the distance between them. With Eq. (7-4), angle $a(BAC)$ can be calculated:

$$\sin \tfrac{1}{2}a\ (BAC) = \frac{BC/2}{100} = \frac{BC}{200} \qquad (7\text{-}4)$$

Many other such uses of a tape are possible. In fact, most area and volume determinations discussed in Chapter 2 can be made in the field with just a tape.

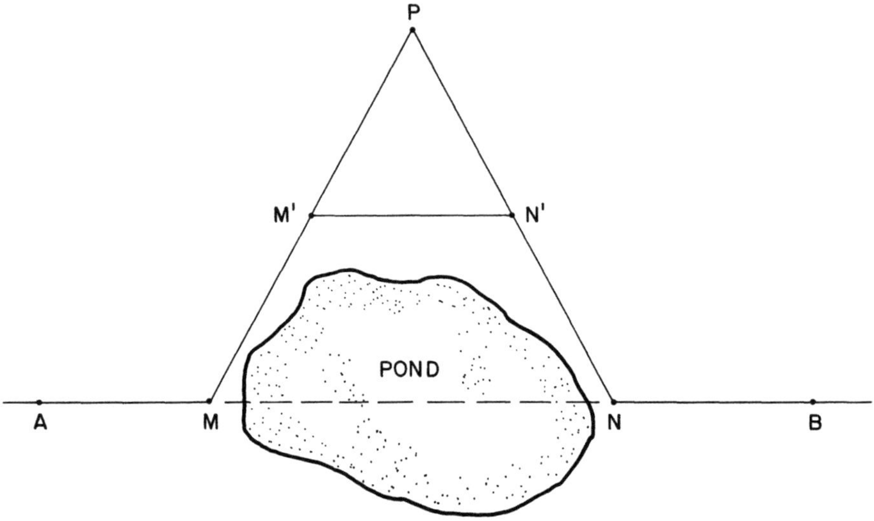

Fig. 7-7. Measuring inaccessible distance *MN* using similar triangles *MPN* and *M' PN'*.

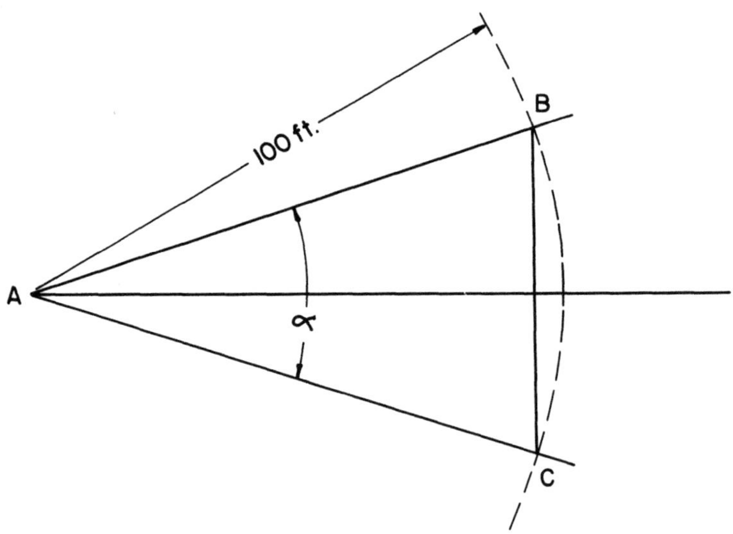

Fig. 7-8. Method of determining the value of an angle that has already been established (laid out) in the field.

Even more complex layout problems can be solved in a fashion similar to many constructions in elementary plane geometry texts. Chapter 8 and Appendix E present additional layout techniques. Many actual field situations require combinations of procedures such as taping, transit work, and leveling.

ANGLE MEASUREMENTS

Just as linear measurements are fundamental to the builder and landscaper, so are angle measurements. These, of course, are commonly made with a transit. In the previous section detailed taping procedures were not discussed, similarly details of transit operation, transit setup, instrument leveling, etc., will not be treated here. Again, surveying texts and transit operation manuals explain the proper and recommended techniques for these. In this section a few common field procedures involving horizontal and vertical angle measurements with a transit will be presented.

Remember from Chapter 1 that an angle is measured *from* the initial side, i.e., the starting point or reference. In actual field work, angles can be measured from any type of permanent or fixed point. Probably the most often used point for beginning an angle measurement is a previously occupied transit station. A sight on such a station is called a *backsight*. However, anything such as a corner of a building, a manhole cover, fire hydrant or property corner pin can be used as a reference.

Horizontal Angles

Although it requires several steps turning a horizontal angle, i.e., one turned about the transit's vertical axis, with a transit is a simple operation. After the instrument is set up over the station and is leveled, the following general steps are followed:

1. *Setting the Plates*. Loosen both the upper and lower clamps and rotate the scales until the 0 on the main scale coincides as closely as possible with the 0 on the vernier. Tighten the upper clamp. Using a magnifying glass and the tangent screw (used to make small changes once the clamps are tightened), "fine tune" the coincidence of the main scale and vernier zeros. Leave the lower clamp loose.
2. *Back Sight*. With the upper clamp still tightened, turn the telescope and aim at the back sight station. Tighten the lower clamp. Then by using the lower tangent screw bring the vertical cross hair *exactly* on station.
3. *Fore Sight*. Loosen the upper clamp. Aim the telescope at the fore sight station and retighten the upper clamp. Using the upper tangent screw bring the vertical cross hair *exactly* on the fore sight station.

4. *Reading the Angle*. Using the vernier (see Chapter 3) measure the angle as carefully and as accurately as the instrument permits. If the project calls for more accuracy than the instrument can provide on a single turn, determine the angle by repeated turnings (see Chapter 3).

Marking Points

Because the optical system has good magnification and allows for excellent accuracy, frequently more than is needed, it should be fully utilized for it requires little extra effort. For example, when establishing a point for an angle determination focus in on a small, thin object exactly marking the point. Ideal and commonly used objects are pencil points, plumb bob string, plumb bob point, nail (surveyor's tack), taping pin (see Fig. 7-9).

If the located points need to be set in asphalt, a nail driven through a folded piece of flagging tape or a nail marked with paint and driven into the asphalt can be used (see Fig. 7-10). On concrete, points are frequently marked by cutting crosses and/or by drilling and setting metal or wooden plugs. In soft ground, short, square (2″ × 2″) wooden stakes commonly called *hubs* with tacks located at the *exact* station position are used (see Fig. 7-11). Hubs are usually marked by a long, thin stick called a *guard stake* (see Fig. 7-11). They are painted a bright color or marked with flagging tape, and have the station data (number, elevation, etc.) written on them.

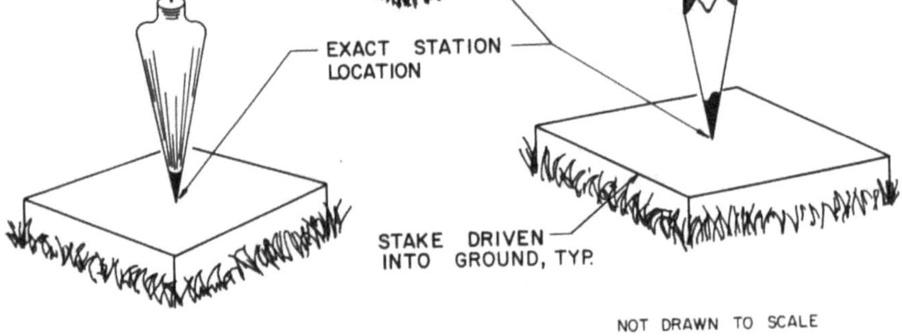

Fig. 7-9. Some common objects used in accurate transit sights and in taping.

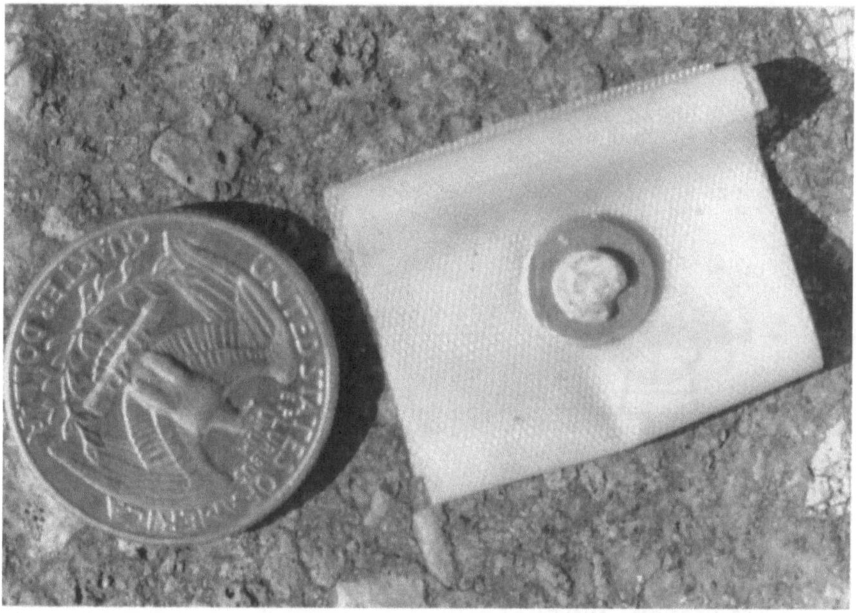

Fig. 7-10. A tack and folded flagging tape used to mark a point in asphalt.

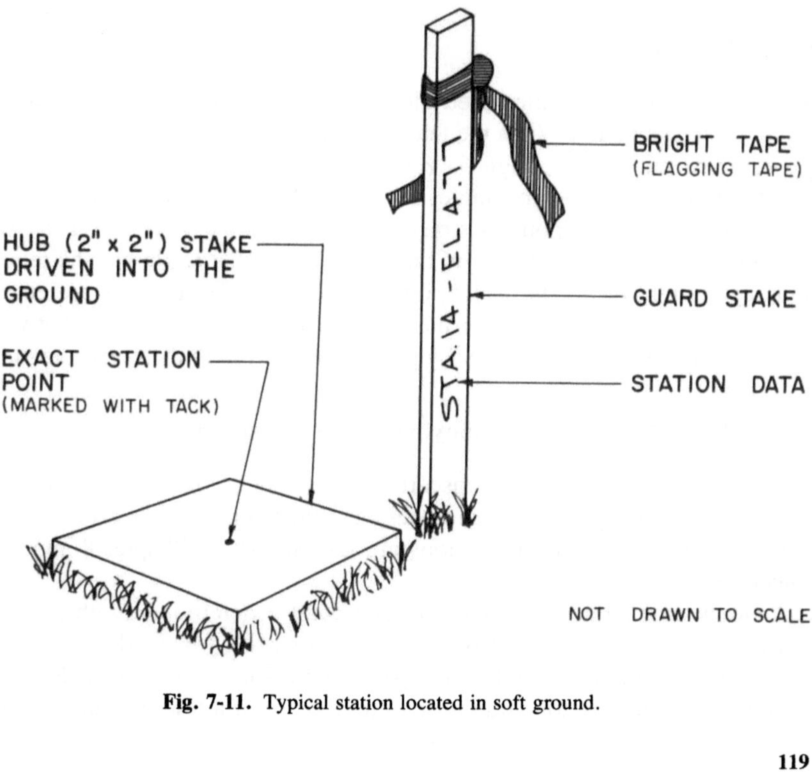

Fig. 7-11. Typical station located in soft ground.

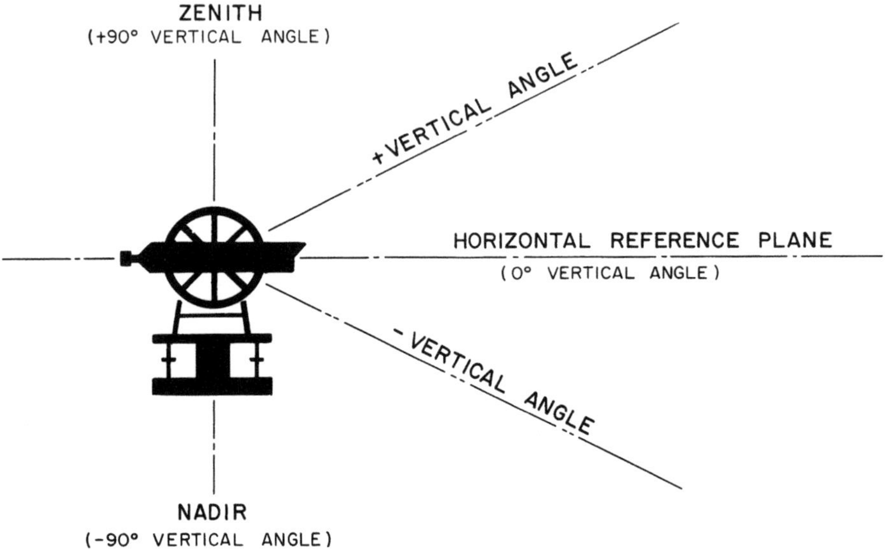

Fig. 7-12. Nomenclature and classification of vertical angles.

Vertical Angles

A *vertical angle* is one measured either up or down from an imaginary horizontal reference plane. When the telescope is level, i.e., at the reference plane, the value of the vertical angle is zero. When the telescope is pointed higher than the reference plane, the angles are said to be positive in value up to +90° when the telescope is pointed straight up the *zenith* (see Fig. 7-12). Angles below the horizontal are negative, and a −90° vertical angle has the telescope pointing straight down the nadir point (see Fig. 7-12). Transits are equipped with a vertical scale that has a vernier, clamp, and tangent screw so that vertical angles can be read with accuracy by procedures similar to the horizontal angles.

Simple Field Techniques Employing Angle Measurements

Because angles are commonly ascertained by use of transits, it is logical to examine a few simple field techniques involving this basic instrument. Additional and more specific illustrations are also shown in Chapter 8 and also in Appendix E.

A simple line from any point *A* to any point *Z* determined with a transit is called a *transit line*. Sighting from *A* to *Z*, however, frequently is not always as simple as a single, level shot in open terrain. Intervening obstacles such as buildings, vegetation, topography, require the use of a few special techniques.

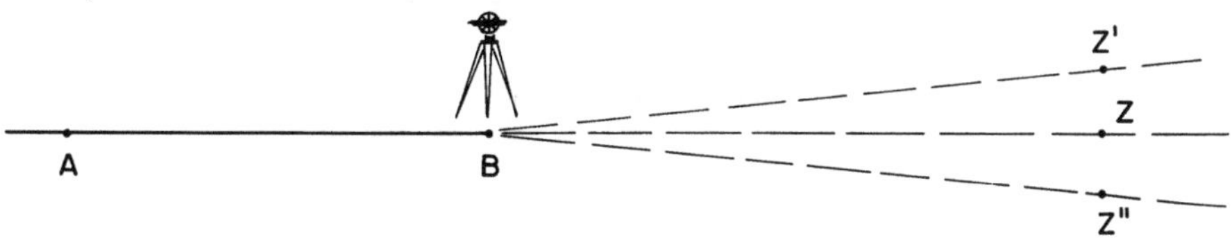

Fig. 7-13. Locating a point by double centering.

Suppose line *AB* needs to be extended to locate point *Z* (see Fig. 7-13). If point *B* is occupied, according to the previous discussion we should simply turn 180°00′ using point *A* as a back sight. However, turning the telescope over about its transverse (horizontal) axis, a procedure called *plunging*, gives more accurate results. if the instrument is in perfect adjustment, line *AB* will be extended to *Z* by this procedure. However, if the instrument is slightly out of adjustment, point *Z* can still be accurately located by a method called *double centering*. Double centering is recommended in most cases as a check even if no instrument maladjustment exists. Follow Fig. 7-13 and the five steps outlined below.

1. Back sight on *A* from *B* with the telescope in the *normal* position.
2. Plunge the telescope. (If the instrument is not in adjustment, *BZ*′ or *BZ*″ will be viewed.) A stake is driven, and the exact location of *BZ*′ or *BZ*″ is marked by a tack.
3. Loosen either the upper or lower clamp and take another back sight on *A*. (The telescope is still in the inverted position from step 2.)
4. Plunge the telescope again. The instrument will now sight a point on the *opposite* side of *Z*. For example, if the first point was marked at *Z*′, the second will be sighted and staked at *Z*″.
5. Point *Z* is halfway between *Z*′ and *Z*″.

It often happens that some obstacle along a transit line has to be passed. There are a number of ways that a transit can be used to accomplish these passes. Figs. 7-14 through 7-18 show some typical procedures.

LEVELING

Like measuring distances and angles, leveling is fundamental to the builder, surveyor, landscaper, geologist, engineer, etc., in the field. Also like other field work, it should be carried out with great care, accuracy, and precision. It might be well to mention here that the tips presented in Chapter 3 for maintaining high-quality work should be reviewed frequently.

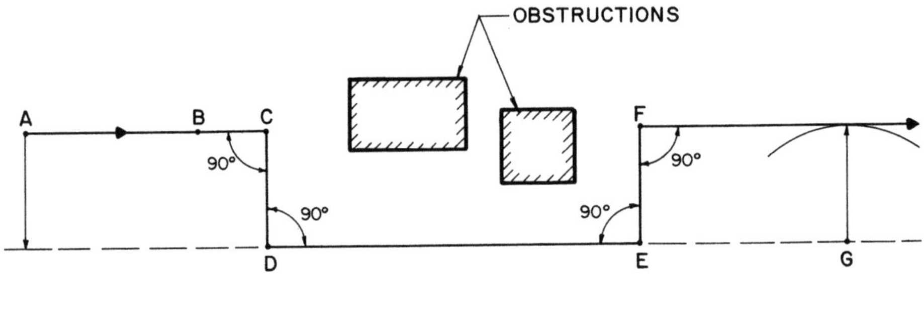

Fig. 7-14. Method of extending line *AB* beyond obstructions using four right angles and, as a check, two tape swings.

Leveling is defined simply as the operation of determining the difference in elevation of two points on the earth's surface. There are two types of leveling that are generally used in the building trades, differential leveling and trigonometric leveling. As might be guessed, there are a multitude of instruments used for leveling. They range from simple line levels, small spirit levels suspended from a string stretched between two points, to hand-held levels, to very precise geodetic levels. The most frequently used instruments are engineer's levels and transits.

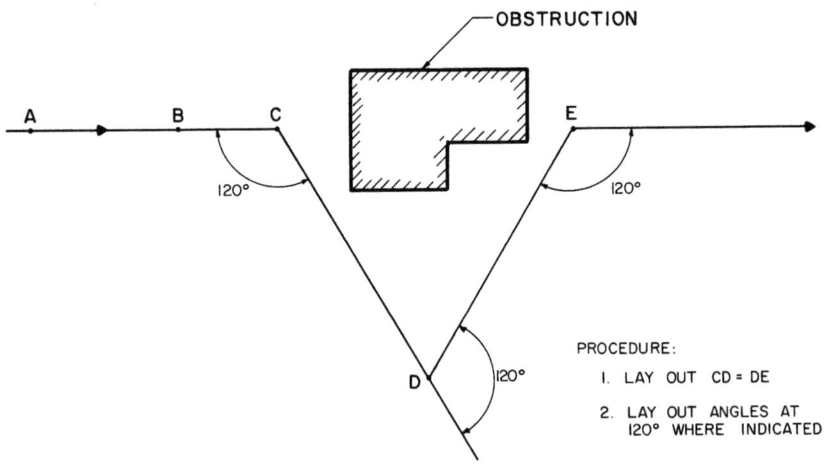

Fig. 7-15. Extending line *AB* past the obstruction using an equilateral triangle.

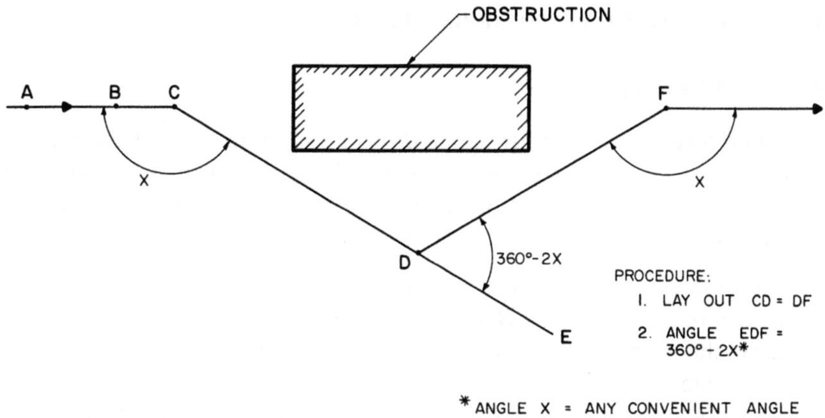

Fig. 7-16. Extending line *AB* beyond obstruction using an isosceles triangle.

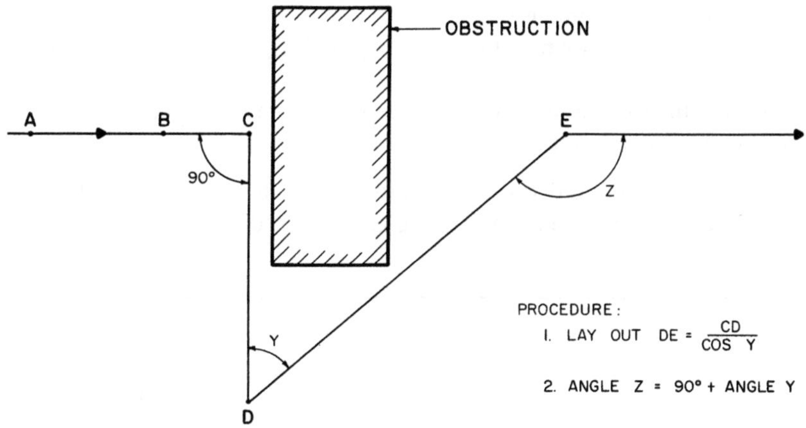

Fig. 7-17. Extending line *AB* past an obstruction using a right triangle.

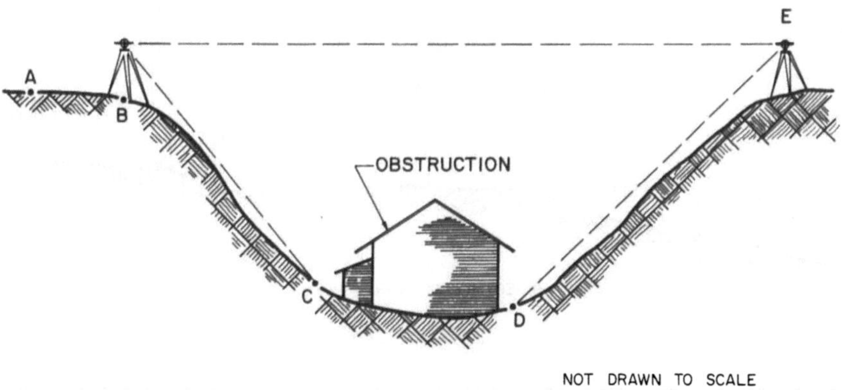

Fig. 7-18. Extending line *ABC* to point *D* hidden by the obstruction by passing over the obstacle to point *E* and then back.

Differential Leveling

Differential leveling, sometimes called *direct leveling*, involves determining the difference in elevation of a known point and the height of the instrument, then the difference in elevation from the instrument height to an unknown point. Fig. 7-19 and the following steps illustrate the principle of differential leveling:

1. Set up the instrument approximately halfway between the known and unknown points, BM_K and BM_U, respectively.
2. Level the instrument and double-check the level bubble several times while taking sights.
3. Back sight on the known station reading the rod as accurately as possible (see Chapter 3).
4. *Add* this back sight reading (B.S.) to the known elevation. This is the height of the instrument (H.I.).
5. Sight on the unknown station (fore sight, F.S.). Do not forget to check the bubble.
6. *Subtract* the rod reading from the fore sight from H.I. This is the elevation of the unknown point.

These six steps can be distilled down to BM_K + B.S. = H.I.; H.I. − F.S. = BM_U. These two relationships are the essence of differential leveling. Note

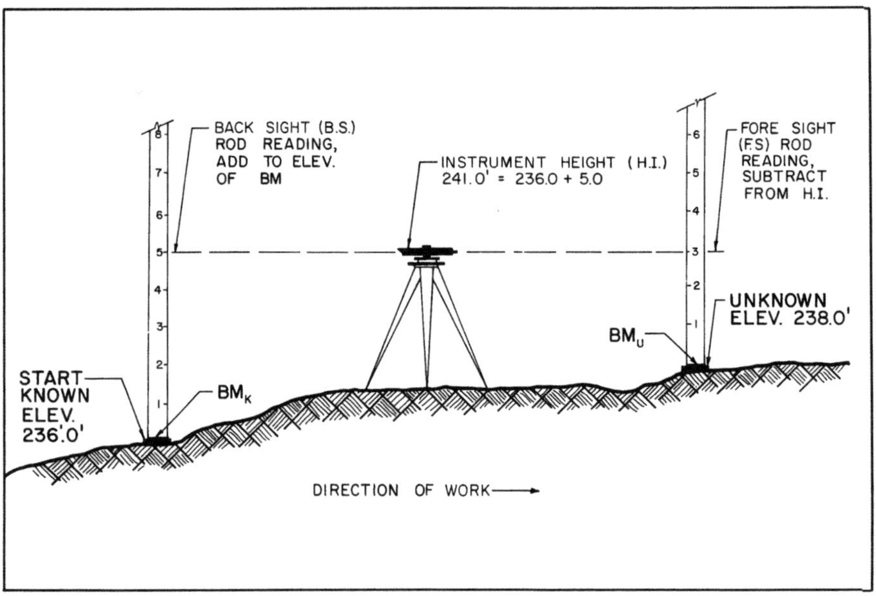

Fig. 7-19. Principle of differential leveling.

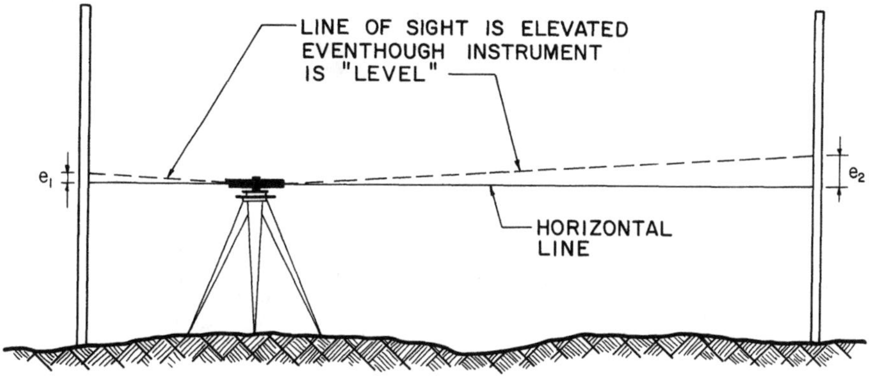

Fig. 7-20. Uneven lengths of back sights and fore sights will cause a small error *e* to be magnified over a long distance.

that step 1 indicates that the instrument should be set up halfway between the known and unknown points. The reason for this is to minimize error caused by an instrument slightly out of adjustment. If the instrument is sighting slightly high or low, even if everything indicates that it is level, a small error in the back sight reading will cause a larger error in the fore sight (see Fig. 7-20). If the same sight, either fore or back, is consistently long, this error will accumulate and become unacceptably large. This source of error can be easily eliminated by always setting up halfway between the points.

Figure 7-19, while showing the principle of differential leveling, implies that the two points must be visible from the station setup. This is not always possible. Fig. 7-21 shows a leveling operation between two points that are not visible from a single set-up. Suppose that Station 14 is the known station, and is, therefore, the starting point. The intermediate points are called *turning points* (T.P.), because traverses will not always lie along a straight line. Notice that the procedure at each setup is identical to what is illustrated in Fig. 7-19.

It is also common practice to determine the elevation of many points from a single setup. This can be particularly useful in topographic mapping, to be discussed later, and in the construction of profiles and cross sections. The object of profiling is to determine the elevation of many points along a given line. The elevations are then plotted graphically to give a visual representation of the line of points.

The usual procedure is to first stake the line in question at some regular interval and at intermediate points if there is a change in slope that will not be shown by the chosen interval. The line of stakes does not have to be straight but can be laid out as needed. The elevation of the points are then determined using a single back sight and multiple fore sights (see Fig. 7-22).

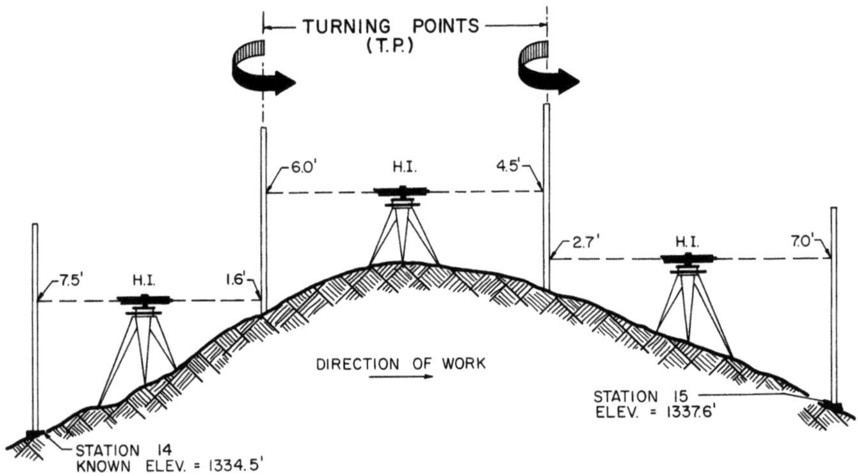

Fig. 7-21. Differential leveling using several turning points.

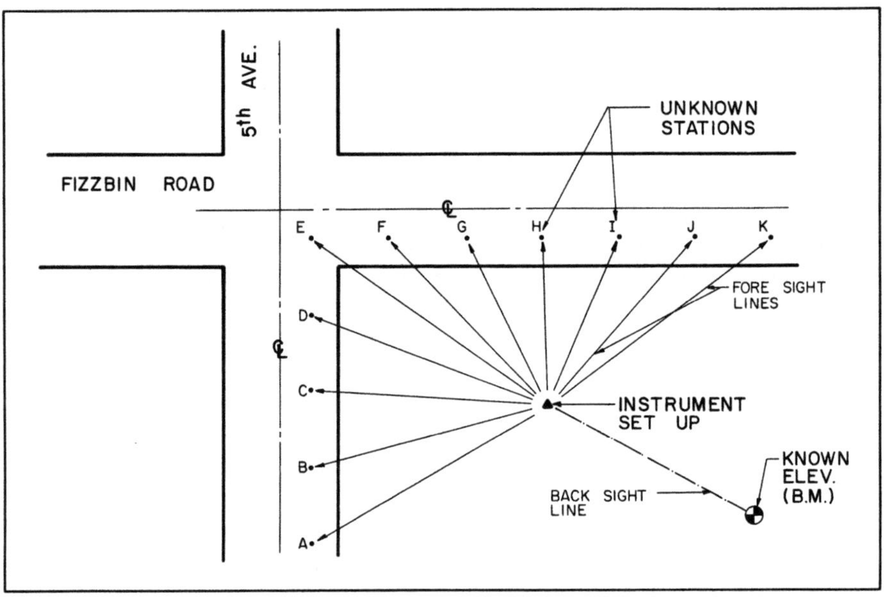

Fig. 7-22. Differential leveling using a single back sight and multiple fore sights.

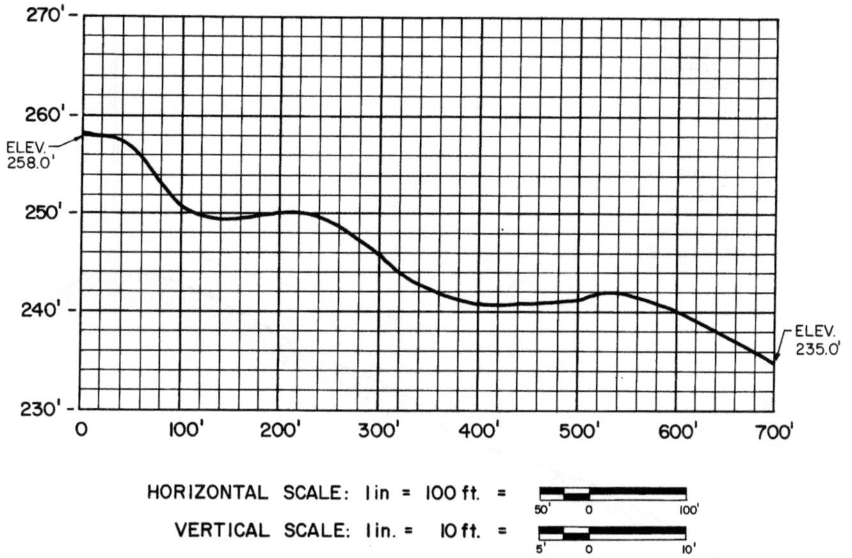

Fig. 7-23. Completed profile showing the vertical scale exaggerated 10× with respect to the horizontal scale.

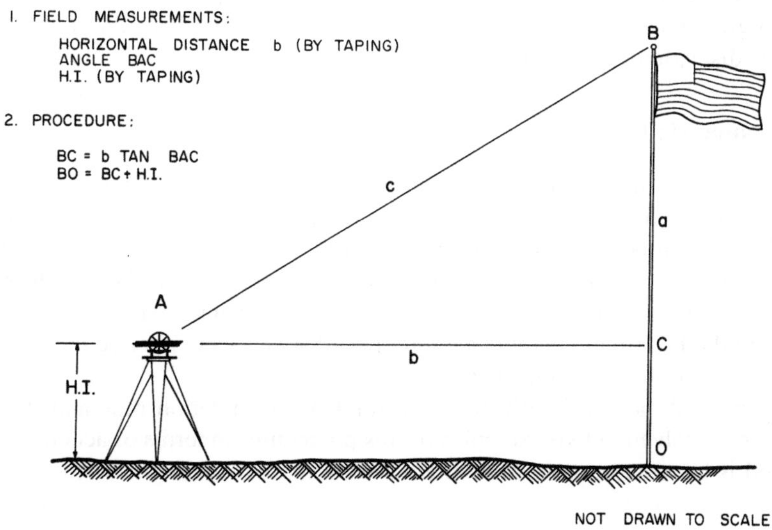

Fig. 7-24. Using trigonometric leveling to measure the elevation of an unknown point.

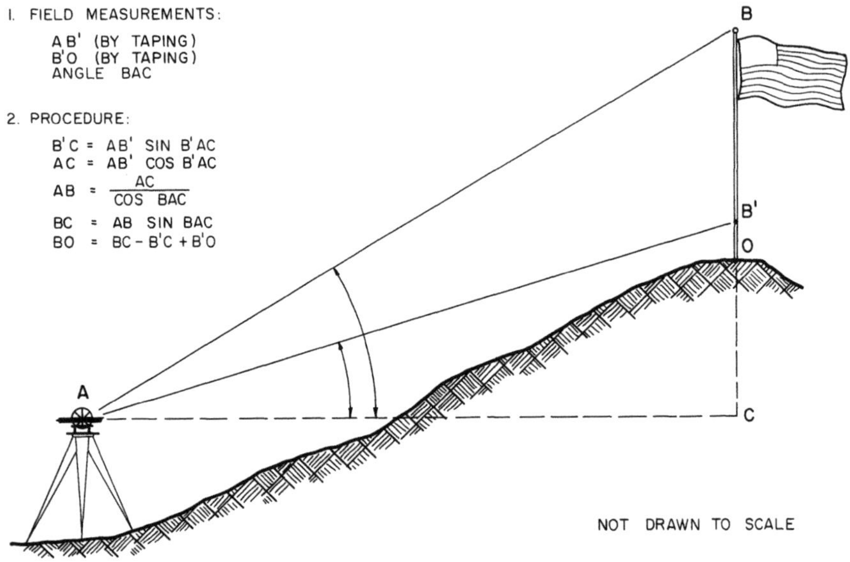

Fig. 7-25. Using trigonometric leveling to determine the elevation of an unknown point atop a hill.

Plotting the field data is done to scale both vertically and horizontally. However, the vertical scale is usually exaggerated with respect to the horizontal so that slight changes in the ground slope can be seen (see Fig. 7-23). Profiling will be discussed in detail in Chapter 9.

Trigonometric Leveling

Sometimes it is not practical or possible to occupy an unknown point to place a rod on it. Examples are trees, transmission towers, water towers, chimneys, etc. In these situations *trigonometric leveling*, which applies the fundamentals of trigonometry, is used to determine elevation differences. This method requires (1) the use of a transit to measure vertical angles and (2) the distance between the instrument and the station. This distance can either be the horizontal distance or the slope distance.

Figures 7-24 and 7-25 illustrate the fundamental relationships and the calculations involved in two examples of this procedure. In terms of accuracy, this method is not as accurate as the direct or differential leveling described above.

8

Project Layout and Planning

Plans drawn up for a particular project almost always give the exact location of the structure within the plot boundaries. The task here in this chapter then becomes to actually locate the structure in the field as dictated by the plans. This requires not only identifying the proper location, but also defining the orientation and elevation of the proposed structure. This, of course, can be done accurately only if some permanent monument can be utilized as a starting point or reference point. Reference point(s) can be items such as property corners located by a registered surveyor or any other existing municipal, utility, or government control mark.

It should be mentioned here that before any field work is done, the following items should be thoroughly checked.

1. Zoning ordinances
2. Setback lines
3. Easements, rights of way
4. Location of underground and/or overhead utilities, e.g., as water, electric, telephone
5. Other deed restrictions, e.g., location of trees, fences, walls

Being aware of any restrictions beforehand can save much time, money, and energy over the span of the entire project.

Logic indicates that in order to locate a structure on a plot of ground, a systematic procedure must be followed from an established reference. Note that although the techniques described here will serve to locate building corners and other points, the accuracy or the correctness of the points should be thoroughly checked and then staked. The procedures for these operations are also described below.

LOCATING A STRUCTURE ON A PLOT

A frequently used and simple method for locating a structure on a plot of ground is called the *baseline offset* method. It consists of locating the corners of a building by right angle offsets from a baseline (see Fig. 8-1). The baseline is defined from some existing and permanent reference(s). Setting up a transit

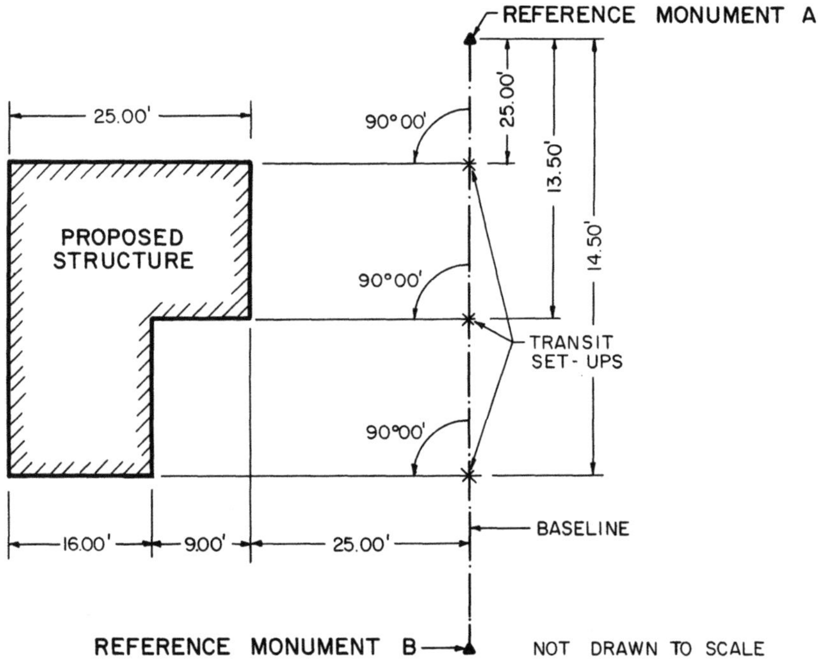

Fig. 8-1. Locating a proposed structure by the baseline offset method.

along the baseline and turning right angles is the most expeditious way of ascertaining the needed offsets. However, it can be done *without* a transit with only a tape using the methods described in the last chapter. This is not recommended, however, if the distances from the baseline are too great.

A second procedure, called the *angle-distance* method, has the advantage of locating many points (corners) from fewer setups (see Fig. 8-2). However, this procedure has a number of limitations that are noteworthy. First, all corners may not be visible from the necessary setup locations. And, second, some sights may have to be quite long which, of course, diminishes both linear and angular accuracy. In the event long sights are unavoidable, turning the angles by repetition is recommended.

The third method is termed the *coordinate* method. This method can be used effectively when large complex projects are planned that require locating many points. For example, in laying out a golf course, a small park, recreation area, or playground where many trees, drinking fountains, shelters, fences, etc., need to be accurately positioned, this method is ideal. The first step is to define and establish two mutually perpendicular baselines located conveniently within or

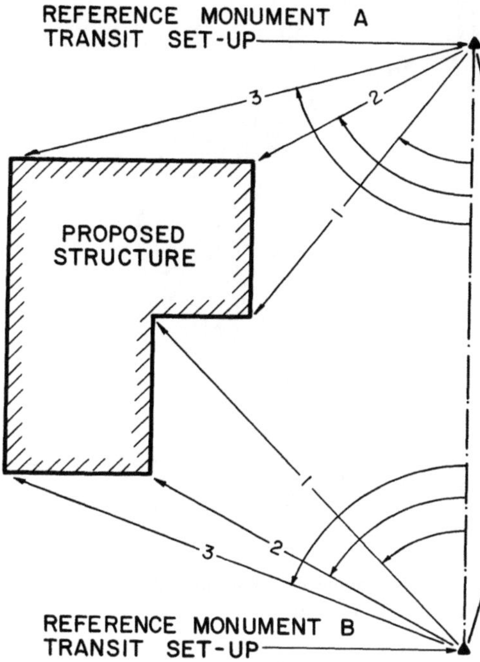

STATION A	BEARING	DISTANCE
1	34° 05'	76.00'
2	58° 30'	49.50'
3	74° 45'	97.75'

STATION B	BEARING	DISTANCE
1	41° 00'	96.04'
2	57° 45'	73.45'
3	67° 45'	104.10'

Fig. 8-2. Locating a proposed structure by the angle-distance method.

along the edge of the project area. The monuments defining the baselines should be permanently set in concrete so that they can be used later to accommodate renovations, repairs, additions, etc. (see Fig. 8-3). Each item within the project area is then located by a set of coordinate distance measurements from the baselines (see Fig. 8-4).

SQUARING UP AND CHECKING FOUNDATION LAYOUTS

Once located on the plot of ground making sure that a foundation layout is square is very important. Accurate layouts make later work much easier and much less expensive. There are a number of ways of squaring up foundations depending on the size, shape, and terrain. Follow Fig. 8-5 which shows the corners of a building located by the perpendicular offset method described above. One option for checking the squareness is to set up the transit at corner *A* and measure an angle of 90°00′ between *B* and *C*. The distances 40.00 feet and 80.00 feet, respectively, are measured. Next, move the transit to *C*, backsight on *A*, turn 90°00′ to *D*, and measure 40.00 feet.

Fig. 8-3. A permanent marker (bench mark) used as a reference for locating building corners, lamp posts, sewers, underground cables, roadways, sidewalks, etc.

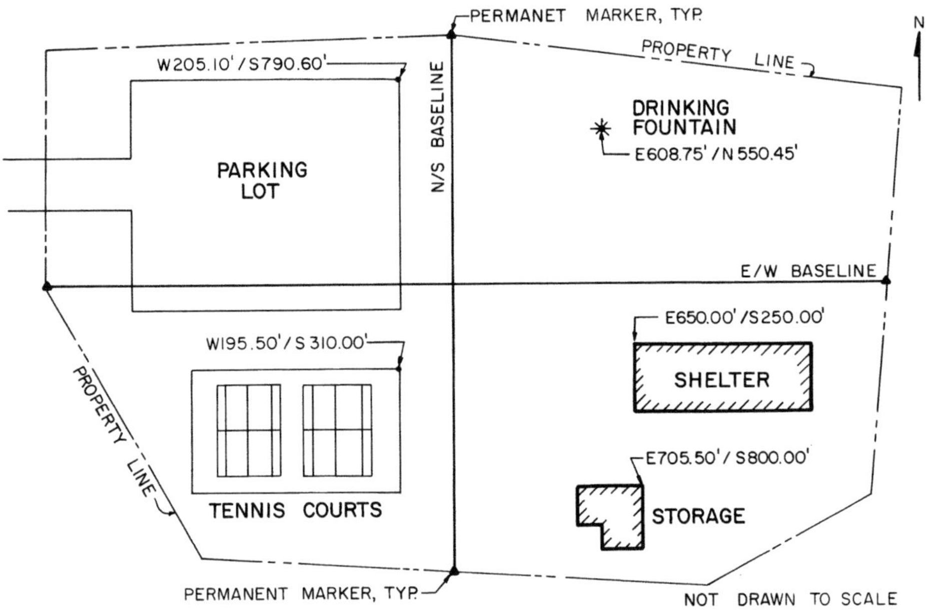

Fig. 8-4. Locating structures by the coordinate method.

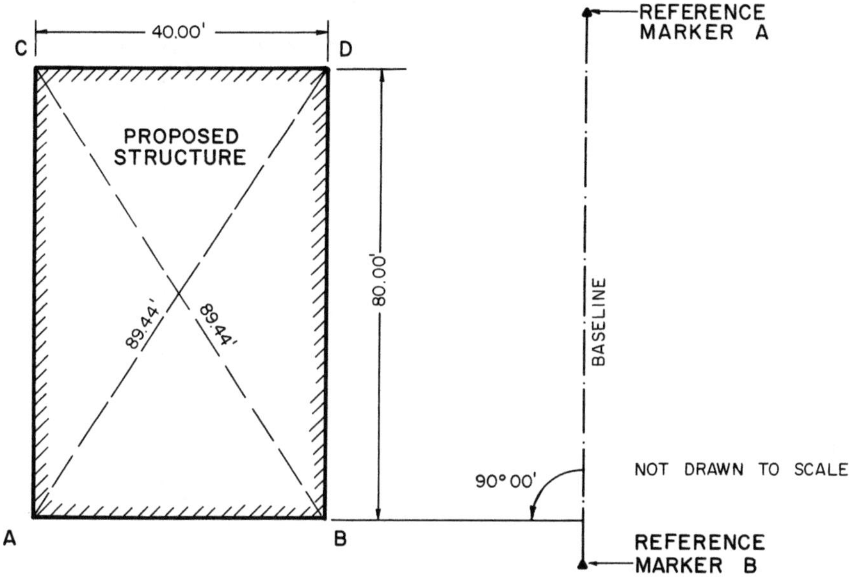

Fig. 8-5. Squaring up a simple foundation by using diagonals.

As a second check on the work, calculate the length of diagonals *AD* and *BC*. In this example, the diagonals should be:

$$AD \text{ or } BC = \sqrt{(AB)^2 + (BD)^2 \text{ or } (AC)^2} \tag{8-1}$$

$$AD \text{ or } BC = 89.44 = \sqrt{(40.00)^2 + (80.00)^2}$$

If a transit is used that can read to seconds, a further check on the work can be made from setups at *A* and *C* by determining acute angles *CAD*, *BAD*, *ACB*, and *BCD*. These can be calculated by trigonometry in the office and then actually checked at the site.

With more complex buildings such as the one shown in Fig. 8-6, smaller rectangles can be defined for various parts of the structure. Diagonal and acute angle checks should be made wherever possible to be sure that not only each part but also the structure as a whole is square.

ESTABLISHING ELEVATIONS AND PRESERVING CORNER LOCATIONS

Once the corners or other points on the structure have been accurately positioned, it is necessary to establish their correct elevation, again as per the plans.

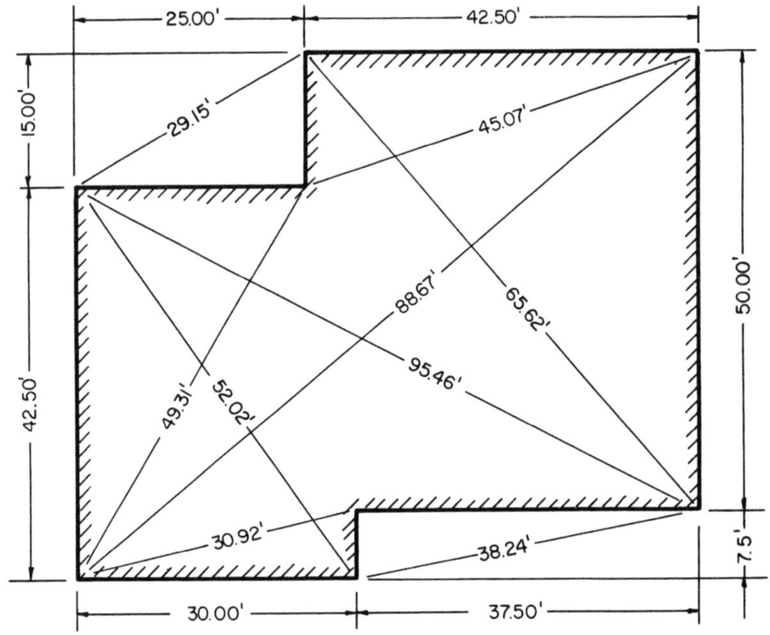

Fig. 8-6. Squaring up a complex foundation by using diagonals. (For illustrative purposes, more diagonals are calculated and shown than would normally be used.)

It is assumed at this point that consideration has already been given to drainage, driveway or sidewalk grades, sewer gradients, cut and fill considerations, general aesthetics, etc. Hence, grades from a reference monument to some specified part of the building, such as the top of the footer, basement floor, first floor, etc., can be established. For this, usually the standard techniques of differential and trigonometric leveling using a transit or a builder's level will provide adequate accuracy. Fig. 8-7 exhibits a few commonly used items whose elevations must be known before work is actually begun.

After the corners of the structure have been accurately located, their position must be marked so that the corners can be relocated after the excavation for the foundation has been completed. The most common way of marking corners is with *batter boards*. They are horizontal boards nailed to vertical stakes and are located three to five feet outside the excavation so they will not be disturbed (see Fig. 8-8). They are located at the building corners *and* at appropriate locations to define and locate all the *inside* corners as well (see Fig. 8-9). The tops of the batter boards must all be at the same elevation, an elevation which bears some relation to the reference marker and the footer, basement floor, etc. Note that the batter board should not be disturbed during construction, and

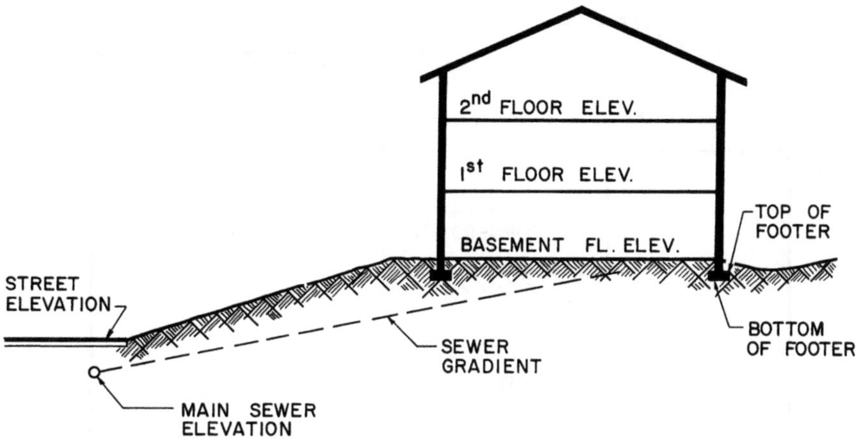

Fig. 8-7. Some features whose elevation must be known before construction is begun.

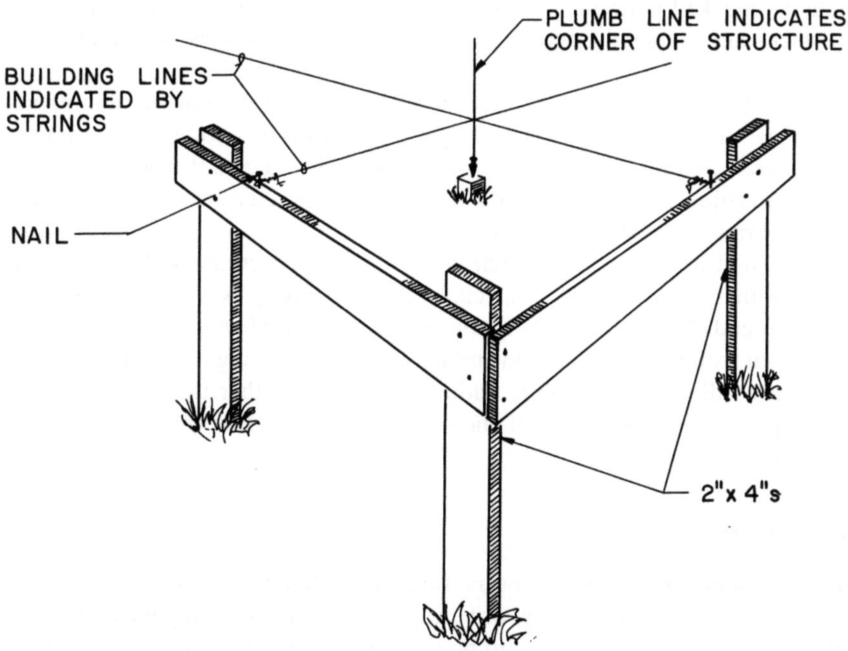

Fig. 8-8. The use of batter boards to mark the corners of a structure.

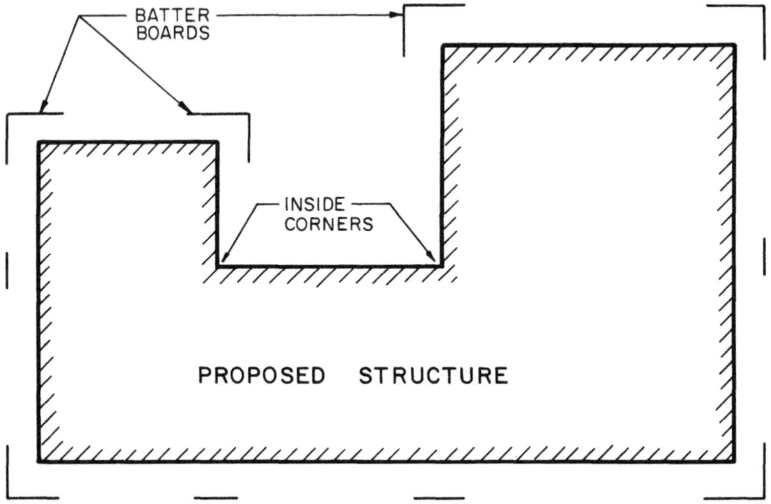

Fig. 8-9. Location of batter boards around a complex foundation. (Note the batter board location for the inside corners.)

should be checked periodically. Hence, they should be quite sturdy. The vertical stakes are frequently two by fours and the cross-piece is commonly a 1″ × 6″.

If batter boards are not used, then it is necessary to lay out the foundation three times: one, before excavation begins; two, after it is completed to square up the footer; and three, after the footer is poured to locate the outside corner of the foundation on the footer. Masonry nails hammered into the footer then serve as permanent corner markers.

This second method of additional layouts is more accurate and *may* be less expensive than batter boards, especially if the building is located on steeply sloping ground. This is particularly true where the batter boards need to be more than several feet above the ground surface. However, laying out a structure two extra times will certainly require additional time. Whether the cost in time is more important than the cost of money is a decision that must be made as needed on the job.

EARTHWORK

From experience and general observation, it is obvious that most structures, i.e., buildings, roads, sidewalks, driveways, require the rearrangement of some earth materials. Some areas need to be excavated (cut), while others need to be embanked (filled). The process of bringing the original land surface to the proper grade is sometimes called *earthwork*, while the actual field work of setting and labeling the stakes is called *slope-staking*.

Some earthwork calculations can become confusing if a few simple concepts are not understood. It is convenient to compute cuts and fills directly from rod readings without referring to actual elevations. This can be accomplished by using what is called grade rod, i.e., what the rod would read if it were held on the finished grade. Follow Fig. 8-10 and the steps below to determine grade rod:

1. Determine H.I. from some bench mark.
2. Find the final grade elevation from the plans.
3. Grade Rod = H.I.—Grade Elevation (in our example, Grade Rod = 100.00′ − 95.00′ = +5.00′).
4. Consider grade rod (e.g., +5.00′) as a back sight in subsequent rod readings. (Note grade rod can be either + or −).
5. Ground elevations are fore sight readings (sometimes called ground rod).

With the value of grade rod established, the amount of cut or fill for any point can be ascertained by reading the rod at various points and using the following equation:

$$\text{Cut } or \text{ fill} = \text{Grade rod} - \text{ground rod}. \tag{8-2}$$

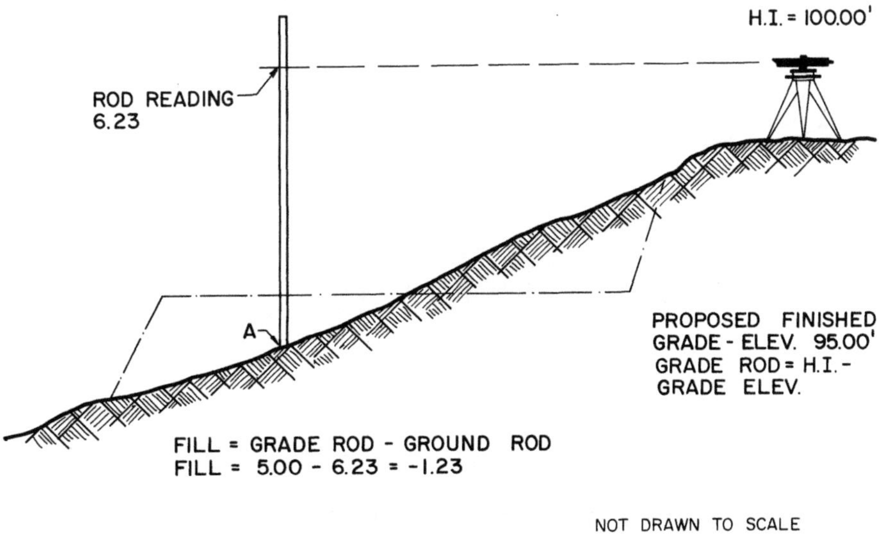

Fig. 8-10. Determination of the amount of fill at point *A* using the grade rod concept.

If the result of sighting on a point is positive, the point is above the intended grade and needs to be cut. If negative, the point is below grade and fill is needed. Again in our example, suppose the rod is placed at point *A* and reads 6.23. Using Eq. (8-2), we get -1.23, which means that the ground elevation is below the intended elevation and has to be filled or embanked by 1.23 feet of material (see Fig. 8-10).

A series of ground rod sightings are then taken over the project area, usually at some regular interval, to ascertain all necessary cut and fill data. It is important to note that when marking cut and fill points in the field with stakes, several different techniques are used. In projects where earthwork is being carried over a wide area such as a parking lot or athletic field, the stakes are set at the proper location *within* the field. The stakes are marked so that they can be easily seen by the equipment operators and read ''Cut'' or ''Fill'' followed by the amount recorded in feet and tenths of a foot. An ''even'' designation indicates neither cut nor fill. In the case of pipe lines or sewers, the stakes are usually set back from the centerline of the cut, usually an even number of feet where they will not be disturbed by the actual excavation work (see Fig. 8-11). Labeling stakes in this situation requires not only the data on the amount of cut, but also must include the amount of setback.

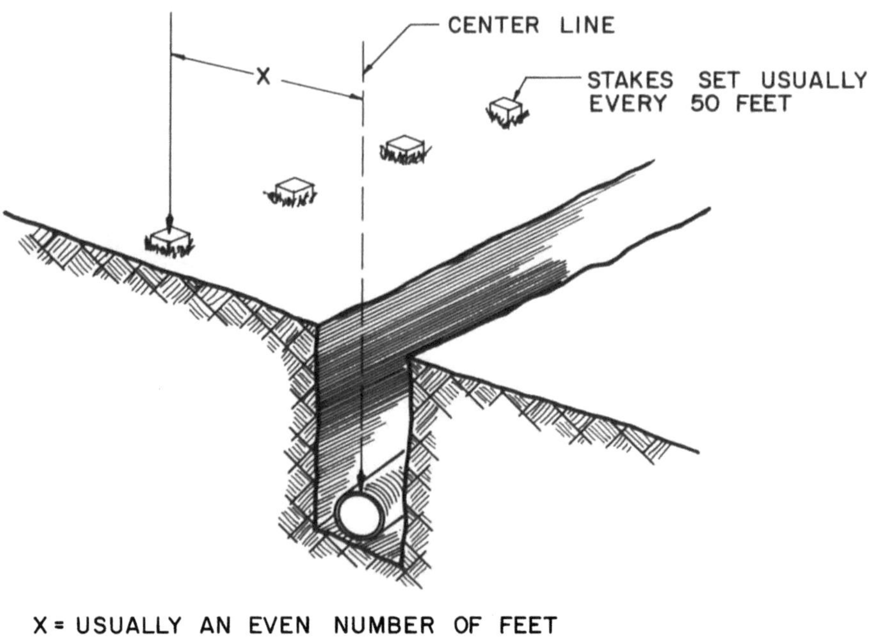

Fig. 8-11. Relationship between centerline and stakes.

Another situation arises when cut or fill side slopes are to be permanently graded to some specified slope. The stakes are labeled and set at points where the cut slope (or fill slope) intersects the original and undisturbed ground. This point is called the *catch point*. When the cut is to be made on level ground, the stakes on either side of the centerline will be equidistant from it. However, when the cut is on sloping ground, in order to maintain the required side slope, the catch points will vary in distance from the centerline. Eq. (8-3) and Fig. 8-12 illustrate this relationship and show how to determine the catch point, i.e., stake location.

$$D = \frac{B}{2} + hZ \tag{8-3}$$

where: D = catch point distance from centerline
B = width of ditch base (or top of embankment)
h = amount of cut or fill
Z = side slope ratio (horizontal to vertical, e.g., 2 : 1 slope, $Z = 2$)

The same procedure is followed in situations where embankments need to be built. A good example is the construction of a small earthen dam for a farm

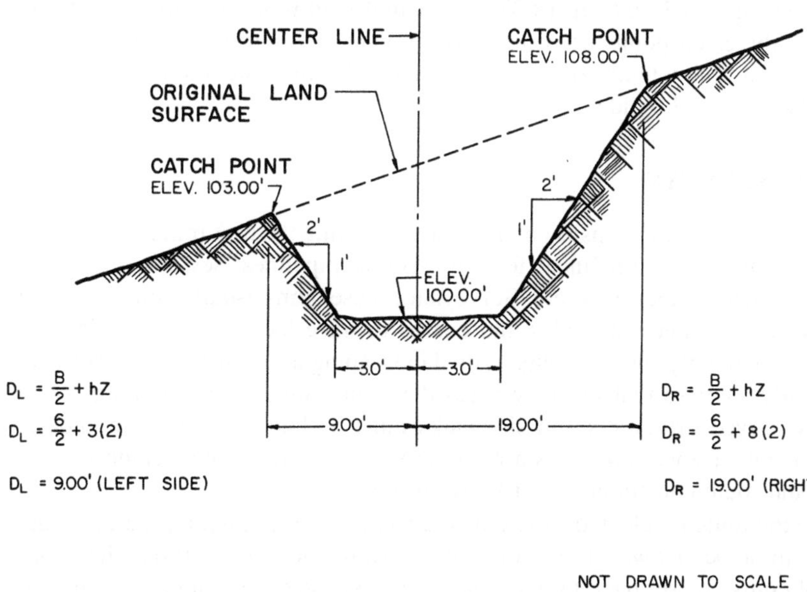

Fig. 8-12. Calculation of catch point distances from the centerline of a cut made on sloping ground.

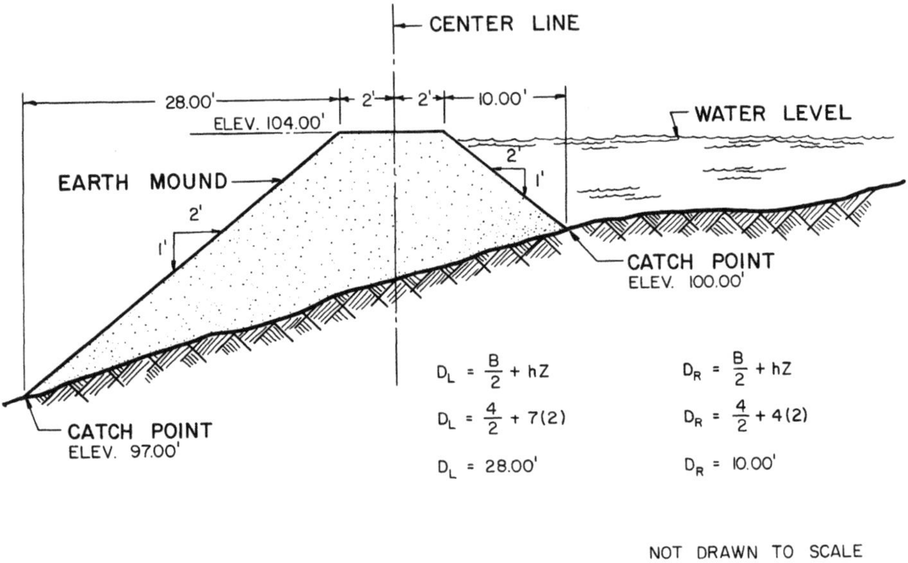

Fig. 8-13. Calculation of catch point distances from the centerline of an embankment made on sloping ground.

pond; see Fig. 8-13 and Eq. (8-3). The actual field procedure for locating the stakes at the catch points does not have to be carried out with great accuracy. Plus or minus a few tenths of a foot is sufficient for most situations encountered by landscapers and builders.

ROUTE SURVEYS

This type of surveying consists of accurately locating the best route, or the route laid out in the plans, for linear features such as pipelines, sewers, driveways, drainage ditches, sidewalks, etc. Surveys for these items usually consist of open traverse using deflection angles and curves to trace the route from point *A* to point *B*. Frequently route layouts in the landscaping and construction trades are sufficiently short so that in many cases the routes are straight, although occasionally curved driveways and sidewalks must be laid out. This presentation will illustrate a sewer layout as a route survey example because grade requirements and route constraints are usually stringent.

Once the route is selected, it is staked every 50 feet, at each turn, *and* at each change in slope. However, because of the subsequent excavation, the stakes must of course be set back to one side. The stake (station) number, the amount of setback, and the amount of cut must be recorded on each stake in some

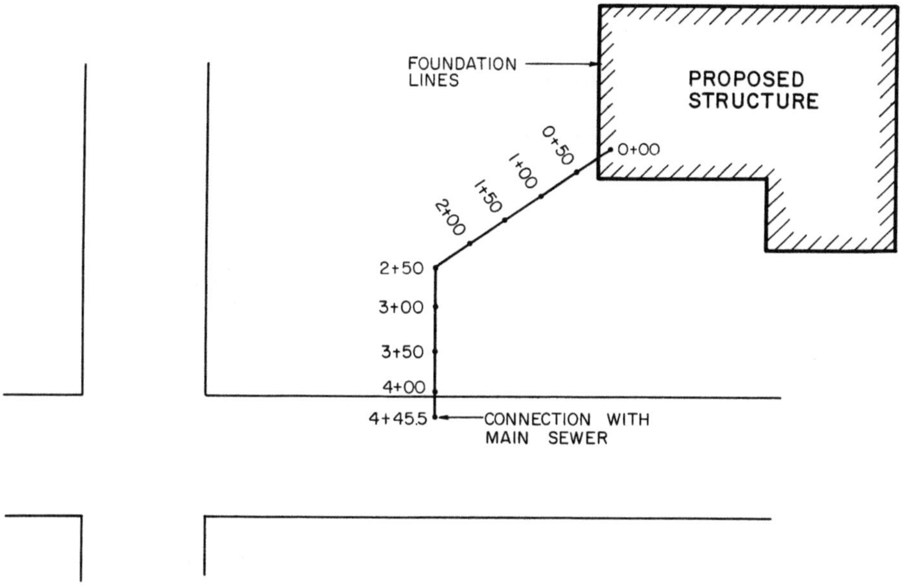

Fig. 8-14. Usual notation for labeling stations (stakes) set every 50 feet for a sewer from a structure to the connection in the street.

orderly and systematic fashion. The stations are started at the higher end and are labeled as shown in Fig. 8-14. If the route is staked every 50 feet, then station 0 + 50 is 50 feet from the beginning, station 1 + 00 is 100 feet, and by the same notation station 1 + 50 is 150 feet from the starting point. A stake labeled 3 + 64.5 is 364.5 feet along the line.

Batter boards are set across the trench at right angles to the centerline. Battens (small strips of wood) are then nailed to each batter board and are used to indicate the centerline and the depth to the invert of the sewer pipe (see Fig. 8-15). Sometimes a string may be stretched between the battens to assist in alignment of the pipe invert along the centerline.

LAYOUT OF CURVES

For most situations in the building and landscaping trades, curves are quite small and can be laid out with a 100-foot tape swung about the center. However, occasionally larger curve layouts are needed, and hence a layout procedure should be well understood. This presentation will concentrate on the layout of a simple circular curve (see Fig. 8-16). Other types of curves are more complex in their layout procedure and moreover are seldom used except in highway and railroad work. Although there are several methods for laying curves, only two

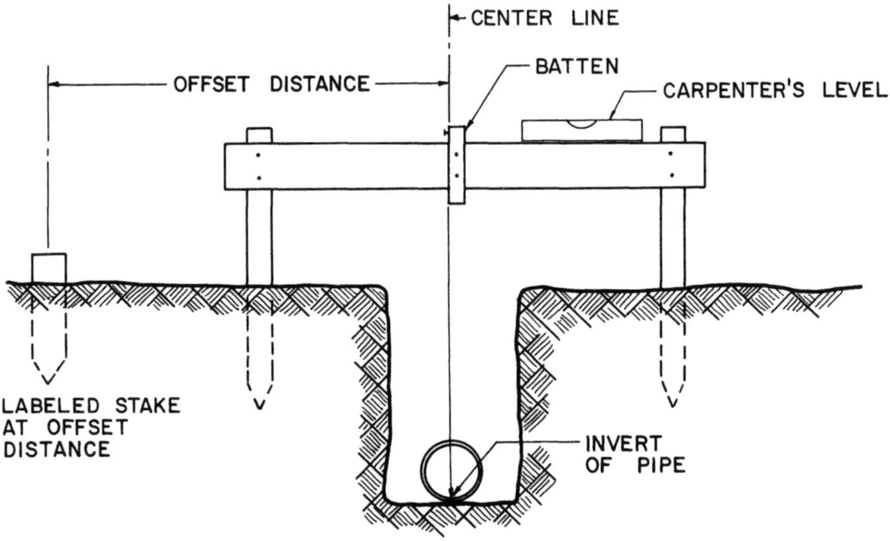

Fig. 8-15. Ideal arrangement of batter board to sewer centerline, pipe invert, and stakes.

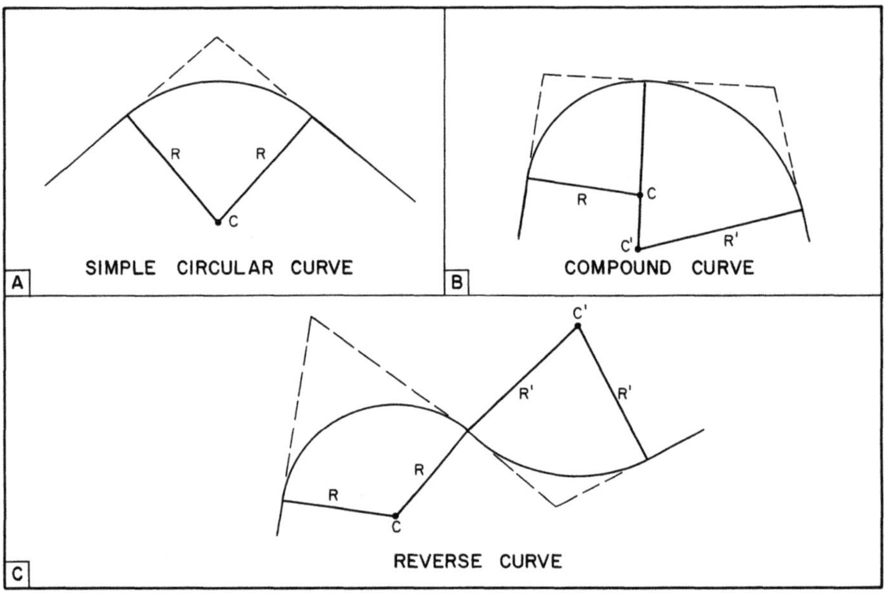

Fig. 8-16. Some common types of curves used in layout work.

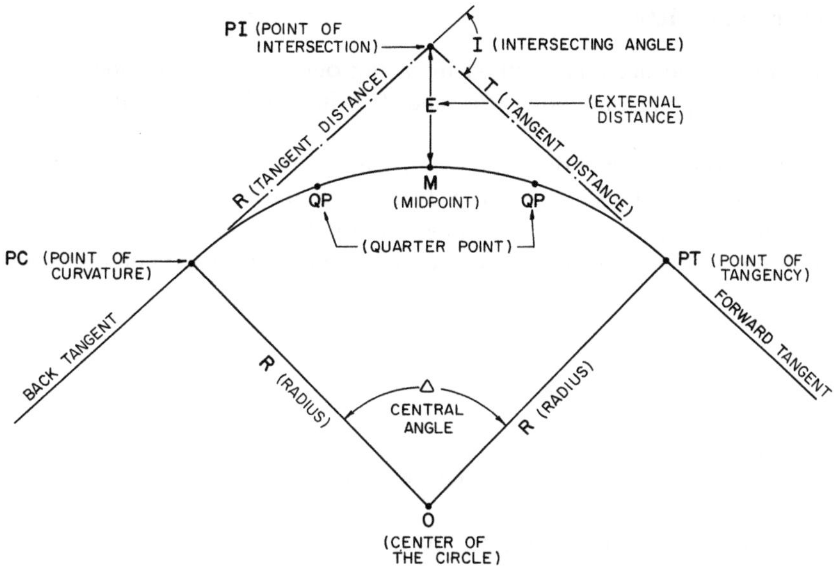

Fig. 8-17. Elements of a simple circular curve.

will be presented here. First, however, note the elements of a circular curve as shown in Fig. 8-17 and described as follows:

Point of Intersection (*PI*)	The point where the back and forward tangents meet
Intersecting Angle (*I*)	The deflection of angle at *PI*
Radius (*R*)	The radius of the circle of which the curve is a part
Point of Curvature (*PC*)	Point where the curve begins
Point of Tangency (*PT*)	The end of the curve
Tangent Distance (*T*)	The distance along the tangents from *PI* to *PC* or *PT*
Central Angle (Δ)	The angle formed by the two radii drawn from the center (*O*) of the circle to *PC* and *PT* (*Note*: $\Delta = I$)
External Distance (*E*)	The distance from *PI* to the midpoint (*M*) on the curve
Quarter Point (*QP*)	Point on the curve midway between *M* and *PT* or *PC*

Quarter Point Method

A very quick and accurate method of laying out a curve is by locating the curve's quarter points, midpoint, *PC*, and *PT* using Tables 4 and 5 in Appendix A.

Table 5 gives the direction *a* and the distance *r* from *PI* to the first and third quarter points (see Figs. 8-17 and 8-18). For a given intersecting angle (or central angle), angle *a* remains the same for *any* curve. But the value of *r* from Table 5 must be divided by the degree of curvature *D* for the curve; see Eq. (8-4):

$$D = \frac{5730}{R} \tag{8-4}$$

where: D = degree of curvature
R = radius of the circle of which the curve is a part

The midpoint on the curve is located by angle a_3; see Fig. 8-16 and Eq. (8-5):

$$a_3 = \tfrac{1}{2}(180 - \Delta) \tag{8-5}$$

where: Δ = central angle
a_3 = angle measured at *PI* from *PT* to *M*

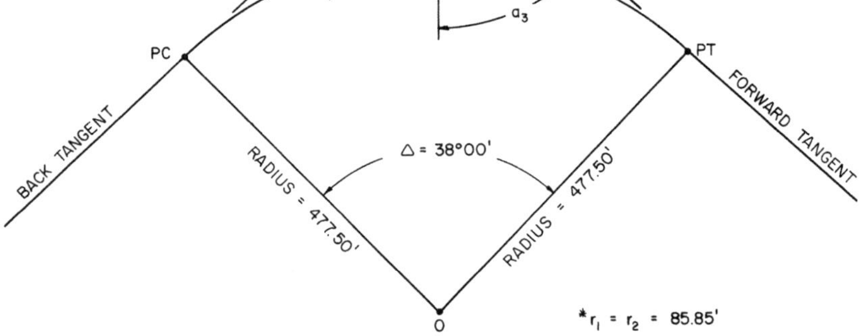

Fig. 8-18. Laying out a curve given the central angle and radius the transit set at *PI*.

The external distance is determined from Table 4, though the value must be *divided* by D. Table 4 also gives the tangent distance T.

The following example, keyed to Fig. 8-18, should help in understanding this method. With the transit set at PI and given $\Delta = 38°00'$ and $R = 477.50$ feet, the element required for curve layout can be determined from Eq. (8-4), $D = 12.00$. From Table 5, a_1 and $a_2 = 4°22'30''$, r_1 and $r_2 = 1030.2 \div 12 = 85.85$ feet. From Table 4, $E = 330.14 \div 12 = 27.51$ feet and $T = 1972.85 \div 164.40$ feet.

Deflection Angle Method

A second procedure for laying out curves uses *deflection angles*, i.e., the angle that a chord makes with a tangent (see Fig. 8-19). Using this method, any number of points can be located and staked on the arc. Note that curves that have relatively small radii need more points and stakes located than do curves with larger radii.

Two relationships, however, need to be understood before this procedure can be employed. First, Eq. (8-6) shows the length of any chord in relation to R and the central angle:

$$C = 2R \sin \frac{\Delta}{2} \tag{8-6}$$

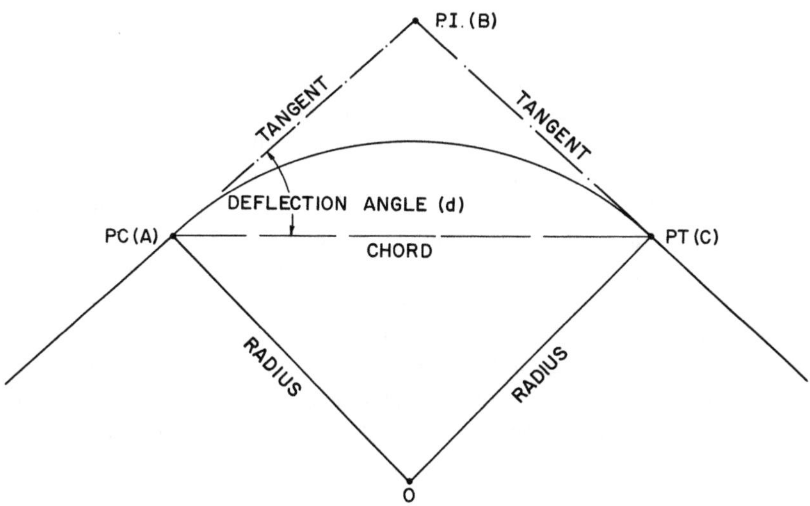

Fig. 8-19. The deflection angle is the angle made by any chord and the tangent. The magnitude of the deflection angle is $\frac{1}{2}$ the chord angle, i.e., $BAC = \frac{1}{2}AOC$.

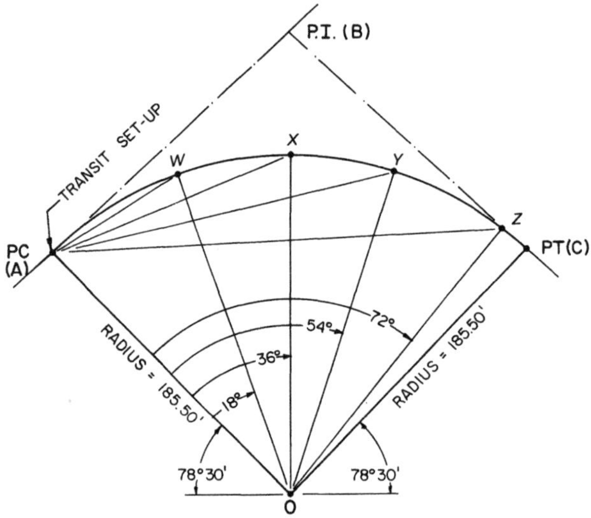

DEFLECTION ANGLES	
BAW	9°00'
BAX	18°00'
BAY	27°00'
BAZ	36°00'
BAC	39°15'

CHORD LENGTHS	
AW	58.04'
AX	114.65'
AY	168.43'
AZ	218.07'
AC	234.73'

NOT DRAWN TO SCALE

Fig. 8-20. Method of curve layout using deflection angles given the radius and the central angle. The transit is set up at *PC*.

where: C = chord length
R = radius
Δ = intersection or central angle

The second necessary relationship states that the deflection angle is always one-half the central angle. In Fig. 8-19, *BAC* is equal to $\frac{1}{2}$ *AOC*. In order to stake out a curve, a number of deflection angles and chord lengths are first computed on paper, usually in the office. Then in the field with the transit set up at *PC*, points on the arc are staked. As an example, suppose the curve in Fig. 8-20 has a radius of 185.50 feet and a central angle of 78°30′. The usual procedure is to divide the central angle into a number of equal arcs with a smaller arc left over at the end (see Fig. 8-20). By using the relationships stated above, the needed points can be calculated, i.e., in our example points *W*, *X*, *Y*, and *Z*.

Other methods of curve layout can be employed depending on the situation such as curve size, obstacles to lines of sight, inaccessible points, and accuracy requirements. However, the method presented above will be sufficient for most situations encountered by a builder and will provide good accuracy.

9
Map Surveying

A *map* is defined as a graphic representation of the earth's surface. In other words by using a series of lines, symbols, colors, and patterns, the features on the surface can be shown on a sheet of paper. It is essential, however, that the objects shown on the map be drawn in proper spatial relationship to one another. Moreover, in some cases not only is their horizontal spatial relationship important, but their vertical distance above or below a reference datum can be of use as well.

A map that shows only distances and directions in a horizontal sense is called a *planimetric map*. Good examples of this type of map are road maps, tax maps, and lot plans. On the other hand, maps that, in addition to distance and direction, show the elevation of objects in relation to a datum (usually sea level) are called *topographic maps*.

CONTROL

In order to accurately place objects with respect to each other, something must be used as a reference, or *control*, as it is known to cartographers. Ideally there should be a number of control points within the map area and/or along the edge of the map area. In situations where the area is large, the points should be situated inside the area, but for a typical lot, or a several acre plot, the property corners will suffice. The control points must be established horizontally for a planimetric map and both horizontally and vertically for a topographic map.

To start, somewhere outside the area there are points whose positions are known, and it is from these that the new control is established. So measurements must be made that tie the known points to the unknown points. These ties must always consist of one of the following:

1. Two distance measurements (see Fig. 9-1)
2. One angle and one distance measurement (see Fig. 9-2)
3. Two angle measurements (see Fig. 9-3)

There are many other combinations that are possible, but the above three, modified as the situation requires, should be sufficient in most situations. It has been

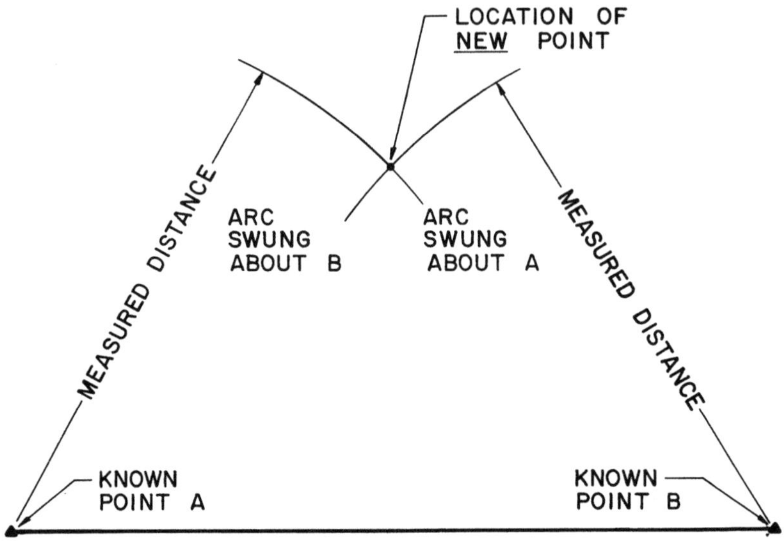

Fig. 9-1. Establishing a new control point by using two distance measurements.

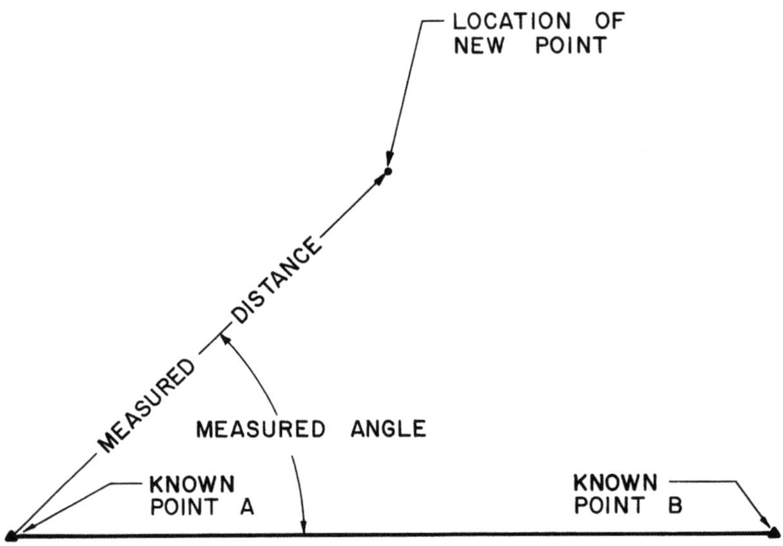

Fig. 9-2. Establishing a new control point by using one angle and one distance measurement, i.e., angle distance method.

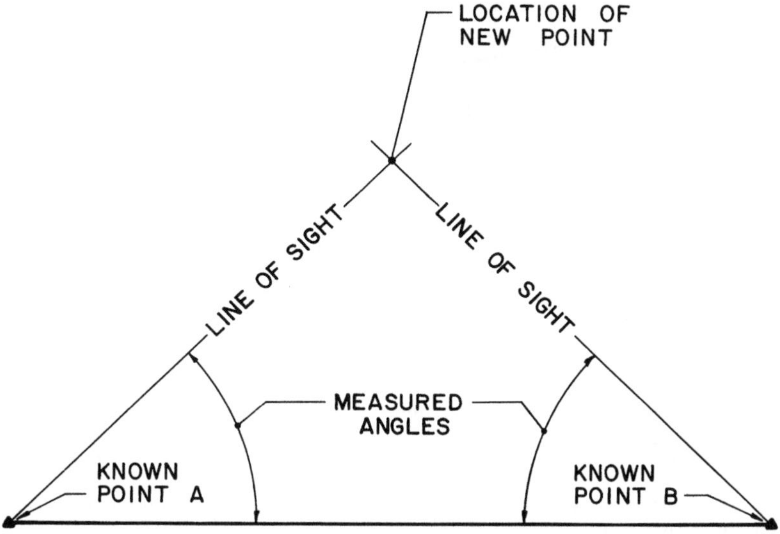

Fig. 9-3. Establishing a new control point by using two angle measurements.

the author's experience that the number two is used most frequently in that all the measurements originate at one point.

The above ties establish only horizontal control of a point. If vertical control is also required, then differential leveling or stadia techniques can be utilized (see Chapter 7).

PLANIMETRIC MAPPING

When locating points for a planimetric map, the easiest and quickest procedure is the angle-distance method (see Fig. 9-4). When building locations are being mapped, it is common practice to locate two corners with the transit and then tape the other corners. In most mapping situations, the distances can be measured by stadia. It is important that note-keeping is done in an orderly fashion, especially if a large number of "rays" are measured from a single point. In most situations a sketch is essential.

Occasionally it is easier to locate features from points already established rather than from the original control point. This is usually done when features cannot be seen from the control (see Fig. 9-5). It must be remembered, however, that any error in locating building *B* will be carried to building *A*, so accuracy is important.

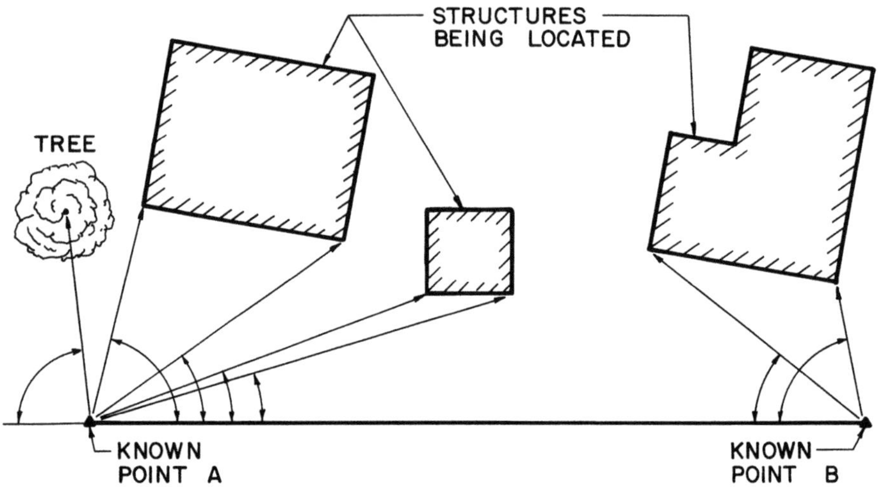

Fig. 9-4. Using the angle-distance method to locate structure corners for a planimetric map. The other remaining corners are located by taping.

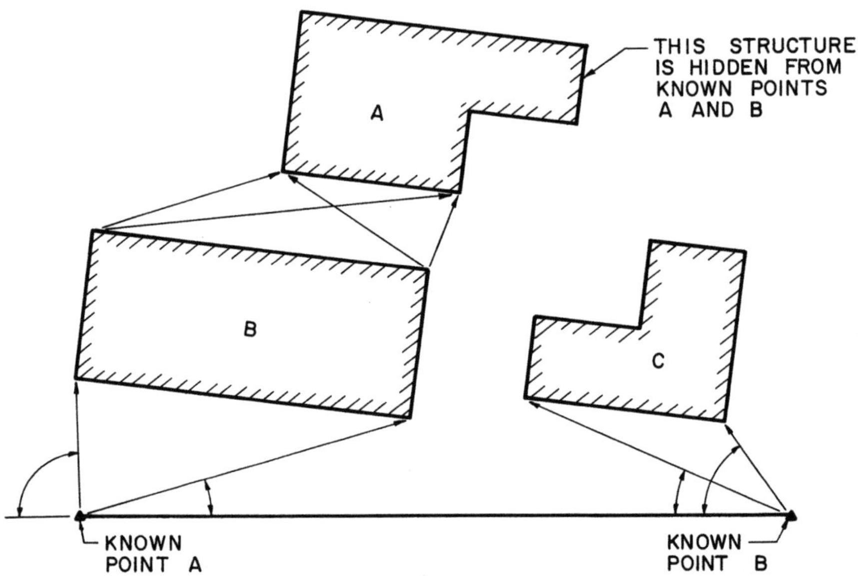

Fig. 9-5. Building *A* is located from the corners of building *B*, the location of which is established by the angle-distance method.

TOPOGRAPHIC MAPPING

Topographic mapping, sometimes called contour mapping, is frequently used in many large landscaping and building projects such as golf courses, parks, real estate developments, etc. Because contour lines identify and delineate all the points in the mapped area that are at a specific elevation, they are useful in determining the shape, steepness, and direction of land slopes. There are two procedures in common usage that identify points for later contouring.

Grid Method

The initial step in the construction of a topographic map is locating a number of points both horizontally and vertically. In relatively small areas such as lots for building sites, the grid method is the easiest to use and most accurate. The area in question is first divided into a series of squares, i.e., a grid is laid out over the area. The size of the grid squares usually varies from 10 feet to as large as 100 feet depending on the lot size, the accuracy needed, and the scale of the final map. The grid intersections are then staked and labeled. The labeling or identification scheme is left to the discretion of the field crew as there is no correct or incorrect method. However, Fig. 9-6 shows a system that is frequently used and is quite simple.

The next step is to ascertain the elevation of each of the grid intersections.

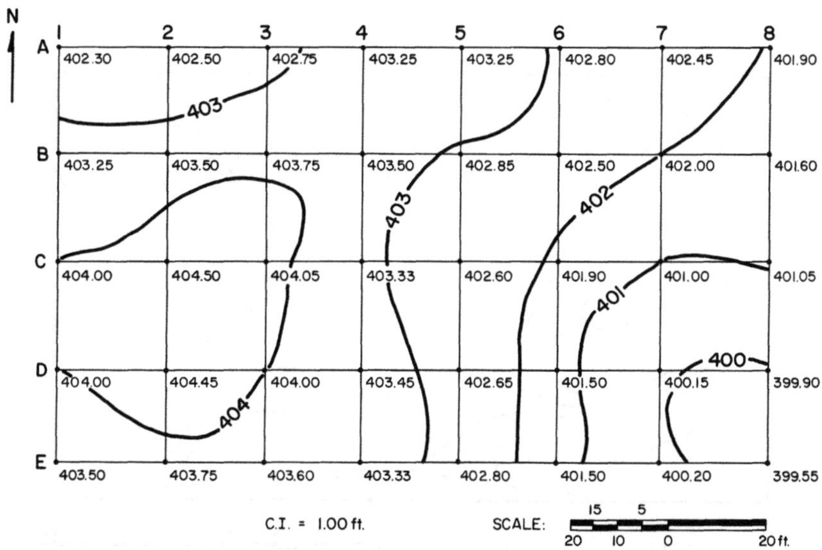

Fig. 9-6. Grid method of contouring. Spot elevations are located at grid intersections. The intersections may be identified by a letter/number coordinate system.

This is done by differential or stadia leveling using a reliable control point as a back sight and the grid intersections as fore sights.

The last step in the procedure is to draw the contour lines. For the plot shown in Fig. 9-6, a one-foot contour interval is used. (*Note:* The *contour interval* is the vertical distance between successive or adjacent contour lines.) The choice of contour interval depends on the relief in the area and the detail that needs to be shown. It should be mentioned here that there are a number of packaged computer programs on the market that will choose a contour interval and draw contour lines. However, unless a great amount of topographic mapping is to be done, drawing the contours "by hand" will suffice.

Governing Point Method

In the grid method of locating spots to serve as data points, the stations were determined by the grid spacing regardless of the topography. Conversely, in this method, which is commonly used in mapping large tracts of land, the data points selected are determined by certain critical features of the topography. Points that are frequently used are: tops of small hills, bends in a stream, the confluence of two streams, the bottom of a depression, and bends in a road.

Once the instrument is set up, it becomes the responsibility of the rodperson(s) to choose points that are important to the accuracy of the final map. In Fig. 9-7 several such points are marked. Notice that points are chosen along

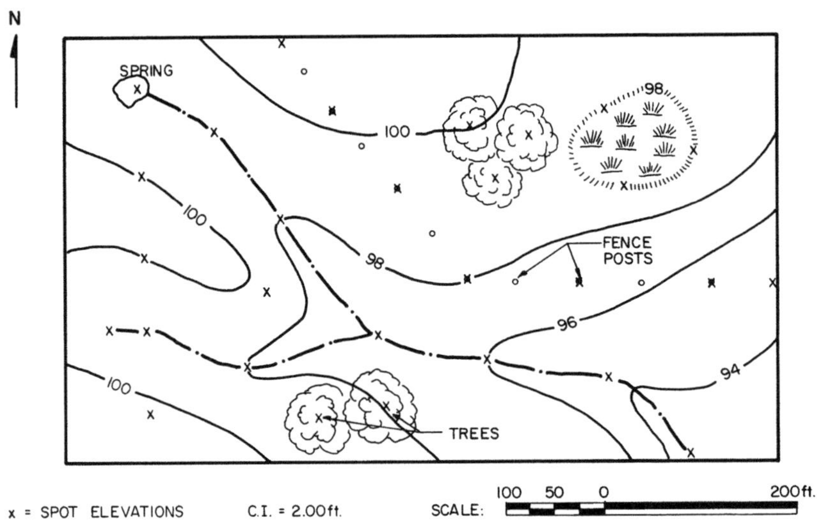

Fig. 9-7. Governing point method for establishing elevation control points. Points are chosen from key spots in the mapped area, rather than at grid intersections.

the streams, at fence posts, around the swamp, at trees, top of a hill, and at a few other spots where control points are sparse. To ensure a good distribution of points, a good rule of thumb is to space the rod stations no farther apart than twice the longest instrument shot.

The horizontal position of the points is most easily ascertained by the angle-distance method (using stadia to determine distances), and the vertical positions by differential leveling or stadia leveling. It is important that the rodperson take good notes to describe the location of each point. His notes must correspond to the notes of the instrument person as well. After 10 or 20 rod sights, the rodperson should return to the instrument to compare notes, check the sight numbering, and to plan the next group of shots.

Once the data points are acquired in sufficient numbers, drawing the contour lines is the next step. The person drawing the contour lines must rely heavily on the notes of the rodperson, especially noting peculiar and unusual features that need to be included on the map. As a reminder, the contour line elevations must be tied to a reliable bench mark.

PROFILES FROM TOPOGRAPHIC MAPS

Profiles are very useful when viewing the topography to get a clearer view of a particular feature. All profiles have two scales. The first scale is a horizontal scale, which is the scale of the map from which the data are being taken. The second scale is the vertical scale. Sometimes it is desirable, or even necessary, to stretch out or exaggerate the vertical scale to allow subtle topographic features to show more clearly. If this is done, the profile is said to have *vertical exaggeration*, or V.E., Eq. (9-1):

$$\text{V.E.} = \frac{\text{horizontal scale}}{\text{vertical scale}} \tag{9-1}$$

To construct a topographic profile, follow the steps listed below while referring to Fig. 9-8:

1. Draw a line (*XY*) on a map along the traverse where you want to make the profile.
2. Examine the contour lines that intersect the profile line looking for the highest and the lowest elevations encountered, i.e., maximum relief.
3. Label the horizontal lines on the profile paper so as to accommodate all of the elevations found in Step 2. Choose a vertical scale to keep the V.E. to a minimum but yet will emphasize the topography.
4. Align the horizontal lines of the profile paper parallel to the profile line *XY*. These may be taped in place to hold them securely while you are working.

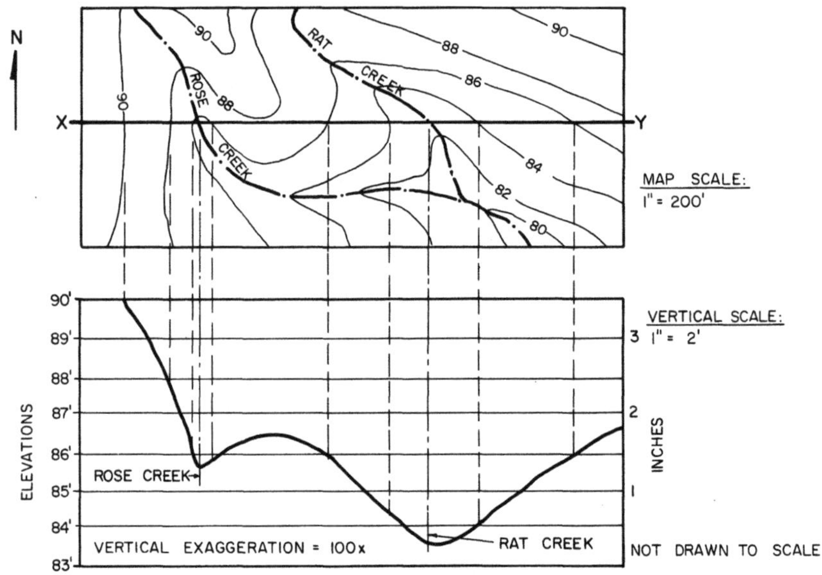

Fig. 9-8. Construction of a topographic profile.

5. Drop a line (dashed and very light to keep from getting confused by too many lines) to the appropriate elevation on the profile paper from each *XY* line/contour line intersection on the map.
6. Using logic and good judgment, connect the points on the profile by a line. This line then represents a cross-sectional view of the land surface along *XY*.
7. Identify on the profile streams, roads, lakes, hilltops, and other features you feel are necessary for completeness.

It should be mentioned here that a little practice may be necessary in choosing the proper V.E. In general, the greater the maximum relief along the profile line, the smaller the vertical exaggeration, and vice versa.

USES OF CONTOUR LINES

Once completed, a topographic map can be a very useful tool for further work. Perhaps the most common use is to *quantitatively* ascertain the magnitude of a slope. This has application not only in designing grades for driveways, sidewalks, cart paths, drainage systems, etc., but also in determining the declivity of existing slopes such as embankments, hill slopes, stream channels, and ditches.

In general terms, a slope can be mathematically expressed by the relationship shown in Eq. (9-2):

$$G = \frac{D}{H} \tag{9-2}$$

where: G = slope gradient between any two points
D = difference in elevation between the two points
H = horizontal distance between the two points

For example, if the difference in elevation between two points is 2.36 feet and the horizontal distance between the points is 112.50 feet, then the gradient (or grade, or slope) is 2.36′/112.50′ = 0.021 or 2.1%.

As the first of several practical examples, suppose the gradient of the stream in Fig. 9-9 must be determined between points *A* and *B*. (*Note:* Stream gradients, due to the extreme length of the channel as compared to the vertical drop, are usually expressed as feet/mile. Stated another way, this shows the number of feet of vertical drop there is in the stream channel for each mile of horizontal travel.) To calculate gradient, follow the steps below while referring to Fig. 9-9:

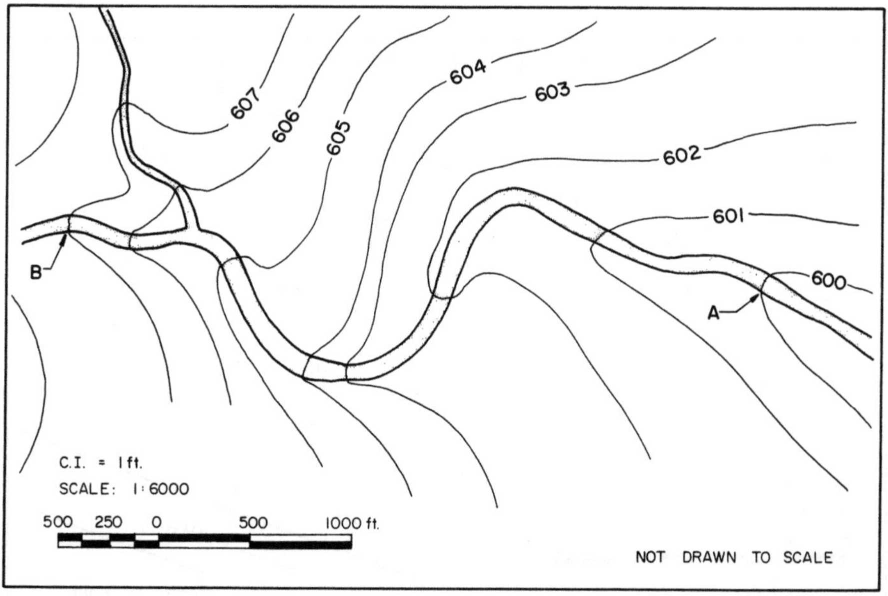

Fig. 9-9. Gradient is the slope of the stream channel in feet/mile between two points, e.g., points *A* and *B*.

1. Locate two points between which you want to ascertain the gradient; points *A* and *B* in Fig. 9-9.
2. Determine the difference in elevation between the two points (7 feet in Fig. 9-9).
3. Determine the horizontal distance between the two points. Note that if the stream has many bends in it, find the distance along the channel, not just the straight-line distance (2.25 miles in Fig. 9-9).
4. Convert your data into "ft/mile" form:

$$\frac{7'}{2.25 \text{ miles}} = \frac{X'}{1 \text{ mile}} = 3.11 \text{ ft/mile}$$

Another problem that frequently arises is to lay out a route at a given grade between two points. As an example, suppose a driveway needs to be built between points *A* and *B* (see Fig. 9-10) with the maximum grade not to exceed 3%. Using Eq. (9-2), it is possible to calculate the length and the location of the driveway between each contour at the required slope.

First, solving Eq. (9-2) for *H*, we get: $H = D \div G$. In this example, *D* is the value of the contour interval, i.e., the vertical distance between adjacent lines (5 feet), and G is the required maximum grade of 3% (0.03); thus H = 5 ÷ 0.03 = 166.67 feet. The procedure starts at point *A*. Swing an arc 166.67

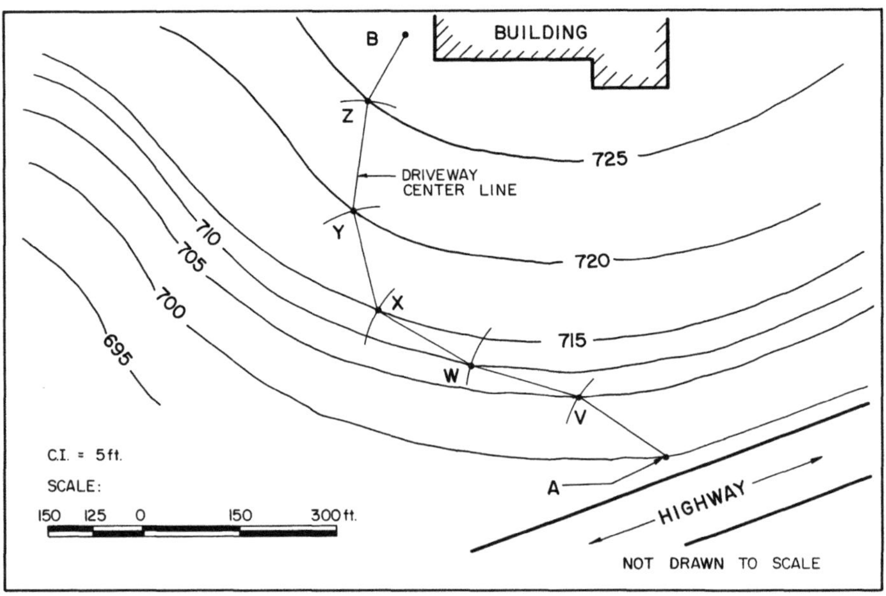

Fig. 9-10. Laying out a driveway between two points at a given grade.

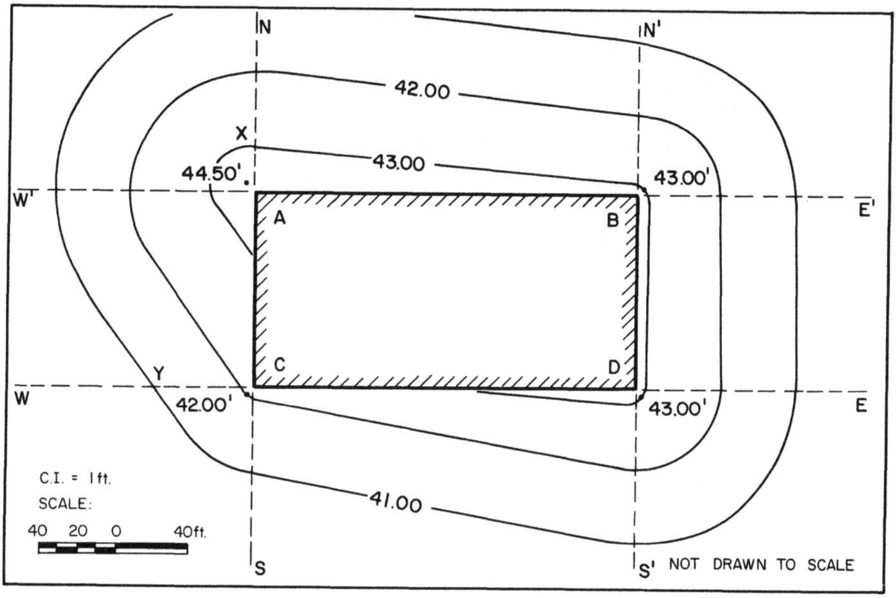

Fig. 9-11. Sloping a finished surface away from a building at a given, prescribed grade.

feet in length about point *A* so that it crosses the next contour line up the slope, i.e., the 705 line. Label that point *V*. The line from *A* to *V* marks the route of the roadway between those two points at a 3% grade. Repeat the procedure at each contour line. Variations in the spacing of the contour lines, which indicate changes in slope angle, will determine the location and direction of the route center line. Note that this route was established on the basis of only one parameter, the percent grade. Other considerations such as amounts of cut and fill, the length of the route, economic restraints, aesthetics, and other factors may well change the location from what was calculated here.

Yet another problem that frequently arises is to slope a finished surface away from a structure at a given grade for drainage purposes. As an example, suppose the finish grade around the structure in Fig. 9-11 is required by the plans to be 3%. The finished grade elevations are shown at the corners. The objective here is to correctly position the slope contours around the building at, say a one foot contour interval and showing the required 3% grade.

Starting at corner *A* (Fig. 9-11) draw line *AN* at right angles to the *AB* face of the building. Note that the difference in elevation between corner *A* and the first contour line down the slope is 0.50 feet. Using $H = D \div G$ from Eq. (9-2), we get D (0.50′) $\div$ G (0.03) = 16.67 feet. Therefore, point *X*, 16.67 feet away from the building along line *AN* is on the 43-foot contour line, and the slope is 3%. At corner *C* the procedure is the same. Draw line *CW* perpendicular to the building. The difference in elevation between *C* and the next contour line down slope is 1.0 foot. Using $H = D \div G$, we get $(1.0') \div (0.03) = 33.33$ feet. Hence, point *Y*, 33.33 feet away from the structure along line *CW*, lies on the 41-foot contour line. The slope is 3%.

The same procedure is followed at all the other corners using all the building-face perpendiculars indicated in Fig. 9-11. The contour lines are drawn as straight lines and the corners curved to indicate that the actual final grading in the field is to be smooth and even.

APPENDIXES

Appendix A
Tables

Table A-1. Natural Sines and Cosines

′	0° Sine	0° Cosine	′	′	0° Sine	0° Cosine	′	′	0° Sine	0° Cosine	′
0	.00000	1	60	21	.00611	.99998	39	41	.01193	.99993	19
1	.00029	1	59	22	.00640	.99998	38	42	.01222	.99993	18
2	.00058	1	58	23	.00669	.99998	37	43	.01251	.99992	17
3	.00087	1	57	24	.00698	.99998	36	44	.01280	.99992	16
4	.00116	1	56	25	.00727	.99997	35	45	.01309	.99991	15
5	.00145	1	55	26	.00756	.99997	34	46	.01338	.99991	14
6	.00175	1	54	27	.00785	.99997	33	47	.01367	.99991	13
7	.00204	1	53	28	.00814	.99997	32	48	.01396	.99990	12
8	.00233	1	52	29	.00844	.99996	31	49	.01425	.99990	11
9	.00262	1	51	30	.00873	.99996	30	50	.01454	.99989	10
10	.00291	1	50	31	.00902	.99996	29	51	.01483	.99989	9
11	.00320	.99999	49	32	.00931	.99996	28	52	.01513	.99989	8
12	.00349	.99999	48	33	.00960	.99995	27	53	.01542	.99988	7
13	.00378	.99999	47	34	.00989	.99995	26	54	.01571	.99988	6
14	.00407	.99999	46	35	.01018	.99995	25	55	.01600	.99987	5
15	.00436	.99999	45	36	.01047	.99995	24	56	.01629	.99987	4
16	.00465	.99999	44	37	.01076	.99994	23	57	.01658	.99986	3
17	.00495	.99999	43	38	.01105	.99994	22	58	.01687	.99986	2
18	.00524	.99999	42	39	.01134	.99994	21	59	.01716	.99985	1
19	.00553	.99998	41	40	.01164	.99993	20	60	.01745	.99985	0
20	.00582	.99998	40								
′	89° Cosine	89° Sine	′	′	89° Cosine	89° Sine	′	′	89° Cosine	89° Sine	′

Table A-1. (*Continued*)

′	1° Sine	1° Cosine	2° Sine	2° Cosine	3° Sine	3° Cosine	4° Sine	4° Cosine	′
0	.01745	.99985	.03490	.99939	.05234	.99863	.06976	.99756	60
1	.01774	.99984	.03519	.99938	.05263	.99861	.07005	.99754	59
2	.01803	.99984	.03548	.99937	.05292	.99860	.07034	.99752	58
3	.01832	.99983	.03577	.99936	.05321	.99858	.07063	.99750	57
4	.01862	.99983	.03606	.99935	.05350	.99857	.07092	.99748	56
5	.01891	.99982	.03635	.99934	.05379	.99855	.07121	.99746	55
6	.01920	.99982	.03664	.99933	.05408	.99854	.07150	.99744	54
7	.01949	.99981	.03693	.99932	.05437	.99852	.07179	.99742	53
8	.01978	.99980	.03723	.99931	.05466	.99851	.07208	.99740	52
9	.02007	.99980	.03752	.99930	.05495	.99849	.07237	.99738	51
10	.02036	.99979	.03781	.99929	.05524	.99847	.07266	.99736	50
11	.02065	.99979	.03810	.99927	.05553	.99846	.07295	.99734	49
12	.02094	.99978	.03839	.99926	.05582	.99844	.07324	.99731	48
13	.02123	.99977	.03868	.99925	.05611	.99842	.07353	.99729	47
14	.02152	.99977	.03897	.99924	.05640	.99841	.07382	.99727	46
15	.02181	.99976	.03926	.99923	.05669	.99839	.07411	.99725	45
16	.02211	.99976	.03955	.99922	.05698	.99838	.07440	.99723	44
17	.02240	.99975	.03984	.99921	.05727	.99836	.07469	.99721	43
18	.02269	.99974	.04013	.99919	.05756	.99834	.07498	.99719	42
19	.02298	.99974	.04042	.99918	.05785	.99833	.07527	.99716	41
20	.02327	.99973	.04071	.99917	.05814	.99831	.07556	.99714	40
21	.02356	.99972	.04100	.99916	.05844	.99829	.07585	.99712	39
22	.02385	.99972	.04129	.99915	.05873	.99827	.07614	.99710	38
23	.02414	.99971	.04159	.99913	.05902	.99826	.07643	.99708	37
24	.02443	.99970	.04188	.99912	.05931	.99824	.07672	.99705	36
25	.02472	.99969	.04217	.99911	.05960	.99822	.07701	.99703	35
26	.02501	.99969	.04246	.99910	.05989	.99821	.07730	.99701	34
27	.02530	.99968	.04275	.99909	.06018	.99819	.07759	.99699	33
28	.02560	.99967	.04304	.99907	.06047	.99817	.07788	.99696	32
29	.02589	.99966	.04333	.99906	.06076	.99815	.07817	.99694	31
30	.02618	.99966	.04362	.99905	.06105	.99813	.07846	.99692	30
31	.02647	.99965	.04391	.99904	.06134	.99812	.07875	.99689	29
32	.02676	.99964	.04420	.99902	.06163	.99810	.07904	.99687	28
33	.02705	.99963	.04449	.99901	.06192	.99808	.07933	.99685	27
34	.02734	.99963	.04478	.99900	.06221	.99806	.07962	.99683	26
35	.02763	.99962	.04507	.99898	.06250	.99804	.07991	.99680	25
36	.02792	.99961	.04536	.99897	.06279	.99803	.08020	.99678	24
37	.02821	.99960	.04565	.99896	.06308	.99801	.08049	.99676	23
38	.02850	.99959	.04594	.99894	.06337	.99799	.08078	.99673	22
39	.02879	.99959	.04623	.99893	.06366	.99797	.08107	.99671	21
40	.02908	.99958	.04653	.99892	.06395	.99795	.08136	.99668	20
41	.02938	.99957	.04682	.99890	.06424	.99793	.08165	.99666	19
42	.02967	.99956	.04711	.99889	.06453	.99792	.08194	.99664	18
43	.02996	.99955	.04740	.99888	.06482	.99790	.08223	.99661	17
44	.03025	.99954	.04769	.99886	.06511	.99788	.08252	.99659	16
45	.03054	.99953	.04798	.99885	.06540	.99786	.08281	.99657	15
46	.03083	.99952	.04827	.99883	.06569	.99784	.08310	.99654	14
47	.03112	.99952	.04856	.99882	.06598	.99782	.08339	.99652	13
48	.03141	.99951	.04885	.99881	.06627	.99780	.08368	.99649	12
49	.03170	.99950	.04914	.99879	.06656	.99778	.08397	.99647	11
50	.03199	.99949	.04943	.99878	.06685	.99776	.08426	.99644	10
51	.03328	.99948	.04972	.99876	.06714	.99774	.08455	.99642	9
52	.03257	.99947	.05001	.99875	.06743	.99772	.08484	.99639	8
53	.03286	.99946	.05030	.99873	.06773	.99770	.08513	.99637	7
54	.03316	.99945	.05059	.99872	.06802	.99768	.08542	.99635	6
55	.03345	.99944	.05088	.99870	.06831	.99766	.08571	.99632	5
56	.03374	.99943	.05117	.99869	.06860	.99764	.08600	.99630	4
57	.03403	.99942	.05146	.99867	.06889	.99762	.08629	.99627	3
58	.03432	.99941	.05175	.99866	.06918	.99760	.08658	.99625	2
59	.03461	.99940	.05205	.99864	.06947	.99758	.08687	.99622	1
60	.03490	.99939	.05234	.99863	.06976	.99756	.08716	.99619	0
′	88° Cosine	88° Sine	87° Cosine	87° Sine	86° Cosine	86° Sine	85° Cosine	85° Sine	′

Table A-1. (*Continued*)

′	5° Sine	5° Cosine	6° Sine	6° Cosine	7° Sine	7° Cosine	8° Sine	8° Cosine	′
0	.08716	.99619	.10453	.99452	.12187	.99255	.13917	.99027	60
1	.08745	.99617	.10482	.99449	.12216	.99251	.13946	.99023	59
2	.08774	.99614	.10511	.99446	.12245	.99248	.13975	.99019	58
3	.08803	.99612	.10540	.99443	.12274	.99244	.14004	.99015	57
4	.08831	.99609	.10569	.99440	.12302	.99240	.14033	.99011	56
5	.08860	.99607	.10597	.99437	.12331	.99237	.14061	.99006	55
6	.08889	.99604	.10626	.99434	.12360	.99233	.14090	.99002	54
7	.08918	.99602	.10655	.99431	.12389	.99230	.14119	.98998	53
8	.08947	.99599	.10684	.99428	.12418	.99226	.14148	.98994	52
9	.08976	.99596	.10713	.99424	.12447	.99222	.14177	.98990	51
10	.09005	.99594	.10742	.99421	.12476	.99219	.14205	.98986	50
11	.09034	.99591	.10771	.99418	.12504	.99215	.14234	.98982	49
12	.09063	.99588	.10800	.99415	.12533	.99211	.14263	.98978	48
13	.09092	.99586	.10829	.99412	.12562	.99208	.14292	.98973	47
14	.09121	.99583	.10858	.99409	.12591	.99204	.14320	.98969	46
15	.09150	.99580	.10887	.99406	.12620	.99200	.14349	.98965	45
16	.09179	.99578	.10916	.99402	.12649	.99197	.14378	.98961	44
17	.09208	.99575	.10945	.99399	.12678	.99193	.14407	.98957	43
18	.09237	.99572	.10973	.99396	.12706	.99189	.14436	.98953	42
19	.09266	.99570	.11002	.99393	.12735	.99186	.14464	.98948	41
20	.09295	.99567	.11031	.99390	.12764	.99182	.14493	.98944	40
21	.09324	.99564	.11060	.99386	.12793	.99178	.14522	.98940	39
22	.09353	.99562	.11089	.99383	.12822	.99175	.14551	.98936	38
23	.09382	.99559	.11118	.99380	.12851	.99171	.14580	.98931	37
24	.09411	.99556	.11147	.99377	.12880	.99167	.14608	.98927	36
25	.09440	.99553	.11176	.99374	.12908	.99163	.14637	.98923	35
26	.09469	.99551	.11205	.99370	.12937	.99160	.14666	.98919	34
27	.09498	.99548	.11234	.99367	.12966	.99156	.14695	.98914	33
28	.09527	.99545	.11263	.99364	.12995	.99152	.14723	.98910	32
29	.09556	.99542	.11291	.99360	.13024	.99148	.14752	.98906	31
30	.09585	.99540	.11320	.99357	.13053	.99144	.14781	.98902	30
31	.09614	.99537	.11349	.99354	.13081	.99141	.14810	.98897	29
32	.09642	.99534	.11378	.99351	.13110	.99137	.14838	.98893	28
33	.09671	.99531	.11407	.99347	.13139	.99133	.14867	.98889	27
34	.09700	.99528	.11436	.99344	.13168	.99129	.14896	.98884	26
35	.09729	.99526	.11465	.99341	.13197	.99125	.14925	.98880	25
36	.09758	.99523	.11494	.99337	.13226	.99122	.14954	.98876	24
37	.09787	.99520	.11523	.99334	.13254	.99118	.14982	.98871	23
38	.09816	.99517	.11552	.99331	.13283	.99114	.15011	.98867	22
39	.09845	.99514	.11580	.99327	.13312	.99110	.15040	.98863	21
40	.09874	.99511	.11609	.99324	.13341	.99106	.15069	.98858	20
41	.09903	.99508	.11638	.99320	.13370	.99102	.15097	.98854	19
42	.09932	.99506	.11667	.99317	.13399	.99098	.15126	.98849	18
43	.09961	.99503	.11696	.99314	.13427	.99094	.15155	.98845	17
44	.09990	.99500	.11725	.99310	.13456	.99091	.15184	.98841	16
45	.10019	.99497	.11754	.99307	.13485	.99087	.15212	.98836	15
46	.10048	.99494	.11783	.99303	.13514	.99083	.15241	.98832	14
47	.10077	.99491	.11812	.99300	.13543	.99079	.15270	.98827	13
48	.10106	.99488	.11840	.99297	.13572	.99075	.15299	.98823	12
49	.10135	.99485	.11869	.99293	.13600	.99071	.15327	.98818	11
50	.10164	.99482	.11898	.99290	.13629	.99067	.15356	.98814	10
51	.10192	.99479	.11927	.99286	.13658	.99063	.15385	.98809	9
52	.10221	.99476	.11956	.99283	.13687	.99059	.15414	.98805	8
53	.10250	.99473	.11985	.99279	.13716	.99055	.15442	.98800	7
54	.10279	.99470	.12014	.99276	.13744	.99051	.15471	.98796	6
55	.10308	.99467	.12043	.99272	.13773	.99047	.15500	.98791	5
56	.10337	.99464	.12071	.99269	.13802	.99043	.15529	.98787	4
57	.10366	.99461	.12100	.99265	.13831	.99039	.15557	.98782	3
58	.10395	.99458	.12129	.99262	.13860	.99035	.15586	.98778	2
59	.10424	.99455	.12158	.99258	.13889	.99031	.15615	.98773	1
60	.10453	.99452	.12187	.99255	.13917	.99027	.15643	.98769	0
′	Cosine 84°	Sine 84°	Cosine 83°	Sine 83°	Cosine 82°	Sine 82°	Cosine 81°	Sine 81°	′

Table A-1. (*Continued*)

′	9° Sine	9° Cosine	10° Sine	10° Cosine	11° Sine	11° Cosine	12° Sine	12° Cosine	′
0	.15643	.98769	.17365	.98481	.19081	.98163	.20791	.97815	60
1	.15672	.98764	.17393	.98476	.19109	.98157	.20820	.97809	59
2	.15701	.98760	.17422	.98471	.19138	.98152	.20848	.97803	58
3	.15730	.98755	.17451	.98466	.19167	.98146	.20877	.97797	57
4	.15758	.98751	.17479	.98461	.19195	.98140	.20905	.97791	56
5	.15787	.98746	.17508	.98455	.19224	.98135	.20933	.97784	55
6	.15816	.98741	.17537	.98450	.19252	.98129	.20962	.97778	54
7	.15845	.98737	.17565	.98445	.19281	.98124	.20990	.97772	53
8	.15873	.98732	.17594	.98440	.19309	.98118	.21019	.97766	52
9	.15902	.98728	.17623	.98435	.19338	.98112	.21047	.97760	51
10	.15931	.98723	.17651	.98430	.19366	.98107	.21076	.97754	50
11	.15959	.98718	.17680	.98425	.19395	.98101	.21104	.97748	49
12	.15988	.98714	.17708	.98420	.19423	.98096	.21132	.97742	48
13	.16017	.98709	.17737	.98414	.19452	.98090	.21161	.97735	47
14	.16046	.98704	.17766	.98409	.19481	.98084	.21189	.97729	46
15	.16074	.98700	.17794	.98404	.19509	.98079	.21218	.97723	45
16	.16103	.98695	.17823	.98399	.19538	.98073	.21246	.97717	44
17	.16132	.98690	.17852	.98394	.19566	.98067	.21275	.97711	43
18	.16160	.98686	.17880	.98389	.19595	.98061	.21303	.97705	42
19	.16189	.98681	.17909	.98383	.19623	.98056	.21331	.97698	41
20	.16218	.98676	.17937	.98378	.19652	.98050	.21360	.97692	40
21	.16246	.98671	.17966	.98373	.19680	.98044	.21388	.97686	39
22	.16275	.98667	.17995	.98368	.19709	.98039	.21417	.97680	38
23	.16304	.98662	.18023	.98362	.19737	.98033	.21445	.97673	37
24	.16333	.98657	.18052	.98357	.19766	.98027	.21474	.97667	36
25	.16361	.98652	.18081	.98352	.19794	.98021	.21502	.97661	35
26	.16390	.98648	.18109	.98347	.19823	.98016	.21530	.97655	34
27	.16419	.98643	.18138	.98341	.19851	.98010	.21559	.97648	33
28	.16447	.98638	.18166	.98336	.19880	.98004	.21587	.97642	32
29	.16476	.98633	.18195	.98331	.19908	.97998	.21616	.97636	31
30	.16505	.98629	.18224	.98325	.19937	.97992	.21644	.97630	30
31	.16533	.98624	.18252	.98320	.19965	.97987	.21672	.97623	29
32	.16562	.98619	.18281	.98315	.19994	.97981	.21701	.97617	28
33	.16591	.98614	.18309	.98310	.20022	.97975	.21729	.97611	27
34	.16620	.98609	.18338	.98304	.20051	.97969	.21758	.97604	26
35	.16648	.98604	.18367	.98299	.20079	.97963	.21786	.97598	25
36	.16677	.98600	.18395	.98294	.20108	.97958	.21814	.97592	24
37	.16706	.98595	.18424	.98288	.20136	.97952	.21843	.97585	23
38	.16734	.98590	.18452	.98283	.20165	.97946	.21871	.97579	22
39	.16763	.98585	.18481	.98277	.20193	.97940	.21899	.97573	21
40	.16792	.98580	.18509	.98272	.20222	.97934	.21928	.97566	20
41	.16820	.98575	.18538	.98267	.20250	.97928	.21956	.97560	19
42	.16849	.98570	.18567	.98261	.20279	.97922	.21985	.97553	18
43	.16878	.98565	.18595	.98256	.20307	.97916	.22013	.97547	17
44	.16906	.98561	.18624	.98250	.20336	.97910	.22041	.97541	16
45	.16935	.98556	.18652	.98245	.20364	.97905	.22070	.97534	15
46	.16964	.98551	.18681	.98240	.20393	.97899	.22098	.97528	14
47	.16992	.98546	.18710	.98234	.20421	.97893	.22126	.97521	13
48	.17021	.98541	.18738	.98229	.20450	.97887	.22155	.97515	12
49	.17050	.98536	.18767	.98223	.20478	.97881	.22183	.97508	11
50	.17078	.98531	.18795	.98218	.20507	.97875	.22212	.97502	10
51	.17107	.98526	.18824	.98212	.20535	.97869	.22240	.97496	9
52	.17136	.98521	.18852	.98207	.20563	.97863	.22268	.97489	8
53	.17164	.98516	.18881	.98201	.20592	.97857	.22297	.97483	7
54	.17193	.98511	.18910	.98196	.20620	.97851	.22325	.97476	6
55	.17222	.98506	.18938	.98190	.20649	.97845	.22353	.97470	5
56	.17250	.98501	.18967	.98185	.20677	.97839	.22382	.97463	4
57	.17279	.98496	.18995	.98179	.20706	.97833	.22410	.97457	3
58	.17308	.98491	.19024	.98174	.20734	.97827	.22438	.97450	2
59	.17336	.98486	.19052	.98168	.20763	.97821	.22467	.97444	1
60	.17365	.98481	.19081	.98163	.20791	.97815	.22495	.97437	0
′	Cosine 80°	Sine 80°	Cosine 79°	Sine 79°	Cosine 78°	Sine 78°	Cosine 77°	Sine 77°	′

Table A-1. (*Continued*)

′	13° Sine	13° Cosine	14° Sine	14° Cosine	15° Sine	15° Cosine	16° Sine	16° Cosine	′
0	.22495	.97437	.24192	.97030	.25882	.96593	.27564	.96126	60
1	.22523	.97430	.24220	.97023	.25910	.96585	.27592	.96118	59
2	.22552	.97424	.24249	.97015	.25938	.96578	.27620	.96110	58
3	.22580	.97417	.24277	.97008	.25966	.96570	.27648	.96102	57
4	.22608	.97411	.24305	.97001	.25994	.96562	.27676	.96094	56
5	.22637	.97404	.24333	.96994	.26022	.96555	.27704	.96086	55
6	.22665	.97398	.24362	.96987	.26050	.96547	.27731	.96078	54
7	.22693	.97391	.24390	.96980	.26079	.96540	.27759	.96070	53
8	.22722	.97384	.24418	.96973	.26107	.96532	.27787	.96062	52
9	.22750	.97378	.24446	.96966	.26135	.96524	.27815	.96054	51
10	.22778	.97371	.24474	.96959	.26163	.96517	.27843	.96046	50
11	.22807	.97365	.24503	.96952	.26191	.96509	.27871	.96037	49
12	.22835	.97358	.24531	.96945	.26219	.96502	.27899	.96029	48
13	.22863	.97351	.24559	.96937	.26247	.96494	.27927	.96021	47
14	.22892	.97345	.24587	.96930	.26275	.96486	.27955	.96013	46
15	.22920	.97338	.24615	.96923	.26303	.96479	.27983	.96005	45
16	.22948	.97331	.24644	.96916	.26331	.96471	.28011	.95997	44
17	.22977	.97325	.24672	.96909	.26359	.96463	.28039	.95989	43
18	.23005	.97318	.24700	.96902	.26387	.96456	.28067	.95981	42
19	.23033	.97311	.24728	.96894	.26415	.96448	.28095	.95972	41
20	.23062	.97304	.24756	.96887	.26443	.96440	.28123	.95964	40
21	.23090	.97298	.24784	.96880	.26471	.96433	.28150	.95956	39
22	.23118	.97291	.24813	.96873	.26500	.96425	.28178	.95948	38
23	.23146	.97284	.24841	.96866	.26528	.96417	.28206	.95940	37
24	.23175	.97278	.24869	.96858	.26556	.96410	.28234	.95931	36
25	.23203	.97271	.24897	.96851	.26584	.96402	.28262	.95923	35
26	.23231	.97264	.24925	.96844	.26612	.96394	.28290	.95915	34
27	.23260	.97257	.24954	.96837	.26640	.96386	.28318	.95907	33
28	.23288	.97251	.24982	.96829	.26668	.96379	.28346	.95898	32
29	.23316	.97244	.25010	.96822	.26696	.96371	.28374	.95890	31
30	.23345	.97237	.25038	.96815	.26724	.96363	.28402	.95882	30
31	.23373	.97230	.25066	.96807	.26752	.96355	.28429	.95874	29
32	.23401	.97223	.25094	.96800	.26780	.96347	.28457	.95865	28
33	.23429	.97217	.25122	.96793	.26808	.96340	.28485	.95857	27
34	.23458	.97210	.25151	.96786	.26836	.96332	.28513	.95849	26
35	.23486	.97203	.25179	.96778	.26864	.96324	.28541	.95841	25
36	.23514	.97196	.25207	.96771	.26892	.96316	.28569	.95832	24
37	.23542	.97189	.25235	.96764	.26920	.96308	.28597	.95824	23
38	.23571	.97182	.25263	.96756	.26948	.96301	.28625	.95816	22
39	.23599	.97176	.25291	.96749	.26976	.96293	.28652	.95807	21
40	.23627	.97169	.25320	.96742	.27004	.96285	.28680	.95799	20
41	.23656	.97162	.25348	.96734	.27032	.96277	.28708	.95791	19
42	.23684	.97155	.25376	.96727	.27060	.96269	.28736	.95782	18
43	.23712	.97148	.25404	.96719	.27088	.96261	.28764	.95774	17
44	.23740	.97141	.25432	.96712	.27116	.96253	.28792	.95766	16
45	.23769	.97134	.25460	.96705	.27144	.96246	.28820	.95757	15
46	.23797	.97127	.25488	.96697	.27172	.96238	.28847	.95749	14
47	.23825	.97120	.25516	.96690	.27200	.96230	.28875	.95740	13
48	.23853	.97113	.25545	.96682	.27228	.96222	.28903	.95732	12
49	.23882	.97106	.25573	.96675	.27256	.96214	.28931	.95724	11
50	.23910	.97100	.25601	.96667	.27284	.96206	.28959	.95715	10
51	.23938	.97093	.25629	.96660	.27312	.96198	.28987	.95707	9
52	.23966	.97086	.25657	.96653	.27340	.96190	.29015	.95698	8
53	.23995	.97079	.25685	.96645	.27368	.96182	.29042	.95690	7
54	.24023	.97072	.25713	.96638	.27396	.96174	.29070	.95681	6
55	.24051	.97065	.25741	.96630	.27424	.96166	.29098	.95673	5
56	.24079	.97058	.25769	.96623	.27452	.96158	.29126	.95664	4
57	.24108	.97051	.25798	.96615	.27480	.96150	.29154	.95656	3
58	.24136	.97044	.25826	.96608	.27508	.96142	.29182	.95647	2
59	.24164	.97037	.25854	.96600	.27536	.96134	.29209	.95639	1
60	.24192	.97030	.25882	.96593	.27564	.96126	.29237	.95630	0
′	Cosine	Sine	Cosine	Sine	Cosine	Sine	Cosine	Sine	′
	76°		75°		74°		73°		

Table A-1. (*Continued*)

′	17° Sine	17° Cosine	18° Sine	18° Cosine	19° Sine	19° Cosine	20° Sine	20° Cosine	′
0	.29237	.95630	.30902	.95106	.32557	.94552	.34202	.93969	60
1	.29265	.95622	.30929	.95097	.32584	.94542	.34229	.93959	59
2	.29293	.95613	.30957	.95088	.32612	.94533	.34257	.93949	58
3	.29321	.95605	.30985	.95079	.32639	.94523	.34284	.93939	57
4	.29348	.95596	.31012	.95070	.32667	.94514	.34311	.93929	56
5	.29376	.95588	.31040	.95061	.32694	.94504	.34339	.93919	55
6	.29404	.95579	.31068	.95052	.32722	.94495	.34366	.93909	54
7	.29432	.95571	.31095	.95043	.32749	.94485	.34393	.93899	53
8	.29460	.95562	.31123	.95033	.32777	.94476	.34421	.93889	52
9	.29487	.95554	.31151	.95024	.32804	.94466	.34448	.93879	51
10	.29515	.95545	.31178	.95015	.32832	.94457	.34475	.93869	50
11	.29543	.95536	.31206	.95006	.32859	.94447	.34503	.93859	49
12	.29571	.95528	.31233	.94997	.32887	.94438	.34530	.93849	48
13	.29599	.95519	.31261	.94988	.32914	.94428	.34557	.93839	47
14	.29626	.95511	.31289	.94979	.32942	.94418	.34584	.93829	46
15	.29654	.95502	.31316	.94970	.32969	.94409	.34612	.93819	45
16	.29682	.95493	.31344	.94961	.32997	.94399	.34639	.93809	44
17	.29710	.95485	.31372	.94952	.33024	.94390	.34666	.93799	43
18	.29737	.95476	.31399	.94943	.33051	.94380	.34694	.93789	42
19	.29765	.95467	.31427	.94933	.33079	.94370	.34721	.93779	41
20	.29793	.95459	.31454	.94924	.33106	.94361	.34748	.93769	40
21	.29821	.95450	.31482	.94915	.33134	.94351	.34775	.93759	39
22	.29849	.95441	.31510	.94906	.33161	.94342	.34803	.93748	38
23	.29876	.95433	.31537	.94897	.33189	.94332	.34830	.93738	37
24	.29904	.95424	.31565	.94888	.33216	.94322	.34857	.93728	36
25	.29932	.95415	.31593	.94878	.33244	.94313	.34884	.93718	35
26	.29960	.95407	.31620	.94869	.33271	.94303	.34912	.93708	34
27	.29987	.95398	.31648	.94860	.33298	.94293	.34939	.93698	33
28	.30015	.95389	.31675	.94851	.33326	.94284	.34966	.93688	32
29	.30043	.95380	.31703	.94842	.33353	.94274	.34993	.93677	31
30	.30071	.95372	.31730	.94832	.33381	.94264	.35021	.93667	30
31	.30098	.95363	.31758	.94823	.33408	.94254	.35048	.93657	29
32	.30126	.95354	.31786	.94814	.33436	.94245	.35075	.93647	28
33	.30154	.95345	.31813	.94805	.33463	.94235	.35102	.93637	27
34	.30182	.95337	.31841	.94795	.33490	.94225	.35130	.93626	26
35	.30209	.95328	.31868	.94786	.33518	.94215	.35157	.93616	25
36	.30237	.95319	.31896	.94777	.33545	.94206	.35184	.93606	24
37	.30265	.95310	.31923	.94768	.33573	.94196	.35211	.93596	23
38	.30292	.95301	.31951	.94758	.33600	.94186	.35239	.93585	22
39	.30320	.95293	.31979	.94749	.33627	.94176	.35266	.93575	21
40	.30348	.95284	.32006	.94740	.33655	.94167	.35293	.93565	20
41	.30376	.95275	.32034	.94730	.33682	.94157	.35320	.93555	19
42	.30403	.95266	.32061	.94721	.33710	.94147	.35347	.93544	18
43	.30431	.95257	.32089	.94712	.33737	.94137	.35375	.93534	17
44	.30459	.95248	.32116	.94702	.33764	.94127	.35402	.93524	16
45	.30486	.95240	.32144	.94693	.33792	.94118	.35429	.93514	15
46	.30514	.95231	.32171	.94684	.33819	.94108	.35456	.93503	14
47	.30542	.95222	.32199	.94674	.33846	.94098	.35484	.93493	13
48	.30570	.95213	.32227	.94665	.33874	.94088	.35511	.93483	12
49	.30597	.95204	.32254	.94656	.33901	.94078	.35538	.93472	11
50	.30625	.95195	.32282	.94646	.33929	.94068	.35565	.93462	10
51	.30653	.95186	.32309	.94637	.33956	.94058	.35592	.93452	9
52	.30680	.95177	.32337	.94627	.33983	.94049	.35619	.93441	8
53	.30708	.95168	.32364	.94618	.34011	.94039	.35647	.93431	7
54	.30736	.95159	.32392	.94609	.34038	.94029	.35674	.93420	6
55	.30763	.95150	.32419	.94599	.34065	.94019	.35701	.93410	5
56	.30791	.95142	.32447	.94590	.34093	.94009	.35728	.93400	4
57	.30819	.95133	.32474	.94580	.34120	.93999	.35755	.93389	3
58	.30846	.95124	.32502	.94571	.34147	.93989	.35782	.93379	2
59	.30874	.95115	.32529	.94561	.34175	.93979	.35810	.93368	1
60	.30902	.95106	.32557	.94552	.34202	.93969	.35837	.93358	0
′	Cosine 72°	Sine 72°	Cosine 71°	Sine 71°	Cosine 70°	Sine 70°	Cosine 69°	Sine 69°	′

Table A-1. (*Continued*)

′	21° Sine	21° Cosine	22° Sine	22° Cosine	23° Sine	23° Cosine	24° Sine	24° Cosine	′
0	.35837	.93358	.37461	.92718	.39073	.92050	.40674	.91355	60
1	.35864	.93348	.37488	.92707	.39100	.92039	.40700	.91343	59
2	.35891	.93337	.37515	.92697	.39127	.92028	.40727	.91331	58
3	.35918	.93327	.37542	.92686	.39153	.92016	.40753	.91319	57
4	.35945	.93316	.37569	.92675	.39180	.92005	.40780	.91307	56
5	.35973	.93306	.37595	.92664	.39207	.91994	.40806	.91295	55
6	.36000	.93295	.37622	.92653	.39234	.91982	.40833	.91283	54
7	.36027	.93285	.37649	.92642	.39260	.91971	.40860	.91272	53
8	.36054	.93274	.37676	.92631	.39287	.91959	.40886	.91260	52
9	.36081	.93264	.37703	.92620	.39314	.91948	.40913	.91248	51
10	.36108	.93253	.37730	.92609	.39341	.91936	.40939	.91236	50
11	.36135	.93243	.37757	.92598	.39367	.91925	.40966	.91224	49
12	.36162	.93232	.37784	.92587	.39394	.91914	.40992	.91212	48
13	.36190	.93222	.37811	.92576	.39421	.91902	.41019	.91200	47
14	.36217	.93211	.37838	.92565	.39448	.91891	.41045	.91188	46
15	.36244	.93201	.37865	.92554	.39474	.91879	.41072	.91176	45
16	.36271	.93190	.37892	.92543	.39501	.91868	.41098	.91164	44
17	.36298	.93180	.37919	.92532	.39528	.91856	.41125	.91152	43
18	.36325	.93169	.37946	.92521	.39555	.91845	.41151	.91140	42
19	.36352	.93159	.37973	.92510	.39581	.91833	.41178	.91128	41
20	.36379	.93148	.37999	.92499	.39608	.91822	.41204	.91116	40
21	.36406	.93137	.38026	.92488	.39635	.91810	.41231	.91104	39
22	.36434	.93127	.38053	.92477	.39661	.91799	.41257	.91092	38
23	.36461	.93116	.38080	.92466	.39688	.91787	.41284	.91080	37
24	.36488	.93106	.38107	.92455	.39715	.91775	.41310	.91068	36
25	.36515	.93095	.38134	.92444	.39741	.91764	.41337	.91056	35
26	.36542	.93084	.38161	.92432	.39768	.91752	.41363	.91044	34
27	.36569	.93074	.38188	.92421	.39795	.91741	.41390	.91032	33
28	.36596	.93063	.38215	.92410	.39822	.91729	.41416	.91020	32
29	.36623	.93052	.38241	.92399	.39848	.91718	.41443	.91008	31
30	.36650	.93042	.38268	.92388	.39875	.91706	.41469	.90996	30
31	.36677	.93031	.38295	.92377	.39902	.91694	.41496	.90984	29
32	.36704	.93020	.38322	.92366	.39928	.91683	.41522	.90972	28
33	.36731	.93010	.38349	.92355	.39955	.91671	.41549	.90960	27
34	.36758	.92999	.38376	.92343	.39982	.91660	.41575	.90948	26
35	.36785	.92988	.38403	.92332	.40008	.91648	.41602	.90936	25
36	.36812	.92978	.38430	.92321	.40035	.91636	.41628	.90924	24
37	.36839	.92967	.38456	.92310	.40062	.91625	.41655	.90911	23
38	.36867	.92956	.38483	.92299	.40088	.91613	.41681	.90899	22
39	.36894	.92945	.38510	.92287	.40115	.91601	.41707	.90887	21
40	.36921	.92935	.38537	.92276	.40141	.91590	.41734	.90875	20
41	.36948	.92924	.38564	.92265	.40168	.91578	.41760	.90863	19
42	.36975	.92913	.38591	.92254	.40195	.91566	.41787	.90851	18
43	.37002	.92902	.38617	.92243	.40221	.91555	.41813	.90839	17
44	.37029	.92892	.38644	.92231	.40248	.91543	.41840	.90826	16
45	.37056	.92881	.38671	.92220	.40275	.91531	.41866	.90814	15
46	.37083	.92870	.38698	.92209	.40301	.91519	.41892	.90802	14
47	.37110	.92859	.38725	.92198	.40328	.91508	.41919	.90790	13
48	.37137	.92849	.38752	.92186	.40355	.91496	.41945	.90778	12
49	.37164	.92838	.38778	.92175	.40381	.91484	.41972	.90766	11
50	.37191	.92827	.38805	.92164	.40408	.91472	.41998	.90753	10
51	.37218	.92816	.38832	.92152	.40434	.91461	.42024	.90741	9
52	.37245	.92805	.38859	.92141	.40461	.91449	.42051	.90729	8
53	.37272	.92794	.38886	.92130	.40488	.91437	.42077	.90717	7
54	.37299	.92784	.38912	.92119	.40514	.91425	.42104	.90704	6
55	.37326	.92773	.38939	.92107	.40541	.91414	.42130	.90692	5
56	.37353	.92762	.38966	.92096	.40567	.91402	.42156	.90680	4
57	.37380	.92751	.38993	.92085	.40594	.91390	.42183	.90668	3
58	.37407	.92740	.39020	.92073	.40621	.91378	.42209	.90655	2
59	.37434	.92729	.39046	.92062	.40647	.91366	.42235	.90643	1
60	.37461	.92718	.39073	.92050	.40674	.91355	.42262	.90631	0
′	68° Cosine	68° Sine	67° Cosine	67° Sine	66° Cosine	66° Sine	65° Cosine	65° Sine	′

Table A-1. (*Continued*)

′	25° Sine	25° Cosine	26° Sine	26° Cosine	27° Sine	27° Cosine	28° Sine	28° Cosine	′
0	.42262	.90631	.43837	.89879	.45399	.89101	.46947	.88295	60
1	.42288	.90618	.43863	.89867	.45425	.89087	.46973	.88281	59
2	.42315	.90606	.43889	.89854	.45451	.89074	.46999	.88267	58
3	.42341	.90594	.43916	.89841	.45477	.89061	.47024	.88254	57
4	.42367	.90582	.43942	.89828	.45503	.89048	.47050	.88240	56
5	.42394	.90569	.43968	.89816	.45529	.89035	.47076	.88226	55
6	.42420	.90557	.43994	.89803	.45554	.89021	.47101	.88213	54
7	.42446	.90545	.44020	.89790	.45580	.89008	.47127	.88199	53
8	.42473	.90532	.44046	.89777	.45606	.88995	.47153	.88185	52
9	.42499	.90520	.44072	.89764	.45632	.88981	.47178	.88172	51
10	.42525	.90507	.44098	.89752	.45658	.88968	.47204	.88158	50
11	.42552	.90495	.44124	.89739	.45684	.88955	.47229	.88144	49
12	.42578	.90483	.44151	.89726	.45710	.88942	.47255	.88130	48
13	.42604	.90470	.44177	.89713	.45736	.88928	.47281	.88117	47
14	.42631	.90458	.44203	.89700	.45762	.88915	.47306	.88103	46
15	.42657	.90446	.44229	.89687	.45787	.88902	.47332	.88089	45
16	.42683	.90433	.44255	.89674	.45813	.88888	.47358	.88075	44
17	.42709	.90421	.44281	.89662	.45839	.88875	.47383	.88062	43
18	.42736	.90408	.44307	.89649	.45865	.88862	.47409	.88048	42
19	.42762	.90396	.44333	.89636	.45891	.88848	.47434	.88034	41
20	.42788	.90383	.44359	.89623	.45917	.88835	.47460	.88020	40
21	.42815	.90371	.44385	.89610	.45942	.88822	.47486	.88006	39
22	.42841	.90358	.44411	.89597	.45968	.88808	.47511	.87993	38
23	.42867	.90346	.44437	.89584	.45994	.88795	.47537	.87979	37
24	.42894	.90334	.44464	.89571	.46020	.88782	.47562	.87965	36
25	.42920	.90321	.44490	.89558	.46046	.88768	.47588	.87951	35
26	.42946	.90309	.44516	.89545	.46072	.88755	.47614	.87937	34
27	.42972	.90296	.44542	.89532	.46097	.88741	.47639	.87923	33
28	.42999	.90284	.44568	.89519	.46123	.88728	.47665	.87909	32
29	.43025	.90271	.44594	.89506	.46149	.88715	.47690	.87896	31
30	.43051	.90259	.44620	.89493	.46175	.88701	.47716	.87882	30
31	.43077	.90246	.44646	.89480	.46201	.88688	.47741	.87868	29
32	.43104	.90233	.44672	.89467	.46226	.88674	.47767	.87854	28
33	.43130	.90221	.44698	.89454	.46252	.88661	.47793	.87840	27
34	.43156	.90208	.44724	.89441	.46278	.88647	.47818	.87826	26
35	.43182	.90196	.44750	.89428	.46304	.88634	.47844	.87812	25
36	.43209	.90183	.44776	.89415	.46330	.88620	.47869	.87798	24
37	.43235	.90171	.44802	.89402	.46355	.88607	.47895	.87784	23
38	.43261	.90158	.44828	.89389	.46381	.88593	.47920	.87770	22
39	.43287	.90146	.44854	.89376	.46407	.88580	.47946	.87756	21
40	.43313	.90133	.44880	.89363	.46433	.88566	.47971	.87743	20
41	.43340	.90120	.44906	.89350	.46458	.88553	.47997	.87729	19
42	.43366	.90108	.44932	.89337	.46484	.88539	.48022	.87715	18
43	.43392	.90095	.44958	.89324	.46510	.88526	.48048	.87701	17
44	.43418	.90082	.44984	.89311	.46536	.88512	.48073	.87687	16
45	.43445	.90070	.45010	.89298	.46561	.88499	.48099	.87673	15
46	.43471	.90057	.45036	.89285	.46587	.88485	.48124	.87659	14
47	.43497	.90045	.45062	.89272	.46613	.88472	.48150	.87645	13
48	.43523	.90032	.45088	.89259	.46639	.88458	.48175	.87631	12
49	.43549	.90019	.45114	.89245	.46664	.88445	.48201	.87617	11
50	.43575	.90007	.45140	.89232	.46690	.88431	.48226	.87603	10
51	.43602	.89994	.45166	.89219	.46716	.88417	.48252	.87589	9
52	.43628	.89981	.45192	.89206	.46742	.88404	.48277	.87575	8
53	.43654	.89968	.45218	.89193	.46767	.88390	.48303	.87561	7
54	.43680	.89956	.45243	.89180	.46793	.88377	.48328	.87546	6
55	.43706	.89943	.45269	.89167	.46819	.88363	.48354	.87532	5
56	.43733	.89930	.45295	.89153	.46844	.88349	.48379	.87518	4
57	.43759	.89918	.45321	.89140	.46870	.88336	.48405	.87504	3
58	.43785	.89905	.45347	.89127	.46896	.88322	.48430	.87490	2
59	.43811	.89892	.45373	.89114	.46921	.88308	.48456	.87476	1
60	.43837	.89879	.45399	.89101	.46947	.88295	.48481	.87462	0
′	64° Cosine	64° Sine	63° Cosine	63° Sine	62° Cosine	62° Sine	61° Cosine	61° Sine	′

Table A-1. (*Continued*)

′	29° Sine	29° Cosine	30° Sine	30° Cosine	31° Sine	31° Cosine	32° Sine	32° Cosine	′
0	.48481	.87462	.50000	.86603	.51504	.85717	.52992	.84805	60
1	.48506	.87448	.50025	.86588	.51529	.85702	.53017	.84789	59
2	.48532	.87434	.50050	.86573	.51554	.85687	.53041	.84774	58
3	.48557	.87420	.50076	.86559	.51579	.85672	.53066	.84759	57
4	.48583	.87406	.50101	.86544	.51604	.85657	.53091	.84743	56
5	.48608	.87391	.50126	.86530	.51628	.85642	.53115	.84728	55
6	.48634	.87377	.50151	.86515	.51653	.85627	.53140	.84712	54
7	.48659	.87363	.50176	.86501	.51678	.85612	.53164	.84697	53
8	.48684	.87349	.50201	.86486	.51703	.85597	.53189	.84681	52
9	.48710	.87335	.50227	.86471	.51728	.85582	.53214	.84666	51
10	.48735	.87321	.50252	.86457	.51753	.85567	.53238	.84650	50
11	.48761	.87306	.50277	.86442	.51778	.85551	.53263	.84635	49
12	.48786	.87292	.50302	.86427	.51803	.85536	.53288	.84619	48
13	.48811	.87278	.50327	.86413	.51828	.85521	.53312	.84604	47
14	.48837	.87264	.50352	.86398	.51852	.85506	.53337	.84588	46
15	.48862	.87250	.50377	.86384	.51877	.85491	.53361	.84573	45
16	.48888	.87235	.50403	.86369	.51902	.85476	.53386	.84557	44
17	.48913	.87221	.50428	.86354	.51927	.85461	.53411	.84542	43
18	.48938	.87207	.50453	.86340	.51952	.85446	.53435	.84526	42
19	.48964	.87193	.50478	.86325	.51977	.85431	.53460	.84511	41
20	.48989	.87178	.50503	.86310	.52002	.85416	.53484	.84495	40
21	.49014	.87164	.50528	.86295	.52026	.85401	.53509	.84480	39
22	.49040	.87150	.50553	.86281	.52051	.85385	.53534	.84464	38
23	.49065	.87136	.50578	.86266	.52076	.85370	.53558	.84448	37
24	.49090	.87121	.50603	.86251	.52101	.85355	.53583	.84433	36
25	.49116	.87107	.50628	.86237	.52126	.85340	.53607	.84417	35
26	.49141	.87093	.50654	.86222	.52151	.85325	.53632	.84402	34
27	.49166	.87079	.50679	.86207	.52175	.85310	.53656	.84386	33
28	.49192	.87064	.50704	.86192	.52200	.85294	.53681	.84370	32
29	.49217	.87050	.50729	.86178	.52225	.85279	.53705	.84355	31
30	.49242	.87036	.50754	.86163	.52250	.85264	.53730	.84339	30
31	.49268	.87021	.50779	.86148	.52275	.85249	.53754	.84324	29
32	.49293	.87007	.50804	.86133	.52299	.85234	.53779	.84308	28
33	.49318	.86993	.50829	.86119	.52324	.85218	.53804	.84292	27
34	.49344	.86978	.50854	.86104	.52349	.85203	.53828	.84277	26
35	.49369	.86964	.50879	.86089	.52374	.85188	.53853	.84261	25
36	.49394	.86949	.50904	.86074	.52399	.85173	.53877	.84245	24
37	.49419	.86935	.50929	.86059	.52423	.85157	.53902	.84230	23
38	.49445	.86921	.50954	.86045	.52448	.85142	.53926	.84214	22
39	.49470	.86906	.50979	.86030	.52473	.85127	.53951	.84198	21
40	.49495	.86892	.51004	.86015	.52498	.85112	.53975	.84182	20
41	.49521	.86878	.51029	.86000	.52522	.85096	.54000	.84167	19
42	.49546	.86863	.51054	.85985	.52547	.85081	.54024	.84151	18
43	.49571	.86849	.51079	.85970	.52572	.85066	.54049	.84135	17
44	.49596	.86834	.51104	.85956	.52597	.85051	.54073	.84120	16
45	.49622	.86820	.51129	.85941	.52621	.85035	.54097	.84104	15
46	.49647	.86805	.51154	.85926	.52646	.85020	.54122	.84088	14
47	.49672	.86791	.51179	.85911	.52671	.85005	.54146	.84072	13
48	.49697	.86777	.51204	.85896	.52696	.84989	.54171	.84057	12
49	.49723	.86762	.51229	.85881	.52720	.84974	.54195	.84041	11
50	.49748	.86748	.51254	.85866	.52745	.84959	.54220	.84025	10
51	.49773	.86733	.51279	.85851	.52770	.84943	.54244	.84009	9
52	.49798	.86719	.51304	.85836	.52794	.84928	.54269	.83994	8
53	.49824	.86704	.51329	.85821	.52819	.84913	.54293	.83978	7
54	.49849	.86690	.51354	.85806	.52844	.84897	.54317	.83962	6
55	.49874	.86675	.51379	.85792	.52869	.84882	.54342	.83946	5
56	.49899	.86661	.51404	.85777	.52893	.84866	.54366	.83930	4
57	.49924	.86646	.51429	.85762	.52918	.84851	.54391	.83915	3
58	.49950	.86632	.51454	.85747	.52943	.84836	.54415	.83899	2
59	.49975	.86617	.51479	.85732	.52967	.84820	.54440	.83883	1
60	.50000	.86603	.51504	.85717	.52992	.84805	.54464	.83867	0
′	Cosine 60°	Sine 60°	Cosine 59°	Sine 59°	Cosine 58°	Sine 58°	Cosine 57°	Sine 57°	′

Table A-1. (*Continued*)

′	33° Sine	33° Cosine	34° Sine	34° Cosine	35° Sine	35° Cosine	36° Sine	36° Cosine	′
0	.54464	.83867	.55919	.82904	.57358	.81915	.58779	.80902	60
1	.54488	.83851	.55943	.82887	.57381	.81899	.58802	.80885	59
2	.54513	.83835	.55968	.82871	.57405	.81882	.58826	.80867	58
3	.54537	.83819	.55992	.82855	.57429	.81865	.58849	.80850	57
4	.54561	.83804	.56016	.82839	.57453	.81848	.58873	.80833	56
5	.54586	.83788	.56040	.82822	.57477	.81832	.58896	.80816	55
6	.54610	.83772	.56064	.82806	.57501	.81815	.58920	.80799	54
7	.54635	.83756	.56088	.82790	.57524	.81798	.58943	.80782	53
8	.54659	.83740	.56112	.82773	.57548	.81782	.58967	.80765	52
9	.54683	.83724	.56136	.82757	.57572	.81765	.58990	.80748	51
10	.54708	.83708	.56160	.82741	.57596	.81748	.59014	.80730	50
11	.54732	83692	.56184	.82724	.57619	.81731	.59037	.80713	49
12	.54756	.83676	.56208	.82708	.57643	.81714	.59061	.80696	48
13	.54781	.83660	.56232	.82692	.57667	.81698	.59084	.80679	47
14	.54805	.83645	.56256	.82675	.57691	.81681	.59108	.80662	46
15	.54829	.83629	.56280	.82659	.57715	.81664	.59131	.80644	45
16	.54854	.83613	.56305	.82643	.57738	.81647	.59154	.80627	44
17	.54878	.83597	.56329	.82626	.57762	.81631	.59178	.80610	43
18	.54902	.83581	.56353	.82610	.57786	.81614	.59201	.80593	42
19	.54927	.83565	.56377	.82593	.57810	.81597	.59225	.80576	41
20	.54951	.83549	.56401	.82577	.57833	.81580	.59248	.80558	40
21	.54975	.83533	.56425	.82561	.57857	.81563	.59272	.80541	39
22	.54999	.83517	.56449	.82544	.57881	.81546	.59295	.80524	38
23	.55024	.83501	.56473	.82528	.57904	.81530	.59318	.80507	37
24	.55048	.83485	.56497	.82511	.57928	.81513	.59342	.80489	36
25	.55072	.83469	.56521	.82495	.57952	.81496	.59365	.80472	35
26	.55097	.83453	.56545	.82478	.57976	.81479	.59389	.80455	34
27	.55121	.83437	.56569	.82462	.57999	.81462	.59412	.80438	33
28	.55145	.83421	.56593	.82446	.58023	.81445	.59436	.80420	32
29	.55169	.83405	.56617	.82429	.58047	.81428	.59459	.80403	31
30	.55194	.83389	.56641	.82413	.58070	.81412	.59482	.80386	30
31	.55218	.83373	.56665	.82396	.58094	.81395	.59506	.80368	29
32	.55242	.83356	.56689	.82380	.58118	.81378	.59529	.80351	28
33	.55266	.83340	.56713	.82363	.58141	.81361	.59552	.80334	27
34	.55291	.83324	.56736	.82347	.58165	.81344	.59576	.80316	26
35	.55315	.83308	.56760	.82330	.58189	.81327	.59599	.80299	25
36	.55339	.83292	.56784	.82314	.58212	.81310	.59622	.80282	24
37	.55363	.83276	.56808	.82297	.58236	.81293	.59646	.80264	23
38	.55388	.83260	.56832	.82281	.58260	.81276	.59669	.80247	22
39	.55412	.83244	.56856	.82264	.58283	.81259	.59693	.80230	21
40	.55436	.83228	.56880	.82248	.58307	.81242	.59716	.80212	20
41	.55460	.83212	.56904	.82231	.58330	.81225	.59739	.80195	19
42	.55484	.83195	.56928	.82214	.58354	.81208	.59763	.80178	18
43	.55509	.83179	.56952	.82198	.58378	.81191	.59786	.80160	17
44	.55533	.83163	.56976	.82181	.58401	.81174	.59809	.80143	16
45	.55557	.83147	.57000	.82165	.58425	.81157	.59832	.80125	15
46	.55581	.83131	.57024	.82148	.58449	.81140	.59856	.80108	14
47	.55605	.83115	.57047	.82132	.58472	.81123	.59879	.80091	13
48	.55630	.83098	.57071	.82115	.58496	.81106	.59902	.80073	12
49	.55654	.83082	.57095	.82098	.58519	.81089	.59926	.80056	11
50	.55678	.83066	.57119	.82082	.58543	.81072	.59949	.80038	10
51	.55702	.83050	.57143	.82065	.58567	.81055	.59972	.80021	9
52	.55726	.83034	.57167	.82048	.58590	.81038	.59995	.80003	8
53	.55750	.83017	.57191	.82032	.58614	.81021	.60019	.79986	7
54	.55775	.83001	.57215	.82015	.58637	.81004	.60042	.79968	6
55	.55799	.82985	.57238	.81999	.58661	.80987	.60065	.79951	5
56	.55823	.82969	.57262	.81982	.58684	.80970	.60089	.79934	4
57	.55847	.82953	.57286	.81965	.58708	.80953	.60112	.79916	3
58	.55871	.82936	.57310	.81949	.58731	.80936	.60135	.79899	2
59	.55895	.82920	.57334	.81932	.58755	.80919	.60158	.79881	1
60	.55919	.82904	.57358	.81915	.58779	.80902	.60182	.79864	0
′	Cosine 56°	Sine 56°	Cosine 55°	Sine 55°	Cosine 54°	Sine 54°	Cosine 53°	Sine 53°	′

Table A-1. (*Continued*)

′	37°		38°		39°		40°		′
	Sine	Cosine	Sine	Cosine	Sine	Cosine	Sine	Cosine	
0	.60182	.79864	.61566	.78801	.62932	.77715	.64279	.76604	60
1	.60205	.79846	.61589	.78783	.62955	.77696	.64301	.76586	59
2	.60228	.79829	.61612	.78765	.62977	.77678	.64323	.76567	58
3	.60251	.79811	.61635	.78747	.63000	.77660	.64346	.76548	57
4	.60274	.79793	.61658	.78729	.63022	.77641	.64368	.76530	56
5	.60298	.79776	.61681	.78711	.63045	.77623	.64390	.76511	55
6	.60321	.79758	.61704	.78694	.63068	.77605	.64412	.76492	54
7	.60344	.79741	.61726	.78676	.63090	.77586	.64435	.76473	53
8	.60367	.79723	.61749	.78658	.63113	.77568	.64457	.76455	52
9	.60390	.79706	.61772	.78640	.63135	.77550	.64479	.76436	51
10	.60414	.79688	.61795	.78622	.63158	.77531	.64501	.76417	50
11	.60437	.79671	.61818	.78604	.63180	.77513	.64524	.76398	49
12	.60460	.79653	.61841	.78586	.63203	.77494	.64546	.76380	48
13	.60483	.79635	.61864	.78568	.63225	.77476	.64568	.76361	47
14	.60506	.79618	.61887	.78550	.63248	.77458	.64590	.76342	46
15	.60529	.79600	.61909	.78532	.63271	.77439	.64612	.76323	45
16	.60553	.79583	.61932	.78514	.63293	.77421	.64635	.76304	44
17	.60576	.79565	.61955	.78496	.63316	.77402	.64657	.76286	43
18	.60599	.79547	.61978	.78478	.63338	.77384	.64679	.76267	42
19	.60622	.79530	.62001	.78460	.63361	.77366	.64701	.76248	41
20	.60645	.79512	.62024	.78442	.63383	.77347	.64723	.76229	40
21	.60668	.79494	.62046	.78424	.63406	.77329	.64746	.76210	39
22	.60691	.79477	.62069	.78405	.63428	.77310	.64768	.76192	38
23	.60714	.79459	.62092	.78387	.63451	.77292	.64790	.76173	37
24	.60738	.79441	.62115	.78369	.63473	.77273	.64812	.76154	36
25	.60761	.79424	.62138	.78351	.63496	.77255	.64834	.76135	35
26	.60784	.79406	.62160	.78333	.63518	.77236	.64856	.76116	34
27	.60807	.79388	.62183	.78315	.63540	.77218	.64878	.76097	33
28	.60830	.79371	.62206	.78297	.63563	.77199	.64901	.76078	32
29	.60853	.79353	.62229	.78279	.63585	.77181	.64923	.76059	31
30	.60876	.79335	.62251	.78261	.63608	.77162	.64945	.76041	30
31	.60899	.79318	.62274	.78243	.63630	.77144	.64967	.76022	29
32	.60922	.79300	.62297	.78225	.63653	.77125	.64989	.76003	28
33	.60945	.79282	.62320	.78206	.63675	.77107	.65011	.75984	27
34	.60968	.79264	.62342	.78188	.63698	.77088	.65033	.75965	26
35	.60991	.79247	.62365	.78170	.63720	.77070	.65055	.75946	25
36	.61015	.79229	.62388	.78152	.63742	.77051	.65077	.75927	24
37	.61038	.79211	.62411	.78134	.63765	.77033	.65100	.75908	23
38	.61061	.79193	.62433	.78116	.63787	.77014	.65122	.75889	22
39	.61084	.79176	.62456	.78098	.63810	.76996	.65144	.75870	21
40	.61107	.79158	.62479	.78079	.63832	.76977	.65166	.75851	20
41	.61130	.79140	.62502	.78061	.63854	.76959	.65188	.75832	19
42	.61153	.79122	.62524	.78043	.63877	.76940	.65210	.75813	18
43	.61176	.79105	.62547	.78025	.63899	.76921	.65232	.75794	17
44	.61199	.79087	.62570	.78007	.63922	.76903	.65254	.75775	16
45	.61222	.79069	.62592	.77988	.63944	.76884	.65276	.75756	15
46	.61245	.79051	.62615	.77970	.63966	.76866	.65298	.75738	14
47	.61268	.79033	.62638	.77952	.63989	.76847	.65320	.75719	13
48	.61291	.79016	.62660	.77934	.64011	.76828	.65342	.75700	12
49	.61314	.78998	.62683	.77916	.64033	.76810	.65364	.75680	11
50	.61337	.78980	.62706	.77897	.64056	.76791	.65386	.75661	10
51	.61360	.78962	.62728	.77879	.64078	.76772	.65408	.75642	9
52	.61383	.78944	.62751	.77861	.64100	.76754	.65430	.75623	8
53	.61406	.78926	.62774	.77843	.64123	.76735	.65452	.75604	7
54	.61429	.78908	.62796	.77824	.64145	.76717	.65474	.75585	6
55	.61451	.78891	.62819	.77806	.64167	.76698	.65496	.75566	5
56	.61474	.78873	.62842	.77788	.64190	.76679	.65518	.75547	4
57	.61497	.78855	.62864	.77769	.64212	.76661	.65540	.75528	3
58	.61520	.78837	.62887	.77751	.64234	.76642	.65562	.75509	2
59	.61543	.78819	.62909	.77733	.64256	.76623	.65584	.75490	1
60	.61566	.78801	.62932	.77715	.64279	.76604	.65606	.75471	0
′	Cosine	Sine	Cosine	Sine	Cosine	Sine	Cosine	Sine	′
	52°		51°		50°		49°		

Table A-1. (*Continued*)

′	41° Sine	41° Cosine	42° Sine	42° Cosine	43° Sine	43° Cosine	44° Sine	44° Cosine	′
0	.65606	.75471	.66913	.74314	.68200	.73135	.69466	.71934	60
1	.65628	.75452	.66935	.74295	.68221	.73116	.69487	.71914	59
2	.65650	.75433	.66956	.74276	.68242	.73096	.69508	.71894	58
3	.65672	.75414	.66978	.74256	.68264	.73076	.69529	.71873	57
4	.65694	.75395	.66999	.74237	.68285	.73056	.69549	.71853	56
5	.65716	.75375	.67021	.74217	.68306	.73036	.69570	.71833	55
6	.65738	.75356	.67043	.74198	.68327	.73016	.69591	.71813	54
7	.65759	.75337	.67064	.74178	.68349	.72996	.69612	.71792	53
8	.65781	.75318	.67086	.74159	.68370	.72976	.69633	.71772	52
9	.65803	.75299	.67107	.74139	.68391	.72957	.69654	.71752	51
10	.65825	.75280	.67129	.74120	.68412	.72937	.69675	.71732	50
11	.65847	.75261	.67151	.74100	.68434	.72917	.69696	.71711	49
12	.65869	.75241	.67172	.74080	.68455	.72897	.69717	.71691	48
13	.65891	.75222	.67194	.74061	.68476	.72877	.69737	.71671	47
14	.65913	.75203	.67215	.74041	.68497	.72857	.69758	.71650	46
15	.65935	.75184	.67237	.74022	.68518	.72837	.69779	.71630	45
16	.65956	.75165	.67258	.74002	.68539	.72817	.69800	.71610	44
17	.65978	.75146	.67280	.73983	.68561	.72797	.69821	.71590	43
18	.66000	.75126	.67301	.73963	.68582	.72777	.69842	.71569	42
19	.66022	.75107	.67323	.73944	.68603	.72757	.69862	.71549	41
20	.66044	.75088	.67344	.73924	.68624	.72737	.69883	.71529	40
21	.66066	.75069	.67366	.73904	.68645	.72717	.69904	.71508	39
22	.66088	.75050	.67387	.73885	.68666	.72697	.69925	.71488	38
23	.66109	.75030	.67409	.73865	.68688	.72677	.69946	.71468	37
24	.66131	.75011	.67430	.73846	.68709	.72657	.69966	.71447	36
25	.66153	.74992	.67452	.73826	.68730	.72637	.69987	.71427	35
26	.66175	.74973	.67473	.73806	.68751	.72617	.70008	.71407	34
27	.66197	.74953	.67495	.73787	.68772	.72597	.70029	.71386	33
28	.66218	.74934	.67516	.73767	.68793	.72577	.70049	.71366	32
29	.66240	.74915	.67538	.73747	.68814	.72557	.70070	.71345	31
30	.66262	.74896	.67559	.73728	.68835	.72537	.70091	.71325	30
31	.66284	.74876	.67580	.73708	.68857	.72517	.70112	.71305	29
32	.66306	.74857	.67602	.73688	.68878	.72497	.70132	.71284	28
33	.66327	.74838	.67623	.73669	.68899	.72477	.70153	.71264	27
34	.66349	.74818	.67645	.73649	.68920	.72457	.70174	.71243	26
35	.66371	.74799	.67666	.73629	.68941	.72437	.70195	.71223	25
36	.66393	.74780	.67688	.73610	.68962	.72417	.70215	.71203	24
37	.66414	.74760	.67709	.73590	.68983	.72397	.70236	.71182	23
38	.66436	.74741	.67730	.73570	.69004	.72377	.70257	.71162	22
39	.66458	.74722	.67752	.73551	.69025	.72357	.70277	.71141	21
40	.66480	.74703	.67773	.73531	.69046	.72337	.70298	.71121	20
41	.66501	.74683	.67795	.73511	.69067	.72317	.70319	.71100	19
42	.66523	.74664	.67816	.73491	.69088	.72297	.70339	.71080	18
43	.66545	.74644	.67837	.73472	.69109	.72277	.70360	.71059	17
44	.66566	.74625	.67859	.73452	.69130	.72257	.70381	.71039	16
45	.66588	.74606	.67880	.73432	.69151	.72236	.70401	.71019	15
46	.66610	.74586	.67901	.73413	.69172	.72216	.70422	.70998	14
47	.66632	.74567	.67923	.73393	.69193	.72196	.70443	.70978	13
48	.66653	.74548	.67944	.73373	.69214	.72176	.70463	.70957	12
49	.66675	.74528	.67965	.73353	.69235	.72156	.70484	.70937	11
50	.66697	.74509	.67987	.73333	.69256	.72136	.70505	.70916	10
51	.66718	.74489	.68008	.73314	.69277	.72116	.70525	.70896	9
52	.66740	.74470	.68029	.73294	.69298	.72095	.70546	.70875	8
53	.66762	.74451	.68051	.73274	.69319	.72075	.70567	.70855	7
54	.66783	.74431	.68072	.73254	.69340	.72055	.70587	.70834	6
55	.66805	.74412	.68093	.73234	.69361	.72035	.70608	.70813	5
56	.66827	.74392	.68115	.73215	.69382	.72015	.70628	.70793	4
57	.66848	.74373	.68136	.73195	.69403	.71995	.70649	.70772	3
58	.66870	.74353	.68157	.73175	.69424	.71974	.70670	.70752	2
59	.66891	.74334	.68179	.73155	.69445	.71954	.70690	.70731	1
60	.66913	.74314	.68200	.73135	.69466	.71934	.70711	.70711	0
′	48° Cosine	48° Sine	47° Cosine	47° Sine	46° Cosine	46° Sine	45° Cosine	45° Sine	′

Table A-2. Natural Tangents and Cotangents

′	0°		1°		2°		3°		′
	TAN.	CO-TAN.	TAN.	CO-TAN.	TAN.	CO-TAN.	TAN.	CO-TAN.	
0	.00000	Infinite.	.01746	57.2900	.03492	28.6363	.05241	19.0811	60
1	.00029	3437.750	.01775	56.3506	.03521	28.3994	.05270	18.9755	59
2	.00058	1718.870	.01804	55.4415	.03550	28.1664	.05299	18.8711	58
3	.00087	1145.920	.01833	54.5613	.03579	27.9372	.05328	18.7678	57
4	.00116	859.436	.01862	53.7086	.03609	27.7117	.05357	18.6656	56
5	.00145	687.549	.01891	52.8821	.03638	27.4899	.05387	18.5645	55
6	.00175	572.957	.01920	52.0807	.03667	27.2715	.05416	18.4645	54
7	.00204	491.106	.01949	51.3032	.03696	27.0566	.05445	18.3655	53
8	.00233	429.718	.01978	50.5485	.03725	26.8450	.05474	18.2677	52
9	.00262	381.971	.02007	49.8157	.03754	26.6367	.05503	18.1708	51
10	.00291	343.774	.02036	49.1039	.03783	26.4316	.05533	18.0750	50
11	.00320	312.521	.02066	48.4121	.03812	26.2296	.05562	17.9802	49
12	.00349	286.478	.02095	47.7395	.03842	26.0307	.05591	17.8863	48
13	.00378	264.441	.02124	47.0853	.03871	25.8348	.05620	17.7934	47
14	.00407	245.552	.02153	46.4489	.03900	25.6418	.05649	17.7015	46
15	.00436	229.182	.02182	45.8294	.03929	25.4517	.05678	17.6106	45
16	.00465	214.858	.02211	45.2261	.03958	25.2644	.05708	17.5205	44
17	.00495	202.219	.02240	44.6386	.03987	25.0798	.05737	17.4314	43
18	.00524	190.984	.02269	44.0661	.04016	24.8978	.05766	17.3432	42
19	.00553	180.932	.02298	43.5081	.04046	24.7185	.05795	17.2558	41
20	.00582	171.885	.02328	42.9641	.04075	24.5418	.05824	17.1693	40
21	.00611	163.700	.02357	42.4335	.04104	24.3675	.05854	17.0837	39
22	.00640	156.259	.02386	41.9158	.04133	24.1957	.05883	16.9990	38
23	.00669	149.465	.02415	41.4106	.04162	24.0263	.05912	16.9150	37
24	.00698	143.237	.02444	40.9174	.04191	23.8593	.05941	16.8319	36
25	.00727	137.507	.02473	40.4358	.04220	23.6945	.05970	16.7496	35
26	.00756	132.219	.02502	39.9655	.04250	23.5321	.05999	16.6681	34
27	.00785	127.321	.02531	39.5059	.04279	23.3718	.06029	16.5874	33
28	.00814	122.774	.02560	39.0568	.04308	23.2137	.06058	16.5075	32
29	.00844	118.540	.02589	38.6177	.04337	23.0577	.06087	16.4283	31
30	.00873	114.589	.02619	38.1885	.04366	22.9038	.06116	16.3499	30
31	.00902	110.892	.02648	37.7686	.04395	22.7519	.06145	16.2722	29
32	.00931	107.426	.02677	37.3579	.04424	22.6020	.06175	16.1952	28
33	.00960	104.171	.02706	36.9560	.04454	22.4541	.06204	16.1190	27
34	.00989	101.107	.02735	36.5627	.04483	22.3081	.06233	16.0435	26
35	.01018	98.2179	.02764	36.1776	.04512	22.1640	.06262	15.9687	25
36	.01047	95.4895	.02793	35.8006	.04541	22.0217	.06291	15.8945	24
37	.01076	92.9085	.02822	35.4313	.04570	21.8813	.06321	15.8211	23
38	.01105	90.4633	.02851	35.0695	.04599	21.7426	.06350	15.7483	22
39	.01135	88.1436	.02881	34.7151	.04628	21.6056	.06379	15.6762	21
40	.01164	85.9398	.02910	34.3678	.04658	21.4704	.06408	15.6048	20
41	.01193	83.8435	.02939	34.0273	.04687	21.3369	.06437	15.5340	19
42	.01222	81.8470	.02968	33.6935	.04716	21.2049	.06467	15.4638	18
43	.01251	79.9434	.02997	33.3662	.04745	21.0747	.06496	15.3943	17
44	.01280	78.1263	.03026	33.0452	.04774	20.9460	.06525	15.3254	16
45	.01309	76.3900	.03055	32.7303	.04803	20.8188	.06554	15.2571	15
46	.01338	74.7292	.03084	32.4213	.04832	20.6932	.06584	15.1893	14
47	.01367	73.1390	.03114	32.1181	.04862	20.5691	.06613	15.1222	13
48	.01396	71.6151	.03143	31.8205	.04891	20.4465	.06642	15.0557	12
49	.01425	70.1533	.03172	31.5284	.04920	20.3253	.06671	14.9898	11
50	.01455	68.7501	.03201	31.2416	.04949	20.2056	.06700	14.9244	10
51	.01484	67.4019	.03230	30.9599	.04978	20.0872	.06730	14.8596	9
52	.01513	66.1055	.03259	30.6833	.05007	19.9702	.06759	14.7954	8
53	.01542	64.8580	.03288	30.4116	.05037	19.8546	.06788	14.7317	7
54	.01571	63.6567	.03317	30.1446	.05066	19.7403	.06817	14.6685	6
55	.01600	62.4992	.03346	29.8823	.05095	19.6273	.06847	14.6059	5
56	.01629	61.3829	.03376	29.6245	.05124	19.5156	.06876	14.5438	4
57	.01658	60.3058	.03405	29.3711	.05153	19.4051	.06905	14.4823	3
58	.01687	59.2659	.03434	29.1220	.05182	19.2959	.06934	14.4212	2
59	.01716	58.2612	.03463	28.8771	.05212	19.1879	.06963	14.3607	1
60	.01746	57.2900	.03492	28.6363	.05241	19.0811	.06993	14.3007	0
′	CO-TAN.	TAN.	CO-TAN.	TAN.	CO-TAN.	TAN.	CO-TAN.	TAN.	′
	89°		88°		87°		86°		

Table A-2. (*Continued*)

′	4° Tan.	4° Co-tan.	5° Tan.	5° Co-tan.	6° Tan.	6° Co-tan.	7° Tan.	7° Co-tan.	′
0	.06993	14.3007	.08749	11.4301	.10510	9.51436	.12278	8.14435	60
1	.07022	14.2411	.08778	11.3919	.10540	9.48781	.12308	8.12481	59
2	.07051	14.1821	.08807	11.3540	.10569	9.46141	.12338	8.10536	58
3	.07080	14.1235	.08837	11.3163	.10599	9.43515	.12367	8.08600	57
4	.07110	14.0655	.08866	11.2789	.10628	9.40904	.12397	8.06674	56
5	.07139	14.0079	.08895	11.2417	.10657	9.38307	.12426	8.04756	55
6	.07168	13.9507	.08925	11.2048	.10687	9.35724	.12456	8.02848	54
7	.07197	13.8940	.08954	11.1681	.10716	9.33154	.12485	8.00948	53
8	.07227	13.8378	.08983	11.1316	.10746	9.30599	.12515	7.99058	52
9	.07256	13.7821	.09013	11.0954	.10775	9.28058	.12544	7.97176	51
10	.07285	13.7267	.09042	11.0594	.10805	9.25530	.12574	7.95302	50
11	.07314	13.6719	.09071	11.0237	.10834	9.23016	.12603	7.93438	49
12	.07344	13.6174	.09101	10.9882	.10863	9.20516	.12633	7.91582	48
13	.07373	13.5634	.09130	10.9529	.10893	9.18028	.12662	7.89734	47
14	.07402	13.5098	.09159	10.9178	.10922	9.15554	.12692	7.87895	46
15	.07431	13.4566	.09189	10.8829	.10952	9.13093	.12722	7.86064	45
16	.07461	13.4039	.09218	10.8483	.10981	9.10646	.12751	7.84242	44
17	.07490	13.3515	.09247	10.8139	.11011	9.08211	.12781	7.82428	43
18	.07519	13.2996	.09277	10.7797	.11040	9.05789	.12810	7.80622	42
19	.07548	13.2480	.09306	10.7457	.11070	9.03379	.12840	7.78825	41
20	.07578	13.1969	.09335	10.7119	.11099	9.00983	.12869	7.77035	40
21	.07607	13.1461	.09365	10.6783	.11128	8.98598	.12899	7.75254	39
22	.07636	13.0958	.09394	10.6450	.11158	8.96227	.12929	7.73480	38
23	.07665	13.0458	.09423	10.6118	.11187	8.93867	.12958	7.71715	37
24	.07695	12.9962	.09453	10.5789	.11217	8.91520	.12988	7.69957	36
25	.07724	12.9469	.09482	10.5462	.11246	8.89185	.13017	7.68208	35
26	.07753	12.8981	.09511	10.5136	.11276	8.86862	.13047	7.66466	34
27	.07782	12.8496	.09541	10.4813	.11305	8.84551	.13076	7.64732	33
28	.07812	12.8014	.09570	10.4491	.11335	8.82252	.13106	7.63005	32
29	.07841	12.7536	.09600	10.4172	.11364	8.79964	.13136	7.61287	31
30	.07870	12.7062	.09629	10.3854	.11394	8.77689	.13165	7.59575	30
31	.07899	12.6591	.09658	10.3538	.11423	8.75425	.13195	7.57872	29
32	.07929	12.6124	.09688	10.3224	.11452	8.73172	.13224	7.56176	28
33	.07958	12.5660	.09717	10.2913	.11482	8.70931	.13254	7.54487	27
34	.07987	12.5199	.09746	10.2602	.11511	8.68701	.13284	7.52806	26
35	.08017	12.4742	.09776	10.2294	.11541	8.66482	.13313	7.51132	25
36	.08046	12.4288	.09805	10.1988	.11570	8.64275	.13343	7.49465	24
37	.08075	12.3838	.09834	10.1683	.11600	8.62078	.13372	7.47806	23
38	.08104	12.3390	.09864	10.1381	.11629	8.59893	.13402	7.46154	22
39	.08134	12.2946	.09893	10.1080	.11659	8.57718	.13432	7.44509	21
40	.08163	12.2505	.09923	10.0780	.11688	8.55555	.13461	7.42871	20
41	.08192	12.2067	.09952	10.0483	.11718	8.53402	.13491	7.41240	19
42	.08221	12.1632	.09981	10.0187	.11747	8.51259	.13521	7.39616	18
43	.08251	12.1201	.10011	9.98931	.11777	8.49128	.13550	7.37999	17
44	.08280	12.0772	.10040	9.96007	.11806	8.47007	.13580	7.36389	16
45	.08309	12.0346	.10069	9.93101	.11836	8.44896	.13609	7.34786	15
46	.08339	11.9923	.10099	9.90211	.11865	8.42795	.13639	7.33190	14
47	.08368	11.9504	.10128	9.87338	.11895	8.40705	.13669	7.31600	13
48	.08397	11.9087	.10158	9.84482	.11924	8.38625	.13698	7.30018	12
49	.08427	11.8673	.10187	9.81641	.11954	8.36555	.13728	7.28442	11
50	.08456	11.8262	.10216	9.78817	.11983	8.34496	.13758	7.26873	10
51	.08485	11.7853	.10246	9.76009	.12013	8.32446	.13787	7.25310	9
52	.08514	11.7448	.10275	9.73217	.12042	8.30406	.13817	7.23754	8
53	.08544	11.7045	.10305	9.70441	.12072	8.28376	.13846	7.22204	7
54	.08573	11.6645	.10334	9.67680	.12101	8.26355	.13876	7.20661	6
55	.08602	11.6248	.10363	9.64935	.12131	8.24345	.13906	7.19125	5
56	.08632	11.5853	.10393	9.62205	.12160	8.22344	.13935	7.17594	4
57	.08661	11.5461	.10422	9.59490	.12190	8.20352	.13965	7.16071	3
58	.08690	11.5072	.10452	9.56791	.12219	8.18370	.13995	7.14553	2
59	.08720	11.4685	.10481	9.54106	.12249	8.16398	.14024	7.13042	1
60	.08749	11.4301	.10510	9.51436	.12278	8.14435	.14054	7.11537	0
′	85° Co-tan.	85° Tan.	84° Co-tan.	84° Tan.	83° Co-tan.	83° Tan.	82° Co-tan.	82° Tan.	′

Table A-2. (*Continued*)

′	8° Tan.	8° Co-tan.	9° Tan.	9° Co-tan.	10° Tan.	10° Co-tan.	11° Tan.	11° Co-tan.	′
0	.14054	7.11537	.15838	6.31375	.17633	5.67128	.19438	5.14455	60
1	.14084	7.10038	.15868	6.30189	.17663	5.66165	.19468	5.13658	59
2	.14113	7.08546	.15898	6.29007	.17693	5.65205	.19498	5.12862	58
3	.14143	7.07059	.15928	6.27829	.17723	5.64248	.19529	5.12069	57
4	.14173	7.05579	.15958	6.26655	.17753	5.63295	.19559	5.11279	56
5	.14202	7.04105	.15988	6.25486	.17783	5.62344	.19589	5.10490	55
6	.14232	7.02637	.16017	6.24321	.17813	5.61397	.19619	5.09704	54
7	.14262	7.01174	.16047	6.23160	.17843	5.60452	.19649	5.08921	53
8	.14291	6.99718	.16077	6.22003	.17873	5.59511	.19680	5.08139	52
9	.14321	6.98268	.16107	6.20851	.17903	5.58573	.19710	5.07360	51
10	.14351	6.96823	.16137	6.19703	.17933	5.57638	.19740	5.06584	50
11	.14381	6.95385	.16167	6.18559	.17963	5.56706	.19770	5.05809	49
12	.14410	6.93952	.16196	6.17419	.17993	5.55777	.19801	5.05037	48
13	.14440	6.92525	.16226	6.16283	.18023	5.54851	.19831	5.04267	47
14	.14470	6.91104	.16256	6.15151	.18053	5.53927	.19861	5.03499	46
15	.14499	6.89688	.16286	6.14023	.18083	5.53007	.19891	5.02734	45
16	.14529	6.88278	.16316	6.12899	.18113	5.52090	.19921	5.01971	44
17	.14559	6.86874	.16346	6.11779	.18143	5.51176	.19952	5.01210	43
18	.14588	6.85475	.16376	6.10664	.18173	5.50264	.19982	5.00451	42
19	.14618	6.84082	.16405	6.09552	.18203	5.49356	.20012	4.99695	41
20	.14648	6.82694	.16435	6.08444	.18233	5.48451	.20042	4.98940	40
21	.14678	6.81312	.16465	6.07340	.18263	5.47548	.20073	4.98188	39
22	.14707	6.79936	.16495	6.06240	.18293	5.46648	.20103	4.97438	38
23	.14737	6.78564	.16525	6.05143	.18323	5.45751	.20133	4.96690	37
24	.14767	6.77199	.16555	6.04051	.18353	5.44857	.20164	4.95945	36
25	.14796	6.75838	.16585	6.02962	.18383	5.43966	.20194	4.95201	35
26	.14826	6.74483	.16615	6.01878	.18414	5.43077	.20224	4.94460	34
27	.14856	6.73133	.16645	6.00797	.18444	5.42192	.20254	4.93721	33
28	.14886	6.71789	.16674	5.99720	.18474	5.41309	.20285	4.92984	32
29	.14915	6.70450	.16704	5.98646	.18504	5.40429	.20315	4.92249	31
30	.14945	6.69116	.16734	5.97576	.18534	5.39552	.20345	4.91516	30
31	.14975	6.67787	.16764	5.96510	.18564	5.38677	.20376	4.90785	29
32	.15005	6.66463	.16794	5.95448	.18594	5.37805	.20406	4.90056	28
33	.15034	6.65144	.16824	5.94390	.18624	5.36936	.20436	4.89330	27
34	.15064	6.63831	.16854	5.93335	.18654	5.36070	.20466	4.83605	26
35	.15094	6.62523	.16884	5.92283	.18684	5.35206	.20497	4.87882	25
36	.15124	6.61219	.16914	5.91235	.18714	5.34345	.20527	4.87162	24
37	.15153	6.59921	.16944	5.90191	.18745	5.33487	.20557	4.86444	23
38	.15183	6.58627	.16974	5.89151	.18775	5.32631	.20588	4.85727	22
39	.15213	6.57339	.17004	5.88114	.18805	5.31778	.20618	4.85013	21
40	.15243	6.56055	.17033	5.87080	.18835	5.30928	.20648	4.84300	20
41	.15272	6.54777	.17063	5.86051	.18865	5.30080	.20679	4.83590	19
42	.15302	6.53503	.17093	5.85024	.18895	5.29235	.20709	4.82882	18
43	.15332	6.52234	.17123	5.84001	.18925	5.28393	.20739	4.82175	17
44	.15362	6.50970	.17153	5.82982	.18955	5.27553	.20770	4.81471	16
45	.15391	6.49710	.17183	5.81966	.18986	5.26715	.20800	4.80769	15
46	.15421	6.48456	.17213	5.80953	.19016	5.25880	.20830	4.80068	14
47	.15451	6.47206	.17243	5.79944	.19046	5.25048	.20861	4.79370	13
48	.15481	6.45961	.17273	5.78938	.19076	5.24218	.20891	4.78673	12
49	.15511	6.44720	.17303	5.77936	.19106	5.23391	.20921	4.77978	11
50	.15540	6.43484	.17333	5.76937	.19136	5.22566	.20952	4.77286	10
51	.15570	6.42253	.17363	5.75941	.19166	5.21744	.20982	4.76595	9
52	.15600	6.41026	.17393	5.74949	.19197	5.20925	.21013	4.75906	8
53	.15630	6.39804	.17423	5.73960	.19227	5.20107	.21043	4.75219	7
54	.15660	6.38587	.17453	5.72974	.19257	5.19293	.21073	4.74534	6
55	.15689	6.37374	.17483	5.71992	.19287	5.18480	.21104	4.73851	5
56	.15719	6.36165	.17513	5.71013	.19317	5.17671	.21134	4.73170	4
57	.15749	6.34961	.17543	5.70037	.19347	5.16863	.21164	4.72490	3
58	.15779	6.33761	.17573	5.69064	.19378	5.16058	.21195	4.71813	2
59	.15809	6.32566	.17603	5.68094	.19408	5.15256	.21225	4.71137	1
60	.15838	6.31375	.17633	5.67128	.19438	5.14455	.21256	4.70463	0
′	Co-tan.	Tan.	Co-tan.	Tan.	Co-tan.	Tan.	Co-tan.	Tan.	′
	81°		80°		79°		78°		

Table A-2. (*Continued*)

′	12° Tan.	12° Co-tan.	13° Tan.	13° Co-tan.	14° Tan.	14° Co-tan.	15° Tan.	15° Co-tan.	′
0	.21256	4.70463	.23087	4.33148	.24933	4.01078	.26795	3.73205	60
1	.21286	4.69791	.23117	4.32573	.24964	4.00582	.26826	3.72771	59
2	.21316	4.69121	.23148	4.32001	.24995	4.00086	.26857	3.72338	58
3	.21347	4.68452	.23179	4.31430	.25026	3.99592	.26888	3.71907	57
4	.21377	4.67786	.23209	4.30860	.25056	3.99099	.26920	3.71476	56
5	.21408	4.67121	.23240	4.30291	.25087	3.98607	.26951	3.71046	55
6	.21438	4.66458	.23271	4.29724	.25118	3.98117	.26982	3.70616	54
7	.21469	4.65797	.23301	4.29159	.25149	3.97627	.27013	3.70188	53
8	.21499	4.65138	.23332	4.28595	.25180	3.97139	.27044	3.69761	52
9	.21529	4.64480	.23363	4.28032	.25211	3.96651	.27076	3.69335	51
10	.21560	4.63825	.23393	4.27471	.25242	3.96165	.27107	3.68909	50
11	.21590	4.63171	.23424	4.26911	.25273	3.95680	.27138	3.68485	49
12	.21621	4.62518	.23455	4.26352	.25304	3.95196	.27169	3.68061	48
13	.21651	4.61868	.23485	4.25795	.25335	3.94713	.27201	3.67638	47
14	.21682	4.61219	.23516	4.25239	.25366	3.94232	.27232	3.67217	46
15	.21712	4.60572	.23547	4.24685	.25397	3.93751	.27263	3.66796	45
16	.21743	4.59927	.23578	4.24132	.25428	3.93271	.27294	3.66376	44
17	.21773	4.59283	.23608	4.23580	.25459	3.92793	.27326	3.65957	43
18	.21804	4.58641	.23639	4.23030	.25490	3.92316	.27357	3.65538	42
19	.21834	4.58001	.23670	4.22481	.25521	3.91839	.27388	3.65121	41
20	.21864	4.57363	.23700	4.21933	.25552	3.91364	.27419	3.64705	40
21	.21895	4.56726	.23731	4.21387	.25583	3.90890	.27451	3.64289	39
22	.21925	4.56091	.23762	4.20842	.25614	3.90417	.27482	3.63874	38
23	.21956	4.55458	.23793	4.20298	.25645	3.89945	.27513	3.63461	37
24	.21986	4.54826	.23823	4.19756	.25676	3.89474	.27545	3.63048	36
25	.22017	4.54196	.23854	4.19215	.25707	3.89004	.27576	3.62636	35
26	.22047	4.53568	.23885	4.18675	.25738	3.88536	.27607	3.62224	34
27	.22078	4.52941	.23916	4.18137	.25769	3.88068	.27638	3.61814	33
28	.22108	4.52316	.23946	4.17600	.25800	3.87601	.27670	3.61405	32
29	.22139	4.51693	.23977	4.17064	.25831	3.87136	.27701	3.60996	31
30	.22169	4.51071	.24008	4.16530	.25862	3.86671	.27732	3.60588	30
31	.22200	4.50451	.24039	4.15997	.25893	3.86208	.27764	3.60181	29
32	.22231	4.49832	.24069	4.15465	.25924	3.85745	.27795	3.59775	28
33	.22261	4.49215	.24100	4.14934	.25955	3.85284	.27826	3.59370	27
34	.22292	4.48600	.24131	4.14405	.25986	3.84824	.27858	3.58966	26
35	.22322	4.47986	.24162	4.13877	.26017	3.84364	.27889	3.58562	25
36	.22353	4.47374	.24193	4.13350	.26048	3.83906	.27920	3.58160	24
37	.22383	4.46764	.24223	4.12825	.26079	3.83449	.27952	3.57758	23
38	.22414	4.46155	.24254	4.12301	.26110	3.82992	.27983	3.57357	22
39	.22444	4.45548	.24285	4.11778	.26141	3.82537	.28015	3.56957	21
40	.22475	4.44942	.24316	4.11256	.26172	3.82083	.28046	3.56557	20
41	.22505	4.44338	.24347	4.10736	.26203	3.81630	.28077	3.56159	19
42	.22536	4.43735	.24377	4.10216	.26235	3.81177	.28109	3.55761	18
43	.22567	4.43134	.24408	4.09699	.26266	3.80726	.28140	3.55364	17
44	.22597	4.42534	.24439	4.09182	.26297	3.80276	.28172	3.54968	16
45	.22628	4.41936	.24470	4.08666	.26328	3.79827	.28203	3.54573	15
46	.22658	4.41340	.24501	4.08152	.26359	3.79378	.28234	3.54179	14
47	.22689	4.40745	.24532	4.07639	.26390	3.78931	.28266	3.53785	13
48	.22719	4.40152	.24562	4.07127	.26421	3.78485	.28297	3.53393	12
49	.22750	4.39560	.24593	4.06616	.26452	3.78040	.28329	3.53001	11
50	.22781	4.38969	.24624	4.06107	.26483	3.77595	.28360	3.52609	10
51	.22811	4.38381	.24655	4.05599	.26515	3.77152	.28391	3.52219	9
52	.22842	4.37793	.24686	4.05092	.26546	3.76709	.28423	3.51829	8
53	.22872	4.37207	.24717	4.04586	.26577	3.76268	.28454	3.51441	7
54	.22903	4.36623	.24747	4.04081	.26608	3.75828	.28486	3.51053	6
55	.22934	4.36040	.24778	4.03578	.26639	3.75388	.28517	3.50666	5
56	.22964	4.35459	.24809	4.03075	.26670	3.74950	.28549	3.50279	4
57	.22995	4.34879	.24840	4.02574	.26701	3.74512	.28580	3.49894	3
58	.23026	4.34300	.24871	4.02074	.26733	3.74075	.28612	3.49509	2
59	.23056	4.33723	.24902	4.01576	.26764	3.73640	.28643	3.49125	1
60	.23087	4.33148	.24933	4.01078	.26795	3.73205	.28675	3.48741	0
′	Co-tan.	Tan.	Co-tan.	Tan.	Co-tan.	Tan.	Co-tan.	Tan.	′
	77°		76°		75°		74°		

Table A-2. (*Continued*)

′	16° Tan.	16° Co-tan.	17° Tan.	17° Co-tan.	18° Tan.	18° Co-tan.	19° Tan.	19° Co-tan.	′
0	.28675	3.48741	.30573	3.27085	.32492	3.07768	.34433	2.90421	60
1	.28706	3.48359	.30605	3.26745	.32524	3.07464	.34465	2.90147	59
2	.28738	3.47977	.30637	3.26406	.32556	3.07160	.34498	2.89873	58
3	.28769	3.47596	.30669	3.26067	.32588	3.06857	.34530	2.89600	57
4	.28800	3.47216	.30700	3.25729	.32621	3.06554	.34563	2.89327	56
5	.28832	3.46837	.30732	3.25392	.32653	3.06252	.34596	2.89055	55
6	.28864	3.46458	.30764	3.25055	.32685	3.05950	.34628	2.88783	54
7	.28895	3.46080	.30796	3.24719	.32717	3.05649	.34661	2.88511	53
8	.28927	3.45703	.30828	3.24383	.32749	3.05349	.34693	2.88240	52
9	.28958	3.45327	.30860	3.24049	.32782	3.05049	.34726	2.87970	51
10	.28990	3.44951	.30891	3.23714	.32814	3.04749	.34758	2.87700	50
11	.29021	3.44576	.30923	3.23381	.32846	3.04450	.34791	2.87430	49
12	.29053	3.44202	.30955	3.23048	.32878	3.04152	.34824	2.87161	48
13	.29084	3.43829	.30987	3.22715	.32911	3.03854	.34856	2.86892	47
14	.29116	3.43456	.31019	3.22384	.32943	3.03556	.34889	2.86624	46
15	.29147	3.43084	.31051	3.22053	.32975	3.03260	.34922	2.86356	45
16	.29179	3.42713	.31083	3.21722	.33007	3.02963	.34954	2.86089	44
17	.29210	3.42343	.31115	3.21392	.33040	3.02667	.34987	2.85822	43
18	.29242	3.41973	.31147	3.21063	.33072	3.02372	.35019	2.85555	42
19	.29274	3.41604	.31178	3.20734	.33104	3.02077	.35052	2.85289	41
20	.29305	3.41236	.31210	3.20406	.33136	3.01783	.35085	2.85023	40
21	.29337	3.40869	.31242	3.20079	.33160	3.01489	.35117	2.84758	39
22	.29368	3.40502	.31274	3.19752	.33201	3.01196	.35150	2.84494	38
23	.29400	3.40136	.31306	3.19426	.33233	3.00903	.35183	2.84229	37
24	.29432	3.39771	.31338	3.19100	.33266	3.00611	.35216	2.83965	36
25	.29463	3.39406	.31370	3.18775	.33298	3.00319	.35248	2.83702	35
26	.29495	3.39042	.31402	3.18451	.33330	3.00028	.35281	2.83439	34
27	.29526	3.38679	.31434	3.18127	.33363	2.99738	.35314	2.83176	33
28	.29558	3.38317	.31466	3.17804	.33395	2.99447	.35346	2.82914	32
29	.29590	3.37955	.31498	3.17481	.33427	2.99158	.35379	2.82653	31
30	.29621	3.37594	.31530	3.17159	.33460	2.98868	.35412	2.82391	30
31	.29653	3.37234	.31562	3.16838	.33492	2.98580	.35445	2.82130	29
32	.29685	3.36875	.31594	3.16517	.33524	2.98292	.35477	2.81870	28
33	.29716	3.36516	.31626	3.16197	.33557	2.98004	.35510	2.81610	27
34	.29748	3.36158	.31658	3.15877	.33589	2.97717	.35543	2.81350	26
35	.29780	3.35800	.31690	3.15558	.33621	2.97430	.35576	2.81091	25
36	.29811	3.35443	.31722	3.15240	.33654	2.97144	.35608	2.80833	24
37	.29843	3.35087	.31754	3.14922	.33686	2.96858	.35641	2.80574	23
38	.29875	3.34732	.31786	3.14605	.33718	2.96573	.35674	2.80316	22
39	.29906	3.34377	.31818	3.14288	.33751	2.96288	.35707	2.80059	21
40	.29938	3.34023	.31850	3.13972	.33783	2.96004	.35740	2.79802	20
41	.29970	3.33670	.31882	3.13656	.33816	2.95721	.35772	2.79545	19
42	.30001	3.33317	.31914	3.13341	.33848	2.95437	.35805	2.79289	18
43	.30033	3.32965	.31946	3.13027	.33881	2.95155	.35838	2.79033	17
44	.30065	3.32614	.31978	3.12713	.33913	2.94872	.35871	2.78778	16
45	.30097	3.32264	.32010	3.12400	.33945	2.94590	.35904	2.78523	15
46	.30128	3.31914	.32042	3.12087	.33978	2.94309	.35937	2.78269	14
47	.30160	3.31565	.32074	3.11775	.34010	2.94028	.35969	2.78014	13
48	.30192	3.31216	.32106	3.11464	.34043	2.93748	.36002	2.77761	12
49	.30224	3.30868	.32139	3.11153	.34075	2.93468	.36035	2.77507	11
50	.30255	3.30521	.32171	3.10842	.34108	2.93189	.36068	2.77254	10
51	.30287	3.30174	.32203	3.10532	.34140	2.92910	.36101	2.77002	9
52	.30319	3.29829	.32235	3.10223	.34173	2.92632	.36134	2.76750	8
53	.30351	3.29483	.32267	3.09914	.34205	2.92354	.36167	2.76498	7
54	.30382	3.29139	.32299	3.09606	.34238	2.92076	.36199	2.76247	6
55	.30414	3.28795	.32331	3.09298	.34270	2.91799	.36232	2.75996	5
56	.30446	3.28452	.32363	3.08991	.34303	2.91523	.36265	2.75746	4
57	.30478	3.28109	.32396	3.08685	.34335	2.91246	.36298	2.75496	3
58	.30509	3.27767	.32428	3.08379	.34368	2.90971	.36331	2.75246	2
59	.30541	3.27426	.32460	3.08073	.34400	2.90696	.36364	2.74997	1
60	.30573	3.27085	.32492	3.07768	.34433	2.90421	.36397	2.74748	0
′	Co-tan.	Tan.	Co-tan.	Tan.	Co-tan.	Tan.	Co-tan.	Tan.	′
	73°		72°		71°		70°		

Table A-2. (*Continued*)

′	20° Tan.	20° Co-tan.	21° Tan.	21° Co-tan.	22° Tan.	22° Co-tan.	23° Tan.	23° Co-tan.	′
0	.36397	2.74748	.38386	2.60509	.40403	2.47509	.42447	2.35585	60
1	.36430	2.74499	.38420	2.60283	.40436	2.47302	.42482	2.35395	59
2	.36463	2.74251	.38453	2.60057	.40470	2.47095	.42516	2.35205	58
3	.36496	2.74004	.38487	2.59831	.40504	2.46888	.42551	2.35015	57
4	.36529	2.73756	.38520	2.59606	.40538	2.46682	.42585	2.34825	56
5	.36562	2.73509	.38553	2.59381	.40572	2.46476	.42619	2.34636	55
6	.36595	2.73263	.38587	2.59156	.40606	2.46270	.42654	2.34447	54
7	.36628	2.73017	.38620	2.58932	.40640	2.46065	.42688	2.34258	53
8	.36661	2.72771	.38654	2.58708	.40674	2.45860	.42722	2.34069	52
9	.36694	2.72526	.38687	2.58484	.40707	2.45655	.42757	2.33881	51
10	.36727	2.72281	.38721	2.58261	.40741	2.45451	.42791	2.33693	50
11	.36760	2.72036	.38754	2.58038	.40775	2.45246	.42826	2.33505	49
12	.36793	2.71792	.38787	2.57815	.40809	2.45043	.42860	2.33317	48
13	.36826	2.71548	.38821	2.57593	.40843	2.44839	.42894	2.33130	47
14	.36859	2.71305	.38854	2.57371	.40877	2.44636	.42929	2.32943	46
15	.36892	2.71062	.38888	2.57150	.40911	2.44433	.42963	2.32756	45
16	.36925	2.70819	.38921	2.56928	.40945	2.44230	.42998	2.32570	44
17	.36958	2.70577	.38955	2.56707	.40979	2.44027	.43032	2.32383	43
18	.36991	2.70335	.38988	2.56487	.41013	2.43825	.43067	2.32197	42
19	.37024	2.70094	.39022	2.56266	.41047	2.43623	.43101	2.32012	41
20	.37057	2.69853	.39055	2.56046	.41081	2.43422	.43136	2.31826	40
21	.37090	2.69612	.39089	2.55827	.41115	2.43220	.43170	2.31641	39
22	.37124	2.69371	.39122	2.55608	.41149	2.43019	.43205	2.31456	38
23	.37157	2.69131	.39156	2.55389	.41183	2.42819	.43239	2.31271	37
24	.37190	2.68892	.39190	2.55170	.41217	2.42618	.43274	2.31086	36
25	.37223	2.68653	.39223	2.54952	.41251	2.42418	.43308	2.30902	35
26	.37256	2.68414	.39257	2.54734	.41285	2.42218	.43343	2.30718	34
27	.37289	2.68175	.39290	2.54516	.41319	2.42019	.43378	2.30534	33
28	.37322	2.67937	.39324	2.54299	.41353	2.41819	.43412	2.30351	32
29	.37355	2.67700	.39357	2.54082	.41387	2.41620	.43447	2.30167	31
30	.37388	2.67462	.39391	2.53865	.41421	2.41421	.43481	2.29984	30
31	.37422	2.67225	.39425	2.53648	.41455	2.41223	.43516	2.29801	29
32	.37455	2.66989	.39458	2.53432	.41490	2.41025	.43550	2.29619	28
33	.37488	2.66752	.39492	2.53217	.41524	2.40827	.43585	2.29437	27
34	.37521	2.66516	.39526	2.53001	.41558	2.40629	.43620	2.29254	26
35	.37554	2.66281	.39559	2.52786	.41592	2.40432	.43654	2.29073	25
36	.37588	2.66046	.39593	2.52571	.41626	2.40235	.43689	2.28891	24
37	.37621	2.65811	.39626	2.52357	.41660	2.40038	.43724	2.28710	23
38	.37654	2.65576	.39660	2.52142	.41694	2.39841	.43758	2.28528	22
39	.37687	2.65342	.39694	2.51929	.41728	2.39645	.43793	2.28348	21
40	.37720	2.65109	.39727	2.51715	.41763	2.39449	.43828	2.28167	20
41	.37754	2.64875	.39761	2.51502	.41797	2.39253	.43862	2.27987	19
42	.37787	2.64642	.39795	2.51289	.41831	2.39058	.43897	2.27806	18
43	.37820	2.64410	.39829	2.51076	.41865	2.38862	.43932	2.27626	17
44	.37853	2.64177	.39862	2.50864	.41899	2.38668	.43966	2.27447	16
45	.37887	2.63945	.39896	2.50652	.41933	2.38473	.44001	2.27267	15
46	.37920	2.63714	.39930	2.50440	.41968	2.38279	.44036	2.27088	14
47	.37953	2.63483	.39963	2.50229	.42002	2.38084	.44071	2.26909	13
48	.37986	2.63252	.39997	2.50018	.42036	2.37891	.44105	2.26730	12
49	.38020	2.63021	.40031	2.49807	.42070	2.37697	.44140	2.26552	11
50	.38053	2.62791	.40065	2.49597	.42105	2.37504	.44175	2.26374	10
51	.38086	2.62561	.40098	2.49386	.42139	2.37311	.44210	2.26196	9
52	.38120	2.62332	.40132	2.49177	.42173	2.37118	.44244	2.26018	8
53	.38153	2.62103	.40166	2.48967	.42207	2.36925	.44279	2.25840	7
54	.38186	2.61874	.40200	2.48758	.42242	2.36733	.44314	2.25663	6
55	.38220	2.61646	.40234	2.48549	.42276	2.36541	.44349	2.25486	5
56	.38253	2.61418	.40267	2.48340	.42310	2.36349	.44384	2.25309	4
57	.38286	2.61190	.40301	2.48132	.42345	2.36158	.44418	2.25132	3
58	.38320	2.60963	.40335	2.47924	.42379	2.35967	.44453	2.24956	2
59	.38353	2.60736	.40369	2.47716	.42413	2.35776	.44488	2.24780	1
60	.38386	2.60509	.40403	2.47509	.42447	2.35585	.44523	2.24604	0
′	Co-tan. 69°	Tan. 69°	Co-tan. 68°	Tan. 68°	Co-tan. 67°	Tan. 67°	Co-tan. 66°	Tan. 66°	′

Table A-2. (*Continued*)

′	24°		25°		26°		27°		
	Tan.	Co-tan.	Tan.	Co-tan.	Tan.	Co-tan.	Tan.	Co-tan.	
0	.44523	2.24604	.46631	2.14451	.48773	2.05030	.50953	1.96261	60
1	.44558	2.24428	.46666	2.14288	.48809	2.04879	.50989	1.96120	59
2	.44593	2.24252	.46702	2.14125	.48845	2.04728	.51026	1.95979	58
3	.44627	2.24077	.46737	2.13963	.48881	2.04577	.51063	1.95838	57
4	.44662	2.23902	.46772	2.13801	.48917	2.04426	.51099	1.95698	56
5	.44697	2.23727	.46808	2.13639	.48953	2.04276	.51136	1.95557	55
6	.44732	2.23553	.46843	2.13477	.48989	2.04125	.51173	1.95417	54
7	.44767	2.23378	.46879	2.13316	.49026	2.03975	.51209	1.95277	53
8	.44802	2.23204	.46914	2.13154	.49062	2.03825	.51246	1.95137	52
9	.44837	2.23030	.46950	2.12993	.49098	2.03675	.51283	1.94997	51
10	.44872	2.22857	.46985	2.12832	.49134	2.03526	.51319	1.94858	50
11	.44907	2.22683	.47021	2.12671	.49170	2.03376	.51356	1.94718	49
12	.44942	2.22510	.47056	2.12511	.49206	2.03227	.51393	1.94579	48
13	.44977	2.22337	.47092	2.12350	.49242	2.03078	.51430	1.94440	47
14	.45012	2.22164	.47128	2.12190	.49278	2.02929	.51467	1.94301	46
15	.45047	2.21992	.47163	2.12030	.49315	2.02780	.51503	1.94162	45
16	.45082	2.21819	.47199	2.11871	.49351	2.02631	.51540	1.94023	44
17	.45117	2.21647	.47234	2.11711	.49387	2.02483	.51577	1.93885	43
18	.45152	2.21475	.47270	2.11552	.49423	2.02335	.51614	1.93746	42
19	.45187	2.21304	.47305	2.11392	.49459	2.02187	.51651	1.93608	41
20	.45222	2.21132	.47341	2.11233	.49495	2.02039	.51688	1.93470	40
21	.45257	2.20961	.47377	2.11075	.49532	2.01891	.51724	1.93332	39
22	.45292	2.20790	.47412	2.10916	.49568	2.01743	.51761	1.93195	38
23	.45327	2.20619	.47448	2.10758	.49604	2.01596	.51798	1.93057	37
24	.45362	2.20449	.47483	2.10600	.49640	2.01449	.51835	1.92920	36
25	.45397	2.20278	.47519	2.10442	.49677	2.01302	.51872	1.92782	35
26	.45432	2.20108	.47555	2.10284	.49713	2.01155	.51909	1.92645	34
27	.45467	2.19938	.47590	2.10126	.49749	2.01008	.51946	1.92508	33
28	.45502	2.19769	.47626	2.09969	.49786	2.00862	.51983	1.92371	32
29	.45537	2.19599	.47662	2.09811	.49822	2.00715	.52020	1.92235	31
30	.45573	2.19430	.47698	2.09654	.49858	2.00569	.52057	1.92098	30
31	.45608	2.19261	.47733	2.09498	.49894	2.00423	.52094	1.91962	29
32	.45643	2.19092	.47769	2.09341	.49931	2.00277	.52131	1.91826	28
33	.45678	2.18923	.47805	2.09184	.49967	2.00131	.52168	1.91690	27
34	.45713	2.18755	.47840	2.09028	.50004	1.99986	.52205	1.91554	26
35	.45748	2.18587	.47876	2.08872	.50040	1.99841	.52242	1.91418	25
36	.45784	2.18419	.47912	2.08716	.50076	1.99695	.52279	1.91282	24
37	.45819	2.18251	.47948	2.08560	.50113	1.99550	.52316	1.91147	23
38	.45854	2.18084	.47984	2.08405	.50149	1.99406	.52353	1.91012	22
39	.45889	2.17916	.48019	2.08250	.50185	1.99261	.52390	1.90876	21
40	.45924	2.17749	.48055	2.08094	.50222	1.99116	.52427	1.90741	20
41	.45960	2.17582	.48091	2.07939	.50258	1.98972	.52464	1.90607	19
42	.45995	2.17416	.48127	2.07785	.50295	1.98828	.52501	1.90472	18
43	.46030	2.17249	.48163	2.07630	.50331	1.98684	.52538	1.90337	17
44	.46065	2.17083	.48198	2.07476	.50368	1.98540	.52575	1.90203	16
45	.46101	2.16917	.48234	2.07321	.50404	1.98396	.52613	1.90069	15
46	.46136	2.16751	.48270	2.07167	.50441	1.98253	.52650	1.89935	14
47	.46171	2.16585	.48306	2.07014	.50477	1.98110	.52687	1.89801	13
48	.46206	2.16420	.48342	2.06860	.50514	1.97966	.52724	1.89667	12
49	.46242	2.16255	.48378	2.06706	.50550	1.97823	.52761	1.89533	11
50	.46277	2.16090	.48414	2.06553	.50587	1.97680	.52798	1.89400	10
51	.46312	2.15925	.48450	2.06400	.50623	1.97538	.52836	1.89266	9
52	.46348	2.15760	.48486	2.06247	.50660	1.97395	.52873	1.89133	8
53	.46383	2.15596	.48521	2.06094	.50696	1.97253	.52910	1.89000	7
54	.46418	2.15432	.48557	2.05942	.50733	1.97111	.52947	1.88867	6
55	.46454	2.15268	.48593	2.05790	.50769	1.96969	.52984	1.88734	5
56	.46489	2.15104	.48629	2.05637	.50806	1.96827	.53022	1.88602	4
57	.46525	2.14940	.48665	2.05485	.50843	1.96685	.53059	1.88469	3
58	.46560	2.14777	.48701	2.05333	.50879	1.96544	.53096	1.88337	2
59	.46595	2.14614	.48737	2.05182	.50916	1.96402	.53134	1.88205	1
60	.46631	2.14451	.48773	2.05030	.50953	1.96261	.53171	1.88073	0
′	Co-tan.	Tan.	Co-tan.	Tan.	Co-tan.	Tan.	Co-tan.	Tan.	′
	65°		64°		63°		62°		

Table A-2. (*Continued*)

	28° Tan.	28° Co-tan.	29° Tan.	29° Co-tan.	30° Tan.	30° Co-tan.	31° Tan.	31° Co-tan.	
0	.53171	1.88073	.55431	1.80405	.57735	1.73205	.60086	1.66428	60
1	.53208	1.87941	.55469	1.80281	.57774	1.73089	.60126	1.66318	59
2	.53246	1.87809	.55507	1.80158	.57813	1.72973	.60165	1. 6209	58
3	.53283	1.87677	.55545	1.80034	.57851	1.72857	.60205	1.66099	57
4	.53320	1.87546	.55583	1.79911	.57890	1.72741	.60245	1.65990	56
5	.53358	1.87415	.55621	1.79788	.57929	1.72625	.60284	1.65881	55
6	.53395	1.87283	.55659	1.79665	.57968	1.72509	.60324	1.65772	54
7	.53432	1.87152	.55697	1.79542	.58007	1.72393	.60364	1.65668	53
8	.53470	1.87021	.55736	1.79419	.58046	1.72278	.60403	1.65534	52
9	.53507	1.86891	.55774	1.79296	.58085	1.72163	.60443	1.65445	51
10	.53545	1.86760	.55812	1.79174	.58124	1.72047	.60483	1.65337	50
11	.53582	1.86630	.55850	1.79051	.58162	1.71932	.60522	1.65228	49
12	.53620	1.86499	.55888	1.78929	.58201	1.71817	.60562	1.65120	48
13	.53657	1.86369	.55926	1.78807	.58240	1.71702	.60602	1.65011	47
14	.53694	1.86239	.55964	1.78685	.58279	1.71588	.60642	1.64903	46
15	.53732	1.86109	.56003	1.78563	.58318	1.71473	.60681	1.64795	45
16	.53769	1.85979	.56041	1.78441	.58357	1.71358	.60721	1.64687	44
17	.53807	1.85850	.56079	1.78319	.58396	1.71244	.60761	1.64579	43
18	.53844	1.85720	.56117	1.78198	.58435	1.71129	.60801	1.64471	42
19	.53882	1.85591	.56156	1.78077	.58474	1.71015	.60841	1.64363	41
20	.53920	1.85462	.56194	1.77955	.58513	1.70901	.60881	1.64256	40
21	.53957	1.85333	.56232	1.77834	.58552	1.70787	.60921	1.64148	39
22	.53995	1.85204	.56270	1.77713	.58591	1.70673	.60960	1.64041	38
23	.54032	1.85075	.56309	1.77592	.58631	1.70560	.61000	1.63934	37
24	.54070	1.84946	.56347	1.77471	.58670	1.70446	.61040	1.63826	36
25	.54107	1.84818	.56385	1.77351	.58709	1.70332	.61080	1.63719	35
26	.54145	1.84689	.56424	1.77230	.58748	1.70219	.61120	1.63612	34
27	.54183	1.84561	.56462	1.77110	.58787	1.70106	.61160	1.63505	33
28	.54220	1.84433	.56500	1.76990	.58826	1.69992	.61200	1.63398	32
29	.54258	1.84305	.56539	1.76869	.58865	1.69879	.61240	1.63292	31
30	.54296	1.84177	.56577	1.76749	.58904	1.69766	.61280	1.63185	30
31	.54333	1.84049	.56616	1.76630	.58944	1.69653	.61320	1.63079	29
32	.54371	1.83922	.56654	1.76510	.58983	1.69541	.61360	1.62972	28
33	.54409	1.83794	.56693	1.76390	.59022	1.69428	.61400	1.62866	27
34	.54446	1.83667	.56731	1.76271	.59061	1.69316	.61440	1.62760	26
35	.54484	1.83540	.56769	1.76151	.59101	1.69203	.61480	1.62654	25
36	.54522	1.83413	.56808	1.76032	.59140	1.69091	.61520	1.62548	24
37	.54560	1.83286	.56846	1.75913	.59179	1.68979	.61561	1.62442	23
38	.54597	1.83159	.56885	1.75794	.59218	1.68866	.61601	1.62336	22
39	.54635	1.83033	.56923	1.75675	.59258	1.68754	.61641	1.62230	21
40	.54673	1.82906	.56962	1.75556	.59297	1.68643	.61681	1.62125	20
41	.54711	1.82780	.57000	1.75437	.59336	1.68531	.61721	1.62019	19
42	.54748	1.82654	.57039	1.75319	.59376	1.68419	.61761	1.61914	18
43	.54786	1.82528	.57078	1.75200	.59415	1.68308	.61801	1.61808	17
44	.54824	1.82402	.57116	1.75082	.59454	1.68196	.61842	1.61703	16
45	.54862	1.82276	.57155	1.74964	.59494	1.68085	.61882	1.61598	15
46	.54900	1.82150	.57193	1.74846	.59533	1.67974	.61922	1.61493	14
47	.54938	1.82025	.57232	1.74728	.59573	1.67863	.61962	1.61388	13
48	.54975	1.81899	.57271	1.74610	.59612	1.67752	.62003	1.61283	12
49	.55013	1.81774	.57309	1.74492	.59651	1.67641	.62043	1.61179	11
50	.55051	1.81649	.57348	1.74375	.59691	1.67530	.62083	1.61074	10
51	.55089	1.81524	.57386	1.74257	.59730	1.67419	.62124	1.60970	9
52	.55127	1.81399	.57425	1.74140	.59770	1.67309	.62164	1.60865	8
53	.55165	1.81274	.57464	1.74022	.59809	1.67198	.62204	1.60761	7
54	.55203	1.81150	.57503	1.73905	.59849	1.67088	.62245	1.60657	6
55	.55241	1.81025	.57541	1.73788	.59888	1.66978	.62285	1.60553	5
56	.55279	1.80901	.57580	1.73671	.59928	1.66867	.62325	1.60449	4
57	.55317	1.80777	.57619	1.73555	.59967	1.66757	.62366	1.60345	3
58	.55355	1.80653	.57657	1.73438	.60007	1.66647	.62406	1.60241	2
59	.55393	1.80529	.57696	1.73321	.60046	1.66538	.62446	1.60137	1
60	.55431	1.80405	.57735	1.73205	.60086	1.66428	.62487	1.60033	0
	61° Co-tan.	61° Tan.	60° Co-tan.	60° Tan.	59° Co-tan.	59° Tan.	58° Co-tan.	58° Tan.	

Table A-2. (*Continued*)

′	32° Tan.	32° Co-tan.	33° Tan.	33° Co-tan.	34° Tan.	34° Co-tan.	35° Tan.	35° Co-tan.	′
0	.62487	1.60033	.64941	1.53986	.67451	1.48256	.70021	1.42815	60
1	.62527	1.59930	.64982	1.53888	.67493	1.48163	.70064	1.42726	59
2	.62568	1.59826	.65023	1.53791	.67536	1.48070	.70107	1.42638	58
3	.62608	1.59723	.65065	1.53693	.67578	1.47977	.70151	1.42550	57
4	.62649	1.59620	.65106	1.53595	.67620	1.47885	.70194	1.42462	56
5	.62689	1.59517	.65148	1.53497	.67663	1.47792	.70238	1.42374	55
6	.62730	1.59414	.65189	1.53400	.67705	1.47699	.70281	1.42286	54
7	.62770	1.59311	.65231	1.53302	.67748	1.47607	.70325	1.42198	53
8	.62811	1.59208	.65272	1.53205	.67790	1.47514	.70368	1.42110	52
9	.62852	1.59105	.65314	1.53107	.67832	1.47422	.70412	1.42022	51
10	.62892	1.59002	.65355	1.53010	.67875	1.47330	.70455	1.41934	50
11	.62933	1.58900	.65397	1.52913	.67917	1.47238	.70499	1.41847	49
12	.62973	1.58797	.65438	1.52816	.67960	1.47146	.70542	1.41759	48
13	.63014	1.58695	.65480	1.52719	.68002	1.47053	.70586	1.41672	47
14	.63055	1.58593	.65521	1.52622	.68045	1.46962	.70629	1.41584	46
15	.63095	1.58490	.65563	1.52525	.68088	1.46870	.70673	1.41497	45
16	.63136	1.58388	.65604	1.52429	.68130	1.46778	.70717	1.41409	44
17	.63177	1.58286	.65646	1.52332	.68173	1.46686	.70760	1.41322	43
18	.63217	1.58184	.65688	1.52235	.68215	1.46595	.70804	1.41235	42
19	.63258	1.58083	.65729	1.52139	.68258	1.46503	.70848	1.41148	41
20	.63299	1.57981	.65771	1.52043	.68301	1.46411	.70891	1.41061	40
21	.63340	1.57879	.65813	1.51946	.68343	1.46320	.70935	1.40974	39
22	.63380	1.57778	.65854	1.51850	.68386	1.46229	.70979	1.40887	38
23	.63421	1.57676	.65896	1.51754	.68429	1.46137	.71023	1.40800	37
24	.63462	1.57575	.65938	1.51658	.68471	1.46046	.71066	1.40714	36
25	.63503	1.57474	.65980	1.51562	.68514	1.45955	.71110	1.40627	35
26	.63544	1.57372	.66021	1.51466	.68557	1.45864	.71154	1.40540	34
27	.63584	1.57271	.66063	1.51370	.68600	1.45773	.71198	1.40454	33
28	.63625	1.57170	.66105	1.51275	.68642	1.45682	.71242	1.40367	32
29	.63666	1.57069	.66147	1.51179	.68685	1.45592	.71285	1.40281	31
30	.63707	1.56969	.66189	1.51084	.68728	1.45501	.71329	1.40195	30
31	.63748	1.56868	.66230	1.50988	.68771	1.45410	.71373	1.40109	29
32	.63789	1.56767	.66272	1.50893	.68814	1.45320	.71417	1.40022	28
33	.63830	1.56667	.66314	1.50797	.68857	1.45229	.71461	1.39936	27
34	.63871	1.56566	.66356	1.50702	.68900	1.45139	.71505	1.39850	26
35	.63912	1.56466	.66398	1.50607	.68942	1.45049	.71549	1.39764	25
36	.63953	1.56366	.66440	1.50512	.68985	1.44958	.71593	1.39679	24
37	.63994	1.56265	.66482	1.50417	.69028	1.44868	.71637	1.39593	23
38	.64035	1.56165	.66524	1.50322	.69071	1.44778	.71681	1.39507	22
39	.64076	1.56065	.66566	1.50228	.69114	1.44688	.71725	1.39421	21
40	.64117	1.55966	.66608	1.50133	.69157	1.44598	.71769	1.39336	20
41	.64158	1.55866	.66650	1.50038	.69200	1.44508	.71813	1.39250	19
42	.64199	1.55766	.66692	1.49944	.69243	1.44418	.71857	1.39165	18
43	.64240	1.55666	.66734	1.49849	.69286	1.44329	.71901	1.39079	17
44	.64281	1.55567	.66776	1.49755	.69329	1.44239	.71946	1.38994	16
45	.64322	1.55467	.66818	1.49661	.69372	1.44149	.71990	1.38909	15
46	.64363	1.55368	.66860	1.49566	.69416	1.44060	.72034	1.38824	14
47	.64404	1.55269	.66902	1.49472	.69459	1.43970	.72078	1.38738	13
48	.64446	1.55170	.66944	1.49378	.69502	1.43881	.72122	1.38653	12
49	.64487	1.55071	.66986	1.49284	.69545	1.43792	.72166	1.38568	11
50	.64528	1.54972	.67028	1.49190	.69588	1.43703	.72211	1.38484	10
51	.64569	1.54873	.67071	1.49097	.69631	1.43614	.72255	1.38399	9
52	.64610	1.54774	.67113	1.49003	.69675	1.43525	.72299	1.38314	8
53	.64652	1.54675	.67155	1.48909	.69718	1.43436	.72344	1.38229	7
54	.64693	1.54576	.67197	1.48816	.69761	1.43347	.72388	1.38145	6
55	.64734	1.54478	.67239	1.48722	.69804	1.43258	.72432	1.38060	5
56	.64775	1.54379	.67282	1.48629	.69847	1.43169	.72477	1.37976	4
57	.64817	1.54281	.67324	1.48536	.69891	1.43080	.72521	1.37891	3
58	.64858	1.54183	.67366	1.48442	.69934	1.42992	.72565	1.37807	2
59	.64899	1.54085	.67409	1.48349	.69977	1.42903	.72610	1.37722	1
60	.64941	1.53986	.67451	1.48256	.70021	1.42815	.72654	1.37638	0
′	Co-tan. 57°	Tan. 57°	Co-tan 56°	Tan. 56°	Co-tan. 55°	Tan. 55°	Co-tan. 54°	Tan. 54°	′

Table A-2. (*Continued*)

′	36° Tan.	36° Co-tan.	37° Tan.	37° Co-tan.	38° Tan.	38° Co-tan.	39° Tan.	39° Co-tan.	′
0	.72654	1.37638	.75355	1.32704	.78129	1.27994	.80978	1.23490	60
1	.72699	1.37554	.75401	1.32624	.78175	1.27917	.81027	1.23416	59
2	.72743	1.37470	.75447	1.32544	.78222	1.27841	.81075	1.23343	58
3	.72788	1.37386	.75492	1.32464	.78269	1.27764	.81123	1.23270	57
4	.72832	1.37302	.75538	1.32384	.78316	1.27688	.81171	1.23196	56
5	.72877	1.37218	.75584	1.32304	.78363	1.27611	.81220	1.23123	55
6	.72921	1.37134	.75629	1.32224	.78410	1.27535	.81268	1.23050	54
7	.72966	1.37050	.75675	1.32144	.78457	1.27458	.81316	1.22977	53
8	.73010	1.36967	.75721	1.32064	.78504	1.27382	.81364	1.22904	52
9	.73055	1.36883	.75767	1.31984	.78551	1.27306	.81413	1.22831	51
10	.73100	1.36800	.75812	1.31904	.78598	1.27230	.81461	1.22758	50
11	.73144	1.36716	.75858	1.31825	.78645	1.27153	.81510	1.22685	49
12	.73189	1.36633	.75904	1.31745	.78692	1.27077	.81558	1.22612	48
13	.73234	1.36549	.75950	1.31666	.78739	1.27001	.81606	1.22539	47
14	.73278	1.36466	.75996	1.31586	.78786	1.26925	.81655	1.22467	46
15	.73323	1.36383	.76042	1.31507	.78834	1.26849	.81703	1.22394	45
16	.73368	1.36300	.76088	1.31427	.78881	1.26774	.81752	1.22321	44
17	.73413	1.36217	.76134	1.31348	.78928	1.26698	.81800	1.22249	43
18	.73457	1.36133	.76180	1.31269	.78975	1.26622	.81849	1.22176	42
19	.73502	1.36051	.76226	1.31190	.79022	1.26546	.81898	1.22104	41
20	.73547	1.35968	.76272	1.31110	.79070	1.26471	.81946	1.22031	40
21	.73592	1.35885	.76318	1.31031	.79117	1.26395	.81995	1.21959	39
22	.73637	1.35802	.76364	1.30952	.79164	1.26319	.82044	1.21886	38
23	.73681	1.35719	.76410	1.30873	.79212	1.26244	.82092	1.21814	37
24	.73726	1.35637	.76456	1.30795	.79259	1.26169	.82141	1.21742	36
25	.73771	1.35554	.76502	1.30716	.79306	1.26093	.82190	1.21670	35
26	.73816	1.35472	.76548	1.30637	.79354	1.26018	.82238	1.21598	34
27	.73861	1.35389	.76594	1.30558	.79401	1.25943	.82287	1.21526	33
28	.73906	1.35307	.76640	1.30480	.79449	1.25867	.82336	1.21454	32
29	.73951	1.35224	.76686	1.30401	.79496	1.25792	.82385	1.21382	31
30	.73996	1.35142	.76733	1.30323	.79544	1.25717	.82434	1.21310	30
31	.74041	1.35060	.76779	1.30244	.79591	1.25642	.82483	1.21238	29
32	.74086	1.34978	.76825	1.30166	.79639	1.25567	.82531	1.21166	28
33	.74131	1.34896	.76871	1.30087	.79686	1.25492	.82580	1.21094	27
34	.74176	1.34814	.76918	1.30009	.79734	1.25417	.82629	1.21023	26
35	.74221	1.34732	.76964	1.29931	.79781	1.25343	.82678	1.20951	25
36	.74267	1.34650	.77010	1.29853	.79829	1.25268	.82727	1.20879	24
37	.74312	1.34568	.77057	1.29775	.79877	1.25193	.82776	1.20808	23
38	.74357	1.34487	.77103	1.29696	.79924	1.25118	.82825	1.20736	22
39	.74402	1.34405	.77149	1.29618	.79972	1.25044	.82874	1.20665	21
40	.74447	1.34323	.77196	1.29541	.80020	1.24969	.82923	1.20593	20
41	.74492	1.34242	.77242	1.29463	.80067	1.24895	.82972	1.20522	19
42	.74538	1.34160	.77289	1.29385	.80115	1.24820	.83022	1.20451	18
43	.74583	1.34079	.77335	1.29307	.80163	1.24746	.83071	1.20379	17
44	.74628	1.33998	.77382	1.29229	.80211	1.24672	.83120	1.20308	16
45	.74674	1.33916	.77428	1.29152	.80258	1.24597	.83169	1.20237	15
46	.74719	1.33835	.77475	1.29074	.80306	1.24523	.83218	1.20166	14
47	.74764	1.33754	.77521	1.28997	.80354	1.24449	.83268	1.20095	13
48	.74810	1.33673	.77568	1.28919	.80402	1.24375	.83317	1.20024	12
49	.74855	1.33592	.77615	1.28842	.80450	1.24301	.83366	1.19953	11
50	.74900	1.33511	.77661	1.28764	.80498	1.24227	.83415	1.19882	10
51	.74946	1.33430	.77708	1.28687	.80546	1.24153	.83465	1.19811	9
52	.74991	1.33349	.77754	1.28610	.80594	1.24079	.83514	1.19740	8
53	.75037	1.33268	.77801	1.28533	.80642	1.24005	.83564	1.19669	7
54	.75082	1.33187	.77848	1.28456	.80690	1.23931	.83613	1.19599	6
55	.75128	1.33107	.77895	1.28379	.80738	1.23858	.83662	1.19528	5
56	.75173	1.33026	.77941	1.28302	.80786	1.23784	.83712	1.19457	4
57	.75219	1.32946	.77988	1.28225	.80834	1.23710	.83761	1.19387	3
58	.75264	1.32865	.78035	1.28148	.80882	1.23637	.83811	1.19316	2
59	.75310	1.32785	.78082	1.28071	.80930	1.23563	.83860	1.19246	1
60	.75355	1.32704	.78129	1.27994	.80978	1.23490	.83910	1.19175	0
′	Co-tan. 53°	Tan. 53°	Co-tan. 52°	Tan. 52°	Co-tan. 51°	Tan. 51°	Co-tan. 50°	Tan 50°	′

Table A-2. (*Continued*)

′	40° Tan.	40° Co-tan.	41° Tan.	41° Co-tan.	42° Tan.	42° Co-tan.	43° Tan.	43° Co-tan.	′
0	.83910	1.19175	.86929	1.15037	.90040	1.11061	.93252	1.07237	60
1	.83960	1.19105	.86980	1.14969	.90093	1.10996	.93306	1.07174	59
2	.84009	1.19035	.87031	1.14902	.90146	1.10931	.93360	1.07112	58
3	.84059	1.18964	.87082	1.14834	.90199	1.10867	.93415	1.07049	57
4	.84108	1.18894	.87133	1.14767	.90251	1.10802	.93469	1.06987	56
5	.84158	1.18824	.87184	1.14699	.90304	1.10737	.93524	1.06925	55
6	.84208	1.18754	.87236	1.14632	.90357	1.10672	.93578	1.06862	54
7	.84258	1.18684	.87287	1.14565	.90410	1.10607	.93633	1.06800	53
8	.84307	1.18614	.87338	1.14498	.90463	1.10543	.93688	1.06738	52
9	.84357	1.18544	.87389	1.14430	.90516	1.10478	.93742	1.06676	51
10	.84407	1.18474	.87441	1.14363	.90569	1.10414	.93797	1.06613	50
11	.84457	1.18404	.87492	1.14296	.90621	1.10349	.93852	1.06551	49
12	.84507	1.18334	.87543	1.14229	.90674	1.10285	.93906	1.06489	48
13	.84556	1.18264	.87595	1.14162	.90727	1.10220	.93961	1.06427	47
14	.84606	1.18194	.87646	1.14095	.90781	1.10156	.94016	1.06365	46
15	.84656	1.18125	.87698	1.14028	.90834	1.10091	.94071	1.06303	45
16	.84706	1.18055	.87749	1.13961	.90887	1.10027	.94125	1.06241	44
17	.84756	1.17986	.87801	1.13894	.90940	1.09963	.94180	1.06179	43
18	.84806	1.17916	.87852	1.13828	.90993	1.09899	.94235	1.06117	42
19	.84856	1.17846	.87904	1.13761	.91046	1.09834	.94290	1.06056	41
20	.84906	1.17777	.87955	1.13694	.91099	1.09770	.94345	1.05994	40
21	.84956	1.17708	.88007	1.13627	.91153	1.09706	.94400	1.05932	39
22	.85006	1.17638	.88059	1.13561	.91206	1.09642	.94455	1.05870	38
23	.85057	1.17569	.88110	1.13494	.91259	1.09578	.94510	1.05809	37
24	.85107	1.17500	.88162	1.13428	.91313	1.09514	.94565	1.05747	36
25	.85157	1.17430	.88214	1.13361	.91366	1.09450	.94620	1.05685	35
26	.85207	1.17361	.88265	1.13295	.91419	1.09386	.94676	1.05624	34
27	.85257	1.17292	.88317	1.13228	.91473	1.09322	.94731	1.05562	33
28	.85307	1.17223	.88369	1.13162	.91526	1.09258	.94786	1.05501	32
29	.85358	1.17154	.88421	1.13096	.91580	1.09195	.94841	1.05439	31
30	.85408	1.17085	.88473	1.13029	.91633	1.09131	.94896	1.05378	30
31	.85458	1.17016	.88524	1.12963	.91687	1.09067	.94952	1.05317	29
32	.85509	1.16947	.88576	1.12897	.91740	1.09003	.95007	1.05255	28
33	.85559	1.16878	.88628	1.12831	.91794	1.08940	.95062	1.05194	27
34	.85609	1.16809	.88680	1.12765	.91847	1.08876	.95118	1.05133	26
35	.85660	1.16741	.88732	1.12699	.91901	1.08813	.95173	1.05072	25
36	.85710	1.16672	.88784	1.12633	.91955	1.08749	.95229	1.05010	24
37	.85761	1.16603	.88836	1.12567	.92008	1.08686	.95284	1.04949	23
38	.85811	1.16535	.88888	1.12501	.92062	1.08622	.95340	1.04888	22
39	.85862	1.16466	.88940	1.12435	.92116	1.08559	.95395	1.04827	21
40	.85912	1.16398	.88992	1.12369	.92170	1.08496	.95451	1.04766	20
41	.85963	1.16329	.89045	1.12303	.92224	1.08432	.95506	1.04705	19
42	.86014	1.16261	.89097	1.12238	.92277	1.08369	.95562	1.04644	18
43	.86064	1.16192	.89149	1.12172	.92331	1.08306	.95618	1.04583	17
44	.86115	1.16124	.89201	1.12106	.92385	1.08243	.95673	1.04522	16
45	.86166	1.16056	.89253	1.12041	.92439	1.08179	.95729	1.04461	15
46	.86216	1.15987	.89306	1.11975	.92493	1.08116	.95785	1.04401	14
47	.86267	1.15919	.89358	1.11909	.92547	1.08053	.95841	1.04340	13
48	.86318	1.15851	.89410	1.11844	.92601	1.07990	.95897	1.04279	12
49	.86368	1.15783	.89463	1.11778	.92655	1.07927	.95952	1.04218	11
50	.86419	1.15715	.89515	1.11713	.92709	1.07864	.96008	1.04158	10
51	.86470	1.15647	.89567	1.11648	.92763	1.07801	.96064	1.04097	9
52	.86521	1.15579	.89620	1.11582	.92817	1.07738	.96120	1.04036	8
53	.86572	1.15511	.89672	1.11517	.92872	1.07676	.96176	1.03976	7
54	.86623	1.15443	.89725	1.11452	.92926	1.07613	.96232	1.03915	6
55	.86674	1.15375	.89777	1.11387	.92980	1.07550	.96288	1.03855	5
56	.86725	1.15308	.89830	1.11321	.93034	1.07487	.96344	1.03794	4
57	.86776	1.15240	.89883	1.11256	.93088	1.07425	.96400	1.03734	3
58	.86827	1.15172	.89935	1.11191	.93143	1.07362	.96457	1.03674	2
59	.86878	1.15104	.89988	1.11126	.93197	1.07299	.96513	1.03613	1
60	.86929	1.15037	.90040	1.11061	.93252	1.07237	.96569	1.03553	0
′	49° Co-tan	49° Tan.	48° Co-tan.	48° Tan.	47° Co-tan.	47° Tan.	46° Co-tan.	46° Tan.	′

Table A-2. (*Continued*)

′	44° Tan.	44° Co-tan.	′	′	44° Tan.	44° Co-tan.	′	′	44° Tan.	44° Co-tan.	′
0	.96569	1.03553	60	21	.97756	1.02295	39	41	.98901	1.01112	19
1	.96625	1.03493	59	22	.97813	1.02236	38	42	.98958	1.01053	18
2	.96681	1.03433	58	23	.97870	1.02176	37	43	.99016	1.00994	17
3	.96738	1.03372	57	24	.97927	1.02117	36	44	.99073	1.00935	16
4	.96794	1.03312	56	25	.97984	1.02057	35	45	.99131	1.00876	15
5	.96850	1.03252	55	26	.98041	1.01998	34	46	.99189	1.00818	14
6	.96907	1.03192	54	27	.98098	1.01939	33	47	.99247	1.00759	13
7	.96963	1.03132	53	28	.98155	1.01879	32	48	.99304	1.00701	12
8	.97020	1.03072	52	29	.98213	1.01820	31	49	.99362	1.00642	11
9	.97076	1.03012	51	30	.98270	1.01761	30	50	.99420	1.00583	10
10	.97133	1.02952	50	31	.98327	1.01702	29	51	.99478	1.00525	9
11	.97189	1.02892	49	32	.98384	1.01642	28	52	.99536	1.00467	8
12	.97246	1.02832	48	33	.98441	1.01583	27	53	.99594	1.00408	7
13	.97302	1.02772	47	34	.98499	1.01524	26	54	.99652	1.00350	6
14	.97359	1.02713	46	35	.98556	1.01465	25	55	.99710	1.00291	5
15	.97416	1.02653	45	36	.98613	1.01406	24	56	.99768	1.00233	4
16	.97472	1.02593	44	37	.98671	1.01347	23	57	.99826	1.00175	3
17	.97529	1.02533	43	38	.98728	1.01288	22	58	.99884	1.00116	2
18	.97586	1.02474	42	39	.98786	1.01229	21	59	.99942	1.00058	1
19	.97643	1.02414	41	40	.98843	1.01170	20	60	1	1	0
20	.97700	1.02355	40								
′	45° Co-tan.	45° Tan.	′	′	45° Co-tan.	45° Tan.	′	′	45° Co-tan.	45° Tan.	′

Table A-3. Stadia Reduction Tables

	0°		1°		2°		3°	
Minutes	Hor. dist.	Diff. elev.	Hor. dist.	Diff. elev.	Hor. dist.	Diff. elev.	Hor. dist.	Diff. elev.
0	100.00	0.00	99.97	1.74	99.88	3.49	99.73	5.23
2	100.00	0.06	99.97	1.80	99.87	3.55	99.72	5.28
4	100.00	0.12	99.97	1.86	99.87	3.60	99.71	5.34
6	100.00	0.17	99.96	1.92	99.87	3.66	99.71	5.40
8	100.00	0.23	99.96	1.98	99.86	3.72	99.70	5.46
10	100.00	0.29	99.96	2.04	99.86	3.78	99.69	5.52
12	100.00	0.35	99.96	2.09	99.85	3.84	99.69	5.57
14	100.00	0.41	99.95	2.15	99.85	3.90	99.68	5.63
16	100.00	0.47	99.95	2.21	99.84	3.95	99.68	5.69
18	100.00	0.52	99.95	2.27	99.84	4.01	99.67	5.75
20	100.00	0.58	99.95	2.33	99.83	4.07	99.66	5.80
22	100.00	0.64	99.94	2.38	99.83	4.13	99.66	5.86
24	100.00	0.70	99.94	2.44	99.82	4.18	99.65	5.92
26	99.99	0.76	99.94	2.50	99.82	4.24	99.64	5.98
28	99.99	0.81	99.93	2.56	99.81	4.30	99.63	6.04
30	99.99	0.87	99.93	2.62	99.81	4.36	99.63	6.09
32	99.99	0.93	99.93	2.67	99.80	4.42	99.62	6.15
34	99.99	0.99	99.93	2.73	99.80	4.48	99.62	6.21
36	99.99	1.05	99.92	2.79	99.79	4.53	99.61	6.27
38	99.99	1.11	99.92	2.85	99.79	4.59	99.60	6.33
40	99.99	1.16	99.92	2.91	99.78	4.65	99.59	6.38
42	99.99	1.22	99.91	2.97	99.78	4.71	99.59	6.44
44	99.98	1.28	99.91	3.02	99.77	4.76	99.58	6.50
46	99.98	1.34	99.90	3.08	99.77	4.82	99.57	6.56
48	99.98	1.40	99.90	3.14	99.76	4.88	99.56	6.61
50	99.98	1.45	99.90	3.20	99.76	4.94	99.56	6.67
52	99.98	1.51	99.89	3.26	99.75	4.99	99.55	6.73
54	99.98	1.57	99.89	3.31	99.74	5.05	99.54	6.78
56	99.97	1.63	99.89	3.37	99.74	5.11	99.53	6.84
58	99.97	1.69	99.88	3.43	99.73	5.17	99.52	6.90
60	99.97	1.74	99.88	3.49	99.73	5.23	99.51	6.96

*These tables are adequate for use in both feet and meters.

Table A-3. *(Continued)*

Minutes	4° Hor. dist.	4° Diff. elev.	5° Hor. dist.	5° Diff. elev.	6° Hor. dist.	6° Diff. elev.	7° Hor. dist.	7° Diff. elev.
0	99.51	6.69	99.24	8.68	98.91	10.40	98.51	12.10
2	99.51	7.02	99.23	8.74	98.90	10.45	98.50	12.15
4	99.50	7.07	99.22	8.80	98.88	10.51	98.48	12.21
6	99.49	7.13	99.21	8.85	98.87	10.57	98.47	12.26
8	99.48	7.19	99.20	8.91	98.86	10.62	98.46	12.32
10	99.47	7.25	99.19	8.97	98.85	10.68	98.44	12.38
12	99.46	7.30	99.18	9.03	98.83	10.74	98.43	12.43
14	99.46	7.36	99.17	9.08	98.82	10.79	98.41	12.49
16	99.45	7.42	99.16	9.14	98.81	10.85	98.40	12.55
18	99.44	7.48	99.15	9.20	98.80	10.91	98.39	12.60
20	99.43	7.53	99.14	9.25	98.78	10.96	98.37	12.66
22	99.42	7.59	99.13	9.31	98.77	11.02	98.36	12.72
24	99.41	7.65	99.11	9.37	98.76	11.08	98.34	12.77
26	99.40	7.71	99.10	9.43	98.74	11.13	98.33	12.83
28	99.39	7.76	99.09	9.48	98.73	11.19	98.31	12.88
30	99.38	7.82	99.08	9.54	98.72	11.25	98.29	12.94
32	99.38	7.88	99.07	9.60	98.71	11.30	98.28	13.00
34	99.37	7.94	99.06	9.65	98.69	11.36	98.27	13.05
36	99.36	7.99	99.05	9.71	98.68	11.42	98.25	13.11
38	99.35	8.05	99.04	9.77	98.67	11.47	98.24	13.17
40	99.34	8.11	99.03	9.83	98.65	11.53	98.22	13.22
42	99.33	8.17	99.01	9.88	98.64	11.59	98.20	13.28
44	99.32	8.22	99.00	9.94	98.63	11.64	98.19	13.33
46	99.31	8.28	98.99	10.00	98.61	11.70	98.17	13.39
48	99.30	8.34	98.98	10.05	98.60	11.76	98.16	13.45
50	99.29	8.40	98.97	10.11	98.58	11.81	98.14	13.50
52	99.28	8.45	98.96	10.17	98.57	11.87	98.13	13.56
54	99.27	8.51	98.94	10.22	98.56	11.93	98.11	13.61
56	99.26	8.57	99.93	10.28	98.54	11.98	98.10	13.67
58	99.25	8.63	98.92	10.34	98.53	12.04	98.08	13.73
60	99.24	8.68	98.91	10.40	98.51	12.10	98.06	13.78

Table A-3. *(Continued)*

Minutes	8° Hor. dist.	8° Diff. elev.	9° Hor. dist.	9° Diff. elev.	10° Hor. dist.	10° Diff. elev.	11° Hor. dist.	11° Diff. elev.
0	98.06	13.78	97.55	15.45	96.98	17.10	96.36	18.73
2	98.05	13.84	97.53	15.51	96.96	17.16	96.34	18.78
4	98.03	13.89	97.52	15.56	96.94	17.21	96.32	18.84
6	98.01	13.95	97.50	15.62	96.92	17.26	96.29	18.89
8	98.00	14.01	97.48	15.67	96.90	17.32	96.27	18.95
10	97.98	14.06	97.46	15.73	96.88	17.37	96.25	19.00
12	97.97	14.12	97.44	15.78	96.86	17.43	96.23	19.05
14	97.95	14.17	97.43	15.84	96.84	17.48	96.21	19.11
16	97.93	14.23	97.41	15.89	96.82	17.54	96.18	19.16
18	97.92	14.28	97.39	15.95	96.80	17.59	96.16	19.21
20	97.90	14.34	97.37	16.00	96.78	17.65	96.14	19.27
22	97.88	14.40	97.35	16.06	96.76	17.70	96.12	19.32
24	97.87	14.45	97.33	16.11	96.74	17.76	96.09	19.38
26	97.85	14.51	97.31	16.17	96.72	17.81	96.07	19.43
28	97.83	14.56	97.29	16.22	96.70	17.86	96.05	19.48
30	97.82	14.62	97.28	16.28	96.68	17.92	96.03	19.54
32	97.80	14.67	97.26	16.33	96.66	17.97	96.00	19.59
34	97.78	14.73	97.24	16.39	96.64	18.03	95.98	19.64
36	97.76	14.79	97.22	16.44	96.62	18.08	95.96	19.70
38	97.75	14.84	97.20	16.50	96.60	18.14	95.93	19.75
40	97.73	14.90	97.18	16.55	96.57	18.19	95.91	19.80
42	97.71	14.95	97.16	16.61	96.55	18.24	95.89	19.86
44	97.69	15.01	97.14	16.66	96.53	18.30	95.86	19.91
46	97.68	15.06	97.12	16.72	96.51	18.35	95.84	19.96
48	97.66	15.12	97.10	16.77	96.49	18.41	95.82	20.02
50	97.64	15.17	97.08	16.83	96.47	18.46	95.79	20.07
52	97.62	15.23	97.06	16.88	96.45	18.51	95.77	20.12
54	97.61	15.28	97.04	16.94	96.42	18.57	95.75	20.18
56	97.59	15.34	97.02	16.99	96.40	18.62	95.72	20.23
58	97.57	15.40	97.00	17.05	96.38	18.68	95.70	20.28
60	97.55	15.45	96.98	17.10	96.36	18.73	95.68	20.34

Table A-3. (*Continued*)

	12°		13°		14°		15°	
Minutes	Hor. dist.	Diff. elev.	Hor. dist.	Diff. elev.	Hor. dist.	Diff. elev.	Hor. dist.	Diff. elev.
0	95.68	20.34	94.94	21.92	94.15	23.47	93.30	25.00
2	95.65	20.39	94.91	21.97	94.12	23.52	93.27	25.05
4	95.63	20.44	94.89	22.02	94.09	23.58	93.24	25.10
6	95.61	20.50	94.86	22.08	94.07	23.63	93.21	25.15
8	95.58	20.55	94.84	22.13	94.04	23.68	93.18	25.20
10	95.56	20.60	94.81	22.18	94.01	23.73	93.16	25.25
12	95.53	20.66	94.79	22.23	93.98	23.78	93.13	25.30
14	95.51	20.71	94.76	22.28	93.95	23.83	93.10	25.35
16	95.49	20.76	94.73	22.34	93.93	23.88	93.07	25.40
18	95.46	20.81	94.71	22.39	93.90	23.93	93.04	25.45
20	95.44	20.87	94.68	22.44	93.87	23.99	93.01	25.50
22	95.41	20.92	94.66	22.49	93.84	24.04	92.98	25.55
24	95.39	20.97	94.63	22.54	93.81	24.09	92.95	25.60
26	95.36	21.03	94.60	22.60	93.79	24.14	92.92	25.65
28	95.34	21.08	94.58	22.65	93.76	24.19	92.89	25.70
30	95.32	21.13	94.55	22.70	93.73	24.24	92.86	25.75
32	95.29	21.18	94.52	22.75	93.70	24.29	92.83	25.80
34	95.27	21.24	94.50	22.80	93.67	24.34	92.80	25.85
36	95.24	21.29	94.47	22.85	93.65	24.39	92.77	25.90
38	95.22	21.34	94.44	22.91	93.62	24.44	92.74	25.95
40	95.19	21.39	94.42	22.96	93.59	24.49	92.71	26.00
42	95.17	21.45	94.39	23.01	93.56	24.55	92.68	26.05
44	95.14	21.50	94.36	23.06	93.53	24.60	92.65	26.10
46	95.12	21.55	94.34	23.11	93.50	24.65	92.62	26.15
48	95.09	21.60	94.31	23.16	93.47	24.70	92.59	26.20
50	95.07	21.66	94.28	23.22	93.45	24.75	92.56	26.25
52	95.04	21.71	94.26	23.27	93.42	24.80	92.53	26.30
54	95.02	21.76	94.23	23.32	93.39	24.85	92.49	26.35
56	94.99	21.81	94.20	23.37	93.36	24.90	92.46	26.40
58	94.97	21.87	94.17	23.42	93.33	24.95	92.43	26.45
60	94.94	21.92	94.15	23.47	93.90	25.00	92.40	26.50

Table A-3. *(Continued)*

	16°		17°		18°		19°	
Minutes	Hor. dist.	Diff. elev.	Hor. dist.	Diff. elev.	Hor. dist.	Diff. elev.	Hor. dist.	Diff. elev.
0	92.40	26.50	91.45	27.96	90.45	29.39	89.40	30.78
2	92.37	26.55	91.42	28.01	90.42	29.44	89.36	30.83
4	92.34	26.59	91.39	28.06	90.38	29.48	89.33	30.87
6	92.31	26.64	91.35	28.10	90.35	29.53	89.29	30.92
8	92.28	26.69	91.32	28.15	90.31	29.58	89.26	30.97
10	92.25	26.74	91.29	28.20	90.28	29.62	89.22	31.01
12	92.22	26.79	91.26	28.25	90.24	29.67	89.18	31.06
14	92.19	26.84	91.22	28.30	90.21	29.72	89.15	31.10
16	92.15	26.89	91.19	28.34	90.18	29.76	89.11	31.15
18	92.12	26.94	91.16	28.39	90.14	29.81	89.08	31.19
20	92.09	26.99	91.12	28.44	90.11	29.86	89.04	31.24
22	92.06	27.04	91.09	28.49	90.07	29.90	89.00	31.28
24	92.03	27.09	91.06	28.54	90.04	29.95	88.96	31.33
26	92.00	27.13	91.02	28.58	90.00	30.00	88.93	31.38
28	91.97	27.18	90.99	28.63	89.97	30.04	88.89	31.42
30	91.93	27.23	90.96	28.68	89.93	30.09	88.86	31.47
32	91.90	27.28	90.92	28.73	89.90	30.14	88.82	31.51
34	91.87	27.33	90.89	28.77	89.86	30.19	88.78	31.56
36	91.84	27.38	90.86	28.82	89.83	30.23	88.75	31.60
38	91.81	27.43	90.82	28.87	89.79	30.28	88.71	31.65
40	91.77	27.48	90.79	28.92	89.76	30.32	88.67	31.69
42	91.74	27.52	90.76	28.96	89.72	30.37	88.64	31.74
44	91.71	27.57	90.72	29.01	89.69	30.41	88.60	31.78
46	91.68	27.62	90.69	29.06	89.65	30.46	88.56	31.83
48	91.65	27.67	90.66	29.11	89.61	30.51	88.53	31.87
50	91.61	27.72	90.62	29.15	89.58	30.55	88.49	31.92
52	91.58	27.77	90.59	29.20	89.54	30.60	88.45	31.96
54	91.55	27.81	90.55	29.25	89.51	30.65	88.41	32.01
56	91.52	27.86	90.52	29.30	89.47	30.69	88.38	32.05
58	91.48	27.91	90.48	29.34	89.44	30.74	88.34	32.09
60	91.45	27.96	90.45	29.39	89.40	30.78	88.30	32.14

Table A-3. (*Continued*)

Minutes	20°		21°		22°		23°	
	Hor. dist.	Diff. elev.	Hor. dist.	Diff. elev.	Hor. dist.	Diff. elev.	Hor. dist.	Diff. elev.
0	88.30	32.14	87.16	33.46	85.97	34.73	84.73	35.97
2	88.26	32.18	87.12	33.50	85.93	34.77	84.69	36.01
4	88.23	32.23	87.08	33.54	85.89	34.82	84.65	36.05
6	88.19	32.27	87.04	33.59	85.85	34.86	84.61	36.09
8	88.15	32.32	87.00	33.63	85.80	34.90	84.57	36.13
10	88.11	32.36	86.96	33.67	85.76	34.94	84.52	36.17
12	88.08	32.41	86.92	33.72	85.72	34.98	84.48	36.21
14	88.04	32.45	86.88	33.76	85.68	35.02	84.44	36.25
16	88.00	32.49	86.84	33.80	85.64	35.07	84.40	36.29
18	87.96	32.54	86.80	33.84	85.60	35.11	84.35	36.33
20	87.93	32.58	86.77	33.89	85.56	35.15	84.31	36.37
22	87.89	32.63	86.73	33.93	85.52	35.19	84.27	36.41
24	87.85	32.67	86.69	33.97	85.48	35.23	84.23	36.45
26	87.81	32.72	86.65	34.01	85.44	35.27	84.18	36.49
28	87.77	32.76	86.61	34.06	85.40	35.31	84.14	36.53
30	87.74	32.80	86.57	34.10	85.36	35.36	84.10	36.57
32	87.70	32.85	86.53	34.14	85.31	35.40	84.06	36.61
34	87.66	32.89	86.49	34.18	85.27	35.44	84.01	36.65
36	87.62	32.93	86.45	34.23	85.23	35.48	83.97	36.69
38	87.58	32.98	86.41	34.27	85.19	35.52	83.93	36.73
40	87.54	33.02	86.37	34.31	85.15	35.56	83.89	36.77
42	87.51	33.07	86.33	34.35	85.11	35.60	83.84	36.80
44	87.47	33.11	86.29	34.40	85.07	35.64	83.80	36.84
46	87.43	33.15	86.25	34.44	85.02	35.68	83.76	36.88
48	87.39	33.20	86.21	34.48	84.98	35.72	83.72	36.92
50	87.35	33.24	86.17	34.52	84.94	35.76	83.67	36.96
52	87.31	33.28	86.13	34.57	84.90	35.80	83.63	37.00
54	87.27	33.33	86.09	34.61	84.86	35.85	83.59	37.04
56	87.24	33.37	86.05	34.65	84.82	35.89	83.54	37.08
58	87.20	33.41	86.01	34.69	84.77	35.93	83.50	37.12
60	87.16	33.46	85.97	34.73	84.73	35.97	83.46	37.16

Table A-3. *(Continued)*

	24°		25°		26°		27°	
Minutes	Hor. dist.	Diff. elev.	Hor. dist.	Diff. elev.	Hor. dist.	Diff. elev.	Hor. dist.	Diff. elev.
0	83.46	37.16	82.14	38.30	80.78	39.40	79.39	40.45
2	83.41	37.20	82.09	38.34	80.74	39.44	79.34	40.49
4	83.37	37.23	82.05	38.38	80.69	39.47	79.30	40.52
6	83.33	37.27	82.01	38.41	80.65	39.51	79.25	40.55
8	83.28	37.31	81.96	38.45	80.60	39.54	79.20	40.59
10	83.24	37.35	81.92	38.49	80.55	39.58	79.15	40.62
12	83.20	37.39	81.87	38.53	80.51	39.61	79.11	40.66
14	83.15	37.43	81.83	38.56	80.46	39.65	79.06	40.69
16	83.11	37.47	81.78	38.60	80.41	39.69	79.01	40.72
18	83.07	37.51	81.74	38.64	80.37	39.72	78.96	40.76
20	83.02	37.54	81.69	38.67	80.32	39.76	78.92	40.79
22	82.98	37.58	81.65	38.71	80.28	39.79	78.87	40.82
24	82.93	37.62	81.60	38.75	80.23	39.83	78.82	40.86
26	82.89	37.66	81.56	38.78	80.18	39.86	78.77	40.89
28	82.85	37.70	81.51	38.82	80.14	39.90	78.73	40.92
30	82.80	37.74	81.47	38.86	80.09	39.93	78.68	40.96
32	82.76	37.77	81.42	38.89	80.04	39.97	78.63	40.99
34	82.72	37.81	81.38	38.93	80.00	40.00	78.58	41.02
36	82.67	37.85	81.33	38.97	79.95	40.04	78.54	41.06
38	82.63	37.89	81.28	39.00	79.90	40.07	78.49	41.09
40	82.58	37.93	81.24	39.04	79.86	40.11	78.44	41.12
42	82.54	37.96	81.19	39.08	79.81	40.14	78.39	41.16
44	82.49	38.00	81.15	39.11	79.76	40.18	78.34	41.19
46	82.45	38.04	81.10	39.15	79.72	40.21	78.30	41.22
48	82.41	38.08	81.06	39.18	79.67	40.24	78.25	41.26
50	82.36	38.11	81.01	39.22	79.62	40.28	78.20	41.29
52	82.32	38.15	80.97	39.26	79.58	40.31	78.15	41.32
54	82.27	38.19	80.92	39.29	79.53	40.35	78.10	41.35
56	82.23	38.23	80.87	39.33	79.48	40.38	78.06	41.39
58	82.18	38.26	80.83	39.36	79.44	40.42	78.01	41.42
60	82.14	38.30	80.78	39.40	79.39	40.45	77.96	41.45

Table A-3. (*Continued*)

Minutes	28°		29°		30°	
	Hor. dist.	Diff. elev.	Hor. dist.	Diff. elev.	Hor. dist.	Diff. elev.
0	77.96	41.45	76.50	42.40	75.00	43.30
2	77.91	41.48	76.45	42.43	74.95	43.33
4	77.86	41.52	76.40	42.46	74.90	43.36
6	77.81	41.55	76.35	42.49	74.85	43.39
8	77.77	41.58	76.30	42.53	74.80	43.42
10	77.72	41.61	76.25	42.56	74.75	43.45
12	77.67	41.65	76.20	42.59	74.70	43.47
14	77.62	41.68	76.15	42.62	74.65	43.50
16	77.57	41.71	76.10	42.65	74.60	43.53
18	77.52	41.74	76.05	42.68	74.55	43.56
20	77.48	41.77	76.00	42.71	74.49	43.59
22	77.42	41.81	75.95	42.74	74.44	43.62
24	77.38	41.84	75.90	42.77	74.39	43.65
26	77.33	41.87	75.85	42.80	74.34	43.67
28	77.28	41.90	75.80	42.83	74.29	43.70
30	77.23	41.93	75.75	42.86	74.24	43.73
32	77.18	41.97	75.70	42.89	74.19	43.76
34	77.13	42.00	75.65	42.92	74.14	43.79
36	77.09	42.03	75.60	42.95	74.09	43.82
38	77.04	42.06	75.55	42.98	74.04	43.84
40	76.99	42.09	75.50	43.01	73.99	46.87
42	76.94	42.12	75.45	43.04	73.93	43.90
44	76.89	42.15	75.40	43.07	73.88	43.93
46	76.84	42.19	75.35	43.10	73.83	43.95
48	76.79	42.22	75.30	43.13	73.78	43.98
50	76.74	42.25	75.25	43.16	73.73	44.01
52	76.69	42.28	75.20	43.18	73.68	44.04
54	76.64	42.31	75.15	43.21	73.63	44.07
56	76.59	42.34	75.10	43.24	73.58	44.09
58	76.55	42.37	75.05	43.27	73.52	44.12
60	76.50	42.40	75.00	43.30	73.47	44.15

Table A-4. Tangents and External Distances for Circular Curves

′	Δ = 0°		Δ = 1°		Δ = 2°		Δ = 3°		′
	T	E	T	E	T	E	T	E	
0	0.000	0.000	50.001	0.218	100.010	0.873	150.034	1.964	0
1	0.833	0.000	50.835	0.226	100.844	0.887	150.868	1.986	1
2	1.667	0.000	51.668	0.233	101.677	0.902	151.702	2.008	2
3	2.500	0.001	52.501	0.241	102.511	0.917	152.536	2.030	3
4	3.333	0.001	53.335	0.248	103.345	0.932	153.370	2.052	4
5	4.167	0.002	54.168	0.256	104.178	0.947	154.204	2.075	5
6	5.000	0.002	55.002	0.264	105.012	0.962	155.038	2.097	6
7	5.833	0.003	55.835	0.272	105.845	0.977	155.872	2.120	7
8	6.667	0.004	56.668	0.280	106.679	0.993	156.706	2.142	8
9	7.500	0.005	57.502	0.288	107.513	1.009	157.540	2.165	9
10	8.333	0.006	58.335	0.297	108.346	1.024	158.374	2.188	10
11	9.167	0.007	59.169	0.305	109.180	1.040	159.208	2.212	11
12	10.000	0.009	60.002	0.314	110.013	1.056	160.042	2.235	12
13	10.833	0.010	60.836	0.323	110.847	1.072	160.876	2.258	13
14	11.667	0.012	61.669	0.332	111.681	1.088	161.710	2.282	14
15	12.500	0.014	62.502	0.341	112.514	1.105	162.544	2.305	15
16	13.333	0.015	63.336	0.350	113.348	1.121	163.378	2.329	16
17	14.167	0.017	64.169	0.359	114.182	1.138	164.212	2.353	17
18	15.000	0.019	65.003	0.368	115.015	1.154	165.046	2.377	18
19	15.834	0.022	65.836	0.378	115.849	1.171	165.880	2.401	19
20	16.667	0.024	66.670	0.388	116.683	1.188	166.714	2.425	20
21	17.500	0.027	67.503	0.398	117.516	1.205	167.548	2.449	21
22	18.333	0.029	68.337	0.407	118.350	1.222	168.382	2.474	22
23	19.167	0.032	69.170	0.417	119.184	1.239	169.216	2.498	23
24	20.000	0.035	70.003	0.427	120.018	1.257	170.050	2.523	24
25	20.834	0.038	70.837	0.438	120.851	1.274	170.884	2.548	25
26	21.667	0.041	71.670	0.448	121.685	1.292	171.718	2.573	26
27	22.500	0.044	72.504	0.459	122.519	1.310	172.552	2.598	27
28	23.333	0.048	73.337	0.469	123.352	1.328	173.386	2.623	28
29	24.167	0.051	74.171	0.480	124.186	1.346	174.220	2.648	29
30	25.000	0.054	75.004	0.491	125.020	1.364	175.054	2.674	30
31	25.834	0.058	75.838	0.502	125.854	1.382	175.889	2.699	31
32	26.667	0.062	76.671	0.513	126.687	1.400	176.723	2.725	32
33	27.500	0.066	77.505	0.524	127.521	1.419	177.557	2.751	33
34	28.333	0.070	78.338	0.536	128.355	1.438	178.391	2.776	34
35	29.167	0.074	79.172	0.547	129.189	1.456	179.225	2.802	35
36	30.000	0.078	80.005	0.559	130.022	1.475	180.059	2.829	36
37	30.834	0.083	80.839	0.570	130.856	1.494	180.893	2.855	37
38	31.667	0.088	81.672	0.582	131.690	1.513	181.728	2.881	38
39	32.500	0.092	82.506	0.594	132.524	1.532	182.562	2.908	39
40	33.334	0.097	83.339	0.606	133.357	1.552	183.396	2.934	40
41	34.167	0.102	84.173	0.618	134.191	1.571	184.230	2.961	41
42	35.000	0.107	85.006	0.630	135.025	1.591	185.064	2.988	42
43	35.834	0.112	85.840	0.643	135.859	1.610	185.898	3.015	43
44	36.667	0.117	86.673	0.655	136.693	1.630	186.733	3.042	44
45	37.501	0.123	87.507	0.668	137.526	1.650	187.567	3.069	45
46	38.333	0.128	88.340	0.681	138.360	1.670	188.401	3.097	46
47	39.167	0.134	89.174	0.694	139.194	1.691	189.235	3.124	47
48	40.001	0.140	90.007	0.707	140.028	1.711	190.070	3.152	48
49	40.834	0.146	90.841	0.720	140.862	1.731	190.904	3.180	49
50	41.667	0.151	91.675	0.733	141.696	1.752	191.738	3.207	50
51	42.501	0.158	92.508	0.747	142.529	1.773	192.572	3.235	51
52	43.333	0.164	93.342	0.760	143.363	1.793	193.407	3.263	52
53	44.168	0.170	94.175	0.774	144.197	1.814	194.241	3.292	53
54	45.001	0.176	95.009	0.788	145.031	1.835	195.075	3.320	54
55	45.834	0.183	95.842	0.802	145.865	1.856	195.910	3.348	55
56	46.667	0.190	96.676	0.815	146.699	1.878	196.744	3.377	56
57	47.501	0.197	97.509	0.830	147.533	1.899	197.578	3.406	57
58	48.333	0.204	98.343	0.844	148.366	1.921	198.413	3.434	58
59	49.168	0.211	99.177	0.858	149.200	1.942	199.247	3.463	59

Table A-4. (*Continued*)

′	Δ = 4°		Δ = 5°		Δ = 6°		Δ = 7°		′
	T	E	T	E	T	E	T	E	
0	200.081	3.492	250.159	5.458	300.274	7.863	350.436	10.707	0
1	200.916	3.522	250.994	5.495	301.110	7.907	351.272	10.758	1
2	201.750	3.551	251.829	5.532	301.946	7.951	352.109	10.809	2
3	202.584	3.580	252.664	5.568	302.781	7.995	352.945	10.861	3
4	203.419	3.610	253.499	5.605	303.617	8.039	353.782	10.912	4
5	204.253	3.640	254.333	5.642	304.453	8.083	354.618	10.964	5
6	205.088	3.669	255.168	5.679	305.288	8.128	355.455	11.015	6
7	205.922	3.699	256.003	5.716	306.124	8.172	356.291	11.067	7
8	206.756	3.729	256.838	5.754	306.960	8.217	357.128	11.119	8
9	207.591	3.759	257.673	5.791	307.796	8.261	357.965	11.171	9
10	208.425	3.790	258.509	5.829	308.631	8.306	358.801	11.224	10
11	209.260	3.820	259.344	5.866	309.467	8.351	359.638	11.276	11
12	210.094	3.851	260.179	5.904	310.303	8.397	360.474	11.328	12
13	210.929	3.881	261.014	5.942	311.139	8.442	361.311	11.381	13
14	211.763	3.912	261.849	5.980	311.974	8.487	362.148	11.434	14
15	212.597	3.943	262.684	6.018	312.810	8.533	362.984	11.486	15
16	213.432	3.974	263.519	6.057	313.646	8.578	363.821	11.539	16
17	214.266	4.005	264.354	6.095	314.482	8.624	364.658	11.593	17
18	215.101	4.036	265.189	6.134	315.318	8.670	365.495	11.646	18
19	215.935	4.068	266.024	6.172	316.154	8.716	366.331	11.699	19
20	216.770	4.099	266.859	6.211	316.989	8.762	367.168	11.753	20
21	217.604	4.131	267.694	6.250	317.825	8.808	368.005	11.806	21
22	218.439	4.162	268.530	6.289	318.661	8.855	368.842	11.860	22
23	219.274	4.194	269.365	6.328	319.497	8.901	369.678	11.914	23
24	220.108	4.226	270.200	6.368	320.333	8.948	370.515	11.968	24
25	220.943	4.258	271.035	6.407	321.169	8.994	371.352	12.022	25
26	221.777	4.291	271.870	6.447	322.005	9.041	372.189	12.076	26
27	222.612	4.323	272.706	6.486	322.841	9.088	373.026	12.130	27
28	223.446	4.356	273.541	6.526	323.677	9.135	373.863	12.185	28
29	224.281	4.388	274.376	6.566	324.513	9.183	374.699	12.239	29
30	225.116	4.421	275.211	6.606	325.349	9.230	375.536	12.294	30
31	225.950	4.454	276.047	6.646	326.185	9.277	376.373	12.349	31
32	226.785	4.486	276.882	6.686	327.021	9.325	377.210	12.404	32
33	227.620	4.519	277.717	6.727	327.857	9.373	378.047	12.459	33
34	228.454	4.553	278.552	6.767	328.693	9.420	378.884	12.514	34
35	229.289	4.586	279.388	6.808	329.529	9.468	379.721	12.569	35
36	230.124	4.620	280.223	6.849	330.365	9.516	380.558	12.624	36
37	230.958	4.653	281.058	6.889	331.201	9.565	381.395	12.680	37
38	231.793	4.687	281.894	6.930	332.038	9.613	382.232	12.736	38
39	232.628	4.721	282.729	6.971	332.874	9.661	383.069	12.791	39
40	233.462	4.754	283.565	7.013	333.710	9.710	383.906	12.847	40
41	234.297	4.788	284.400	7.054	334.546	9.759	384.743	12.903	41
42	235.132	4.823	285.235	7.096	335.382	9.807	385.580	12.959	42
43	235.967	4.857	286.071	7.137	336.218	9.856	386.418	13.016	43
44	236.801	4.891	286.906	7.179	337.055	9.905	387.255	13.072	44
45	237.636	4.926	287.742	7.221	337.891	9.955	388.092	13.129	45
46	238.471	4.961	288.577	7.263	338.727	10.004	388.929	13.185	46
47	239.306	4.995	289.412	7.305	339.563	10.053	389.766	13.242	47
48	240.140	5.030	290.248	7.347	340.400	10.103	390.603	13.299	48
49	240.975	5.065	291.083	7.389	341.236	10.152	391.441	13.356	49
50	241.810	5.100	291.919	7.432	342.072	10.202	392.278	13.413	50
51	242.645	5.136	292.754	7.474	342.909	10.252	393.115	13.470	51
52	243.480	5.171	293.590	7.517	343.745	10.302	393.952	13.528	52
53	244.315	5.207	294.425	7.560	344.581	10.352	394.790	13.585	53
54	245.149	5.242	295.261	7.603	345.418	10.403	395.627	13.643	54
55	245.984	5.278	296.096	7.646	346.254	10.453	396.464	13.700	55
56	246.819	5.314	296.932	7.689	347.090	10.503	397.302	13.758	56
57	247.654	5.350	297.768	7.732	347.927	10.554	398.139	13.816	57
58	248.489	5.386	298.603	7.776	348.763	10.605	398.976	13.874	58
59	249.324	5.422	299.439	7.819	349.600	10.656	399.814	13.933	59

Table A-4. (*Continued*)

′	Δ = 8° T	Δ = 8° E	Δ = 9° T	Δ = 9° E	Δ = 10° T	Δ = 10° E	Δ = 11° T	Δ = 11° E	′
0	400.651	13.991	450.928	17.717	501.273	21.886	551.696	26.500	0
1	401.489	14.050	451.766	17.783	502.113	21.959	552.537	26.581	1
2	402.326	14.108	452.605	17.849	502.953	22.033	553.378	26.661	2
3	403.163	14.167	453.443	17.915	503.792	22.106	554.219	26.742	3
4	404.001	14.226	454.282	17.981	504.632	22.180	555.060	26.823	4
5	404.838	14.285	455.120	18.047	505.472	22.254	555.901	26.904	5
6	405.676	14.344	455.959	18.114	506.312	22.327	556.742	26.986	6
7	406.513	14.403	456.797	18.180	507.152	22.401	557.584	27.067	7
8	407.351	14.462	457.636	18.247	507.991	22.475	558.425	27.149	8
9	408.188	14.522	458.475	18.314	508.831	22.550	559.266	27.230	9
10	409.026	14.581	459.314	18.381	509.671	22.624	560.107	27.312	10
11	409.864	14.641	460.152	18.448	510.511	22.699	560.949	27.394	11
12	410.701	14.701	460.991	18.515	511.351	22.773	561.790	27.476	12
13	411.539	14.761	461.830	18.583	512.191	22.848	562.631	27.558	13
14	412.376	14.821	462.668	18.650	513.031	22.923	563.473	27.641	14
15	413.214	14.881	463.507	18.718	513.871	22.998	564.314	27.723	15
16	414.052	14.941	464.346	18.785	514.711	23.073	565.156	27.806	16
17	414.890	15.002	465.185	18.853	515.551	23.148	565.997	27.888	17
18	415.727	15.062	466.024	18.921	516.391	23.223	566.839	27.971	18
19	416.565	15.123	466.862	18.989	517.232	23.299	567.680	28.054	19
20	417.403	15.184	467.701	19.057	518.072	23.374	568.522	28.137	20
21	418.240	15.245	468.540	19.126	518.912	23.450	569.363	28.220	21
22	419.078	15.306	469.379	19.194	519.752	23.526	570.205	28.303	22
23	419.916	15.367	470.218	19.263	520.592	23.602	571.046	28.387	23
24	420.754	15.428	471.057	19.331	521.432	23.678	571.888	28.470	24
25	421.592	15.490	471.896	19.400	522.273	23.754	572.730	28.554	25
26	422.430	15.551	472.735	19.469	523.113	23.831	573.571	28.638	26
27	423.267	15.613	473.574	19.538	523.953	23.907	574.413	28.722	27
28	424.105	15.675	474.413	19.607	524.794	23.984	575.255	28.806	28
29	424.943	15.737	475.252	19.677	525.634	24.060	576.096	28.890	29
30	425.781	15.799	476.091	19.746	526.474	24.137	576.938	28.974	30
31	426.619	15.861	476.930	19.816	527.315	24.214	577.780	29.058	31
32	427.457	15.923	477.769	19.885	528.155	24.291	578.622	29.143	32
33	428.295	15.986	478.609	19.955	528.995	24.368	579.464	29.228	33
34	429.133	16.048	479.448	20.025	529.836	24.446	580.305	29.312	34
35	429.971	16.111	480.287	20.095	530.676	24.523	581.147	29.397	35
36	430.809	16.174	481.126	20.165	531.517	24.601	581.989	29.482	36
37	431.647	16.236	481.965	20.235	532.357	24.678	582.831	29.567	37
38	432.485	16.299	482.805	20.306	533.198	24.756	583.673	29.653	38
39	433.323	16.363	483.644	20.376	534.038	24.834	584.515	29.738	39
40	434.161	16.426	484.483	20.447	534.879	24.912	585.357	29.824	40
41	435.000	16.489	485.322	20.518	535.720	24.991	586.199	29.909	41
42	435.838	16.553	486.162	20.589	536.560	25.069	587.041	29.995	42
43	436.676	16.616	487.001	20.660	537.401	25.147	587.883	30.081	43
44	437.514	16.680	487.840	20.731	538.242	25.226	588.726	30.167	44
45	438.352	16.744	488.680	20.802	539.082	25.305	589.568	30.253	45
46	439.190	16.808	489.519	20.874	539.923	25.383	590.410	30.339	46
47	440.029	16.872	490.359	20.945	540.764	25.462	591.252	30.426	47
48	440.867	16.936	491.198	21.017	541.605	25.541	592.094	30.512	48
49	441.705	17.001	492.038	21.088	542.445	25.621	592.936	30.599	49
50	442.544	17.065	492.877	21.160	543.286	25.700	593.779	30.686	50
51	443.382	17.130	493.717	21.232	544.127	25.779	594.621	30.773	51
52	444.220	17.195	494.556	21.305	544.968	25.859	595.463	30.860	52
53	445.059	17.259	495.396	21.377	545.809	25.939	596.306	30.947	53
54	445.897	17.324	496.235	21.449	546.650	26.018	597.148	31.034	54
55	446.735	17.390	497.075	21.522	547.491	26.098	597.990	31.121	55
56	447.574	17.455	497.914	21.594	548.332	26.178	598.833	31.209	56
57	448.412	17.520	498.754	21.667	549.173	26.259	599.675	31.296	57
58	449.251	17.586	499.594	21.740	550.014	26.339	600.518	31.384	58
59	450.089	17.651	500.433	21.813	550.855	26.419	601.360	31.472	59

Table A-4. (*Continued*)

′	Δ = 12° T	Δ = 12° E	Δ = 13° T	Δ = 13° E	Δ = 14° T	Δ = 14° E	Δ = 15° T	Δ = 15° E	′
0	602.203	31.560	652.803	37.069	703.504	43.028	754.313	49.440	0
1	603.045	31.648	653.647	37.164	704.350	43.131	755.161	49.551	1
2	603.888	31.736	654.491	37.260	705.195	43.235	756.009	49.662	2
3	604.731	31.825	655.335	37.356	706.041	43.338	756.857	49.773	3
4	605.573	31.913	656.180	37.452	706.887	43.442	757.705	49.884	4
5	606.416	32.002	657.024	37.548	707.733	43.545	758.552	49.995	5
6	607.258	32.091	657.868	37.644	708.580	43.649	759.400	50.107	6
7	608.101	32.180	658.713	37.741	709.426	43.753	760.248	50.218	7
8	608.944	32.269	659.557	37.837	710.272	43.857	761.096	50.330	8
9	609.787	32.358	660.401	37.934	711.118	43.961	761.944	50.441	9
10	610.630	32.447	661.246	38.031	711.964	44.065	762.793	50.553	10
11	611.472	32.536	662.090	38.128	712.810	44.170	763.641	50.665	11
12	612.315	32.626	662.935	38.225	713.657	44.274	764.489	50.777	12
13	613.158	32.716	663.779	38.322	714.503	44.379	765.337	50.889	13
14	614.001	32.805	664.624	38.419	715.349	44.484	766.185	51.002	14
15	614.844	32.895	665.468	38.516	716.195	44.589	767.033	51.114	15
16	615.687	32.985	666.313	38.614	717.042	44.694	767.882	51.227	16
17	616.530	33.075	667.157	38.711	717.888	44.799	768.730	51.340	17
18	617.373	33.166	668.002	38.809	718.735	44.904	769.578	51.453	18
19	618.216	33.256	668.847	38.907	719.581	45.010	770.427	51.566	19
20	619.059	33.346	669.692	39.005	720.428	45.115	771.275	51.679	20
21	619.902	33.437	670.536	39.103	721.274	45.221	772.124	51.792	21
22	620.745	33.528	671.381	39.201	722.121	45.327	772.972	51.905	22
23	621.588	33.619	672.226	39.300	722.967	45.432	773.821	52.019	23
24	622.431	33.710	673.071	39.398	723.814	45.539	774.669	52.133	24
25	623.274	33.801	673.915	39.497	724.661	45.645	775.518	52.246	25
26	624.118	33.892	674.760	39.596	725.507	45.751	776.366	52.360	26
27	624.961	33.983	675.605	39.695	726.354	45.857	777.215	52.474	27
28	625.804	34.075	676.450	39.794	727.201	45.964	778.064	52.588	28
29	626.647	34.167	677.295	39.893	728.047	46.071	778.913	52.703	29
30	627.491	34.258	678.140	39.992	728.894	46.177	779.761	52.817	30
31	628.334	34.350	678.985	40.091	729.741	46.284	780.610	52.931	31
32	629.177	34.442	679.830	40.191	730.588	46.391	781.459	53.046	32
33	630.021	34.534	680.675	40.290	731.435	46.499	782.308	53.161	33
34	630.864	34.627	681.520	40.390	732.282	46.606	783.157	53.276	34
35	631.708	34.719	682.366	40.490	733.129	46.713	784.005	53.391	35
36	632.551	34.811	683.211	40.590	733.976	46.821	784.854	53.506	36
37	633.395	34.904	684.056	40.690	734.823	46.929	785.703	53.621	37
38	634.238	34.997	684.901	40.791	735.670	47.036	786.552	53.737	38
39	635.082	35.090	685.746	40.891	736.517	47.144	787.402	53.852	39
40	635.925	35.183	686.592	40.991	737.364	47.252	788.251	53.968	40
41	636.769	35.276	687.437	41.092	738.211	47.361	789.100	54.084	41
42	637.613	35.369	688.282	41.193	739.058	47.469	789.949	54.200	42
43	638.456	35.462	689.128	41.294	739.906	47.577	790.798	54.316	43
44	639.300	35.556	689.973	41.395	740.753	47.686	791.647	54.432	44
45	640.114	35.649	690.819	41.496	741.600	47.795	792.496	54.548	45
46	640.987	35.743	691.664	41.597	742.447	47.903	793.346	54.665	46
47	641.831	35.837	692.509	41.699	743.295	48.012	794.195	54.781	47
48	642.675	35.931	693.355	41.800	744.142	48.122	795.045	54.898	48
49	643.519	36.025	694.201	41.902	744.990	48.231	795.894	55.015	49
50	644.363	36.119	695.046	42.004	745.837	48.340	796.743	55.132	50
51	645.207	36.214	695.892	42.105	746.684	48.449	797.593	55.249	51
52	646.051	36.308	696.737	42.207	747.532	48.559	798.442	55.366	52
53	646.895	36.403	697.583	42.310	748.379	48.669	799.292	55.483	53
54	647.738	36.498	698.429	42.412	749.227	48.779	800.141	55.601	54
55	648.582	36.593	699.274	42.514	750.075	48.889	800.991	55.718	55
56	649.426	36.688	700.120	42.617	750.922	48.999	801.841	55.836	56
57	650.270	36.783	700.966	42.719	751.770	49.109	802.690	55.954	57
58	651.115	36.878	701.812	42.822	752.618	49.219	803.540	56.072	58
59	651.959	36.973	702.658	42.925	753.465	49.330	804.390	56.190	59

Table A-4. (*Continued*)

′	Δ = 16° T	Δ = 16° E	Δ = 17° T	Δ = 17° E	Δ = 18° T	Δ = 18° E	Δ = 19° T	Δ = 19° E	′
0	805.240	56.308	856.291	63.633	907.476	71.420	958.803	79.670	0
1	806.089	56.426	857.143	63.759	908.330	71.554	959.659	79.812	1
2	806.939	56.545	857.995	63.885	909.184	71.687	960.516	79.953	2
3	807.789	56.663	858.847	64.012	910.039	71.821	961.373	80.095	3
4	808.639	56.782	859.699	64.138	910.893	71.955	962.229	80.237	4
5	809.489	56.901	860.551	64.265	911.748	72.090	963.086	80.379	5
6	810.339	57.020	861.403	64.391	912.602	72.224	963.943	80.521	6
7	811.189	57.139	862.256	64.518	913.457	72.358	964.800	80.663	7
8	812.039	57.258	863.108	64.645	914.311	72.493	965.657	80.806	8
9	812.889	57.377	863.960	64.772	915.166	72.628	966.514	80.948	9
10	813.739	57.497	864.812	64.899	916.020	72.763	967.371	81.091	10
11	814.589	57.617	865.665	65.026	916.875	72.898	968.228	81.233	11
12	815.440	57.736	866.517	65.154	917.730	73.033	969.085	81.376	12
13	816.290	57.856	867.370	65.281	918.584	73.168	969.943	81.519	13
14	817.140	57.976	868.222	65.409	919.439	73.303	970.800	81.663	14
15	817.990	58.096	869.074	65.537	920.294	73.439	971.657	81.806	15
16	818.841	58.216	869.927	65.665	921.149	73.575	972.514	81.949	16
17	819.691	58.337	870.779	65.793	922.004	73.710	973.372	82.093	17
18	820.542	58.457	871.632	65.921	922.859	73.846	974.229	82.236	18
19	821.392	58.578	872.485	66.049	923.714	73.982	975.087	82.380	19
20	822.243	58.699	873.337	66.178	924.569	74.118	975.944	82.524	20
21	823.093	58.820	874.190	66.306	925.424	74.255	976.802	82.668	21
22	823.944	58.941	875.043	66.435	926.279	74.391	977.659	82.812	22
23	824.794	59.062	875.896	66.564	927.134	74.528	978.517	82.957	23
24	825.645	59.183	876.749	66.693	927.989	74.664	979.375	83.101	24
25	826.495	59.304	877.601	66.822	928.844	74.801	980.232	83.246	25
26	827.346	59.426	878.454	66.951	929.699	74.938	981.090	83.390	26
27	828.197	59.548	879.307	67.080	930.555	75.075	981.948	83.535	27
28	829.048	59.669	880.160	67.210	931.410	75.212	982.806	83.680	28
29	829.898	59.791	881.013	67.339	932.266	75.350	983.664	83.825	29
30	830.749	59.913	881.866	67.469	933.121	75.487	984.522	83.971	30
31	831.600	60.035	882.719	67.599	933.976	75.625	985.379	84.116	31
32	832.451	60.158	883.573	67.729	934.832	75.762	986.237	84.261	32
33	833.302	60.280	884.426	67.859	935.687	75.900	987.095	84.407	33
34	834.153	60.403	885.279	67.989	936.543	76.038	987.954	84.553	34
35	835.004	60.525	886.132	68.119	937.399	76.176	988.812	84.699	35
36	835.855	60.648	886.985	68.250	938.254	76.314	989.670	84.845	36
37	836.706	60.771	887.839	68.380	939.110	76.453	990.528	84.991	37
38	837.557	60.894	888.692	68.511	939.966	76.591	991.386	85.137	38
39	838.408	61.017	889.546	68.642	940.822	76.730	992.245	85.283	39
40	839.260	61.140	890.399	68.773	941.677	76.868	993.103	85.430	40
41	840.111	61.264	891.252	68.904	942.533	77.007	993.961	85.577	41
42	840.962	61.387	892.106	69.035	943.389	77.146	994.820	85.723	42
43	841.813	61.511	892.960	69.167	944.245	77.285	995.678	85.870	43
44	842.665	61.635	893.813	69.298	945.101	77.425	996.537	86.017	44
45	843.516	61.759	894.667	69.430	945.957	77.564	997.395	86.165	45
46	844.367	61.883	895.520	69.562	946.813	77.704	998.254	86.312	46
47	845.219	62.007	896.374	69.694	947.669	77.843	999.113	86.459	47
48	846.070	62.132	897.228	69.826	948.525	77.983	999.971	86.607	48
49	846.922	62.256	898.082	69.958	949.382	78.123	1000.830	86.755	49
50	847.774	62.381	898.936	70.090	950.238	78.263	1001.689	86.902	50
51	848.625	62.505	899.789	70.222	951.094	78.403	1002.548	87.050	51
52	849.477	62.630	900.643	70.355	951.950	78.543	1003.407	87.199	52
53	850.328	62.755	901.497	70.488	952.807	78.684	1004.265	87.347	53
54	851.180	62.880	902.351	70.620	953.663	78.824	1005.124	87.495	54
55	852.032	63.005	903.205	70.753	954.520	78.965	1005.983	87.644	55
56	852.884	63.131	904.059	70.886	955.376	79.106	1006.842	87.792	56
57	853.735	63.256	904.913	71.019	956.233	79.247	1007.702	87.941	57
58	854.587	63.382	905.767	71.153	957.089	79.388	1008.561	88.090	58
59	855.439	63.507	906.622	71.286	957.946	79.529	1009.420	88.239	59

Table A-4. (*Continued*)

′	Δ = 20°		Δ = 21°		Δ = 22°		Δ = 23°		′
	T	E	T	E	T	E	T	E	
0	1010.28	88.388	1061.91	97.576	1113.72	107.239	1165.70	117.379	0
1	1011.14	88.537	1062.78	97.733	1114.58	107.404	1166.56	117.552	1
2	1012.00	88.687	1063.64	97.891	1115.45	107.569	1167.43	117.726	2
3	1012.86	88.836	1064.50	98.048	1116.31	107.734	1168.30	117.899	3
4	1013.72	88.986	1065.36	98.206	1117.18	107.900	1169.17	118.073	4
5	1014.58	89.136	1066.22	98.363	1118.04	108.065	1170.04	118.246	5
6	1015.44	89.285	1067.09	98.521	1118.91	108.231	1170.90	118.420	6
7	1016.29	89.436	1067.95	98.679	1119.77	108.397	1171.77	118.594	7
8	1017.15	89.586	1068.81	98.837	1120.64	108.563	1172.64	118.768	8
9	1018.01	89.736	1069.67	98.995	1121.50	108.729	1173.51	118.942	9
10	1018.87	89.887	1070.54	99.154	1122.37	108.896	1174.38	119.116	10
11	1019.73	90.037	1071.40	99.312	1123.23	109.062	1175.24	119.291	11
12	1020.59	90.188	1072.26	99.471	1124.10	109.229	1176.11	119.465	12
13	1021.45	90.339	1073.12	99.629	1124.96	109.395	1176.98	119.640	13
14	1022.31	90.490	1073.99	99.788	1125.83	109.562	1177.85	119.815	14
15	1023.17	90.641	1074.85	99.947	1126.69	109.729	1178.72	119.990	15
16	1024.03	90.792	1075.71	100.106	1127.56	109.896	1179.59	120.165	16
17	1024.89	90.943	1076.57	100.266	1128.43	110.063	1180.46	120.340	17
18	1025.75	91.095	1077.44	100.425	1129.29	110.231	1181.32	120.515	18
19	1026.61	91.246	1078.30	100.584	1130.16	110.398	1182.19	120.691	19
20	1027.47	91.398	1079.16	100.744	1131.02	110.566	1183.06	120.867	20
21	1028.33	91.550	1080.03	100.904	1131.89	110.733	1183.93	121.042	21
22	1029.19	91.702	1080.89	101.064	1132.75	110.901	1184.80	121.218	22
23	1030.05	91.854	1081.75	101.224	1133.62	111.069	1185.67	121.394	23
24	1030.91	92.007	1082.61	101.384	1134.49	111.237	1186.54	121.570	24
25	1031.77	92.159	1083.48	101.544	1135.35	111.406	1187.41	121.747	25
26	1032.63	92.312	1084.34	101.705	1136.22	111.574	1188.28	121.923	26
27	1033.49	92.464	1085.20	101.865	1137.09	111.743	1189.15	122.100	27
28	1034.36	92.617	1086.07	102.026	1137.95	111.911	1190.02	122.276	28
29	1035.22	92.770	1086.93	102.187	1138.82	112.080	1190.88	122.453	29
30	1036.08	92.923	1087.79	102.348	1139.68	112.249	1191.75	122.630	30
31	1036.94	93.076	1088.66	102.509	1140.55	112.418	1192.62	122.807	31
32	1037.80	93.230	1089.52	102.670	1141.42	112.587	1193.49	122.985	32
33	1038.66	93.383	1090.38	102.832	1142.28	112.757	1194.36	123.162	33
34	1039.52	93.537	1091.25	102.993	1143.15	112.926	1195.23	123.340	34
35	1040.38	93.690	1092.11	103.155	1144.02	113.096	1196.10	123.517	35
36	1041.24	93.844	1092.98	103.317	1144.88	113.266	1196.97	123.695	36
87	1042.10	93.998	1093.84	103.478	1145.75	113.435	1197.84	123.873	37
38	1042.96	94.152	1094.70	103.640	1146.62	113.605	1198.71	124.051	38
39	1043.82	94.307	1095.57	103.803	1147.48	113.775	1199.58	124.229	39
40	1044.68	94.461	1096.43	103.965	1148.35	113.946	1200.45	124.407	40
41	1045.55	94.616	1097.29	104.127	1149.22	114.116	1201.32	124.586	41
42	1046.41	94.770	1098.16	104.290	1150.08	114.287	1202.19	124.764	42
43	1047.27	94.925	1099.02	104.453	1150.95	114.457	1203.06	124.943	43
44	1048.13	95.080	1099.89	104.615	1151.82	114.628	1203.93	125.122	44
45	1048.99	95.235	1100.75	104.778	1152.68	114.799	1204.80	125.301	45
46	1049.85	95.340	1101.61	104.942	1153.55	114.970	1205.67	125.480	46
47	1050.71	95.545	1102.48	105.105	1154.42	115.141	1206.54	125.659	47
48	1051.57	95.701	1103.34	105.268	1155.29	115.313	1207.41	125.839	48
49	1052.44	95.856	1104.21	105.432	1156.15	115.484	1208.28	126.018	49
50	1053.30	96.012	1105.07	105.595	1157.02	115.656	1209.15	126.198	50
51	1054.16	96.168	1105.94	105.759	1157.89	115.828	1210.02	126.378	51
52	1055.02	96.324	1106.80	105.923	1158.75	116.000	1210.89	126.558	52
53	1055.88	96.481	1107.66	106.087	1159.62	116.172	1211.76	126.738	53
54	1056.74	96.636	1108.53	106.251	1160.49	116.344	1212.63	126.918	54
55	1057.61	96.792	1109.39	106.415	1161.36	116.516	1213.51	127.098	55
56	1058.47	96.949	1110.26	106.580	1162.22	116.688	1214.38	127.279	56
57	1059.33	97.106	1111.12	106.744	1163.09	116.861	1215.25	127.459	57
58	1060.19	97.262	1111.99	106.909	1163.96	117.034	1216.12	127.640	58
59	1061.05	97.419	1112.85	107.074	1164.83	117.206	1216.99	127.821	59

Table A-4. (*Continued*)

′	Δ = 24°		Δ = 25°		Δ = 26°		Δ = 27°		′
	T	E	T	E	T	E	T	E	
0	1217.86	128.002	1270.22	139.111	1322.78	150.711	1375.55	162.807	0
1	1218.73	128.183	1271.09	139.301	1323.65	150.909	1376.43	163.013	1
2	1219.60	128.364	1271.97	139.490	1324.53	151.107	1377.31	163.219	2
3	1220.48	128.546	1272.84	139.680	1325.41	151.304	1378.19	163.425	3
4	1221.34	128.728	1273.71	139.869	1326.29	151.502	1379.08	163.631	4
5	1222.21	128.909	1274.59	140.059	1327.17	151.700	1379.96	163.838	5
6	1223.09	129.091	1275.46	140.249	1328.04	151.899	1380.84	164.044	6
7	1223.96	129.273	1276.34	140.439	1328.92	152.097	1381.72	164.251	7
8	1224.83	129.455	1277.21	140.630	1329.80	152.295	1382.60	164.458	8
9	1225.70	129.637	1278.09	140.820	1330.69	152.494	1383.48	164.664	9
10	1226.57	129.820	1278.96	141.011	1331.56	152.693	1384.37	164.872	10
11	1227.44	130.002	1279.84	141.201	1332.44	152.892	1385.25	165.079	11
12	1228.31	130.185	1280.71	141.392	1333.31	153.091	1386.13	165.286	12
13	1229.19	130.368	1281.59	141.583	1334.19	153.290	1387.01	165.494	13
14	1230.06	130.551	1282.46	141.774	1335.07	153.489	1387.90	165.701	14
15	1230.93	130.734	1283.34	141.965	1335.95	153.689	1388.78	165.909	15
16	1231.80	130.917	1284.21	142.157	1336.83	153.888	1389.66	166.117	16
17	1232.67	131.100	1285.09	142.348	1337.71	154.088	1390.54	166.325	17
18	1233.55	131.284	1285.96	142.540	1338.59	154.288	1391.42	166.533	18
19	1234.42	131.467	1286.84	142.731	1339.47	154.488	1392.31	166.741	19
20	1235.29	131.651	1287.71	142.923	1340.34	154.688	1393.19	166.950	20
21	1236.16	131.835	1288.59	143.115	1341.22	154.888	1394.07	167.158	21
22	1237.03	132.019	1289.46	143.307	1342.10	155.089	1394.95	167.367	22
23	1237.91	132.203	1290.34	143.500	1342.98	155.289	1395.84	167.576	23
24	1238.78	132.387	1291.22	143.692	1343.86	155.490	1396.72	167.785	24
25	1239.65	132.572	1292.09	143.885	1344.74	155.691	1397.60	167.994	25
26	1240.52	132.756	1292.97	144.077	1345.62	155.892	1398.49	168.203	26
27	1241.40	132.941	1293.84	144.270	1346.50	156.093	1399.37	168.413	27
28	1242.27	133.126	1294.72	144.463	1347.38	156.294	1400.25	168.623	28
29	1243.14	133.311	1295.60	144.656	1348.26	156.495	1401.14	168.832	29
30	1244.01	133.496	1296.47	144.850	1349.14	156.697	1402.02	169.042	30
31	1244.89	133.681	1297.35	145.043	1350.02	156.899	1402.90	169.252	31
32	1245.76	133.866	1298.22	145.237	1350.90	157.101	1403.79	169.462	32
33	1246.63	134.052	1299.10	145.430	1351.78	157.302	1404.67	169.673	33
34	1247.50	134.237	1299.98	145.624	1352.66	157.505	1405.55	169.883	34
35	1248.38	134.423	1300.85	145.818	1353.53	157.707	1406.44	170.094	35
36	1249.25	134.609	1301.73	146.012	1354.41	157.909	1407.32	170.304	36
37	1250.12	134.795	1302.60	146.206	1355.29	158.112	1408.20	170.515	37
38	1251.00	134.981	1303.48	146.401	1356.17	158.314	1409.09	170.726	38
39	1251.87	135.168	1304.36	146.595	1357.05	158.517	1409.97	170.937	39
40	1252.74	135.354	1305.23	146.790	1357.93	158.720	1410.85	171.149	40
41	1253.62	135.540	1306.11	146.985	1358.82	158.923	1411.74	171.360	41
42	1254.49	135.727	1306.99	147.180	1359.70	159.126	1412.62	171.571	42
43	1255.36	135.914	1307.86	147.375	1360.58	159.329	1413.51	171.783	43
44	1256.24	136.101	1308.74	147.570	1361.46	159.533	1414.39	171.995	44
45	1257.11	136.288	1309.62	147.765	1362.34	159.736	1415.27	172.207	45
46	1257.98	136.475	1310.49	147.961	1363.22	159.940	1416.16	172.419	46
47	1258.86	136.663	1311.37	148.156	1364.10	160.144	1417.04	172.631	47
48	1259.73	136.850	1312.25	148.352	1364.98	160.348	1417.93	172.844	48
49	1260.60	137.038	1313.13	148.548	1365.86	160.552	1418.81	173.056	49
50	1261.48	137.226	1314.00	148.744	1366.74	160.756	1419.70	173.269	50
51	1262.35	137.414	1314.88	148.940	1367.62	160.961	1420.58	173.482	51
52	1263.22	137.602	1315.76	149.136	1368.50	161.165	1421.47	173.695	52
53	1264.10	137.790	1316.63	149.332	1369.38	161.370	1422.35	173.908	53
54	1264.97	137.978	1317.51	149.529	1370.26	161.575	1423.23	174.121	54
55	1265.85	138.167	1318.39	149.726	1371.14	161.780	1424.12	174.334	55
56	1266.72	138.355	1319.27	149.923	1372.02	161.985	1425.00	174.548	56
57	1267.59	138.544	1320.14	150.120	1372.91	162.190	1425.89	174.761	57
58	1268.47	138.733	1321.02	150.317	1373.79	162.396	1426.77	174.975	58
59	1269.34	138.922	1321.90	150.514	1374.67	162.601	1427.66	175.189	59

Table A-4. (*Continued*)

′	Δ = 28°		Δ = 29°		Δ = 30°		Δ = 31°		′
	T	E	T	E	T	E	T	E	
0	1428.54	175.403	1481.77	188.505	1535.24	202.118	1588.95	216.247	0
1	1429.43	175.617	1482.66	188.728	1536.13	202.349	1589.85	216.489	1
2	1430.31	175.832	1483.55	188.950	1537.02	202.580	1590.75	216.727	2
3	1431.20	176.046	1484.44	189.173	1537.92	202.812	1591.65	216.967	3
4	1432.09	176.261	1485.33	189.396	1538.81	203.043	1592.54	217.207	4
5	1432.97	176.476	1486.22	189.620	1539.70	203.275	1593.44	217.448	5
6	1433.86	176.690	1487.11	189.843	1540.60	203.507	1594.34	217.688	6
7	1434.74	176.906	1487.99	190.067	1541.49	203.739	1595.24	217.929	7
8	1435.63	177.121	1488.88	190.290	1542.38	203.972	1596.13	218.170	8
9	1436.51	177.336	1489.77	190.514	1543.28	204.204	1597.03	218.411	9
10	1437.40	177.552	1490.66	190.738	1544.17	204.437	1597.93	218.652	10
11	1438.28	177.767	1491.55	190.962	1545.06	204.669	1598.83	218.894	11
12	1439.17	177.983	1492.44	191.186	1545.96	204.902	1599.73	219.135	12
13	1440.06	178.199	1493.33	191.411	1546.85	205.135	1600.63	219.377	13
14	1440.94	178.415	1494.22	191.635	1547.75	205.368	1601.52	219.619	14
15	1441.83	178.631	1495.11	191.860	1548.64	205.601	1602.42	219.861	15
16	1442.71	178.847	1496.00	192.085	1549.54	205.835	1603.32	220.103	16
17	1443.60	179.064	1496.89	192.310	1550.43	206.068	1604.22	220.345	17
18	1444.49	179.280	1497.78	192.535	1551.32	206.302	1605.12	220.587	18
19	1445.37	179.497	1498.67	192.760	1552.22	206.536	1606.02	220.830	19
20	1446.26	179.714	1499.56	192.985	1553.11	206.770	1606.92	221.073	20
21	1447.15	179.931	1500.45	193.211	1554.01	207.004	1607.81	221.315	21
22	1448.03	180.148	1501.35	193.436	1554.90	207.238	1608.71	221.558	22
23	1448.92	180.365	1502.24	193.662	1555.80	207.473	1609.61	221.801	23
24	1449.81	180.583	1503.13	193.888	1556.69	207.707	1610.51	222.045	24
25	1450.69	180.800	1504.02	194.144	1557.59	207.942	1611.41	222.288	25
26	1451.58	181.018	1504.91	194.341	1558.48	208.177	1612.31	222.532	26
27	1452.47	181.236	1505.80	194.567	1559.38	208.412	1613.21	222.775	27
28	1453.35	181.454	1506.69	194.794	1560.27	208.647	1614.11	223.019	28
29	1454.24	181.672	1507.58	195.020	1561.17	208.882	1615.01	223.263	29
30	1455.13	181.890	1508.47	195.247	1562.06	209.117	1615.91	223.507	30
31	1456.01	182.109	1509.36	195.474	1562.96	209.353	1616.81	223.752	31
32	1456.90	182.327	1510.25	195.701	1563.85	209.589	1617.71	223.996	32
33	1457.79	182.546	1511.15	195.928	1564.75	209.825	1618.61	224.240	33
34	1458.68	182.765	1512.04	196.156	1565.64	210.061	1619.51	224.485	34
35	1459.56	182.984	1512.93	196.383	1566.54	210.297	1620.41	224.730	35
36	1460.45	183.203	1513.82	196.611	1567.44	210.533	1621.31	224.975	36
37	1461.34	183.422	1514.71	196.839	1568.33	210.769	1622.21	225.220	37
38	1462.23	183.642	1515.60	197.067	1569.23	211.006	1623.11	225.465	38
39	1463.11	183.861	1516.49	197.295	1570.12	211.243	1624.01	225.711	39
40	1464.00	184.081	1517.39	197.523	1571.02	211.479	1624.91	225.956	40
41	1464.89	184.301	1518.28	197.751	1571.91	211.716	1625.81	226.202	41
42	1465.78	184.521	1519.17	197.980	1572.81	211.954	1626.71	226.448	42
43	1466.66	184.741	1520.06	198.208	1573.71	212.191	1627.61	226.694	43
44	1467.55	184.961	1520.95	198.437	1574.60	212.428	1628.51	226.940	44
45	1468.44	185.182	1521.85	198.666	1575.50	212.666	1629.41	227.186	45
46	1469.33	185.402	1522.74	198.895	1576.40	212.904	1630.31	227.433	46
47	1470.22	185.623	1523.63	199.125	1577.29	213.141	1631.21	227.679	47
48	1471.11	185.844	1524.52	199.354	1578.19	213.380	1632.11	227.926	48
49	1471.99	186.065	1525.42	199.583	1579.09	213.618	1633.01	228.173	49
50	1472.88	186.286	1526.31	199.813	1579.98	213.856	1633.92	228.420	50
51	1473.77	186.507	1527.20	200.043	1580.88	214.094	1634.82	228.667	51
52	1474.66	186.729	1528.09	200.273	1581.78	214.333	1635.72	228.915	52
53	1475.55	186.950	1528.99	200.503	1582.67	214.572	1636.62	229.162	53
54	1476.44	187.172	1529.88	200.733	1583.57	214.811	1637.52	229.410	54
55	1477.33	187.394	1530.77	200.964	1584.47	215.050	1638.42	229.658	55
56	1478.21	187.616	1531.66	201.194	1585.36	215.289	1639.32	229.906	56
57	1479.10	187.838	1532.56	201.425	1586.26	215.528	1640.22	230.154	57
58	1479.99	188.060	1533.45	201.656	1587.16	215.768	1641.13	230.402	58
59	1480.88	188.282	1534.34	201.887	1588.06	216.007	1642.03	230.650	59

Table A-4. (*Continued*)

′	Δ = 32°		Δ = 33°		Δ = 34°		Δ = 35°		′
	T	E	T	E	T	E	T	E	
0	1642.93	230.899	1697.18	246.079	1751.71	261.795	1806.53	278.051	0
1	1643.83	231.148	1698.08	246.337	1752.62	262.061	1807.45	278.327	1
2	1644.73	231.396	1698.99	246.594	1753.53	262.328	1808.36	278.603	2
3	1645.64	231.645	1699.90	246.852	1754.44	262.594	1809.28	278.879	3
4	1646.54	231.894	1700.80	247.110	1755.35	262.861	1810.19	279.155	4
5	1647.44	232.144	1701.71	247.368	1756.26	263.128	1811.11	279.431	5
6	1648.34	232.393	1702.62	247.626	1757.18	263.396	1812.03	279.707	6
7	1649.24	232.642	1703.53	247.885	1758.09	263.663	1812.94	279.983	7
8	1650.15	232.892	1704.43	248.143	1759.00	263.931	1813.86	280.260	8
9	1651.05	233.142	1705.34	248.402	1759.91	264.198	1814.78	280.537	9
10	1651.95	233.392	1706.25	248.661	1760.82	264.466	1815.70	280.814	10
11	1652.85	233.642	1707.15	248.920	1761.74	264.734	1816.61	281.091	11
12	1653.76	233.892	1708.06	249.179	1762.65	265.002	1817.53	281.368	12
13	1654.66	234.143	1708.97	249.439	1763.56	265.271	1818.45	281.646	13
14	1655.56	234.393	1709.88	249.698	1764.47	265.539	1819.36	281.923	14
15	1656.47	234.644	1710.78	249.958	1765.39	265.808	1820.28	282.201	15
16	1657.37	234.895	1711.69	250.217	1766.30	266.077	1821.20	282.479	16
17	1658.27	235.146	1712.60	250.477	1767.21	266.345	1822.12	282.757	17
18	1659.18	235.397	1713.51	250.737	1768.12	266.614	1823.03	283.035	18
19	1660.08	235.648	1714.41	250.997	1769.04	266.884	1823.95	283.314	19
20	1660.98	235.900	1715.32	251.258	1769.95	267.153	1824.87	283.592	20
21	1661.89	236.152	1716.23	251.518	1770.86	267.423	1825.79	283.871	21
22	1662.79	236.403	1717.14	251.779	1771.77	267.692	1826.71	284.149	22
23	1663.69	236.655	1718.05	252.040	1772.69	267.962	1827 62	284.428	23
24	1664.60	236.907	1718.96	252.301	1773.60	268.232	1828.54	284.707	24
25	1665.50	237.159	1719.86	252.562	1774.51	268.502	1829.46	284.987	25
26	1666.40	237.412	1720.77	252.823	1775.43	268.772	1830.38	285.266	26
27	1667.31	237.664	1721.68	253.084	1776.34	269.043	1831.30	285.546	27
28	1668.21	237.917	1722.59	253.346	1777.25	269.313	1832.21	285.825	28
29	1669.12	238.170	1723.50	253.608	1778.17	269.584	1833.13	286.105	29
30	1670.02	238.423	1724.41	253.869	1779.08	269.855	1834.05	286.385	30
31	1670.92	238.676	1725.32	254.131	1779.99	270.126	1834.97	286.665	31
32	1671.83	238.929	1726.22	254.394	1780.91	270.397	1835.89	286.946	32
33	1672.73	239.182	1727.13	254.656	1781.82	270.668	1836.81	287.226	33
34	1673.64	239.436	1728.04	254.918	1782.74	270.940	1837.73	287.507	34
35	1674.54	239.689	1728.95	255.181	1783.65	271.211	1838.65	287.787	35
36	1675.45	239.943	1729.86	255.444	1784.56	271.483	1839.57	288.068	36
37	1676.35	240.197	1730.77	255.707	1785.48	271.755	1840.49	288.350	37
38	1677.26	240.451	1731.68	255.970	1786.39	272.027	1841.40	288.631	38
39	1678.16	240.706	1732.59	256.233	1787.31	272.300	1842.32	288.912	39
40	1679.06	240.960	1733.50	256.496	1788.22	272.572	1843.24	289.194	40
41	1679.97	241.214	1734.41	256.760	1789.14	272.844	1844.16	289.475	41
42	1680.87	241.469	1735.32	257.023	1790.05	273.117	1845.08	289.757	42
43	1681.78	241.724	1736.23	257.287	1790.97	273.390	1846.00	290.039	43
44	1682.69	241.979	1737.14	257.551	1791.88	273.663	1846.92	290.321	44
45	1683.59	242.234	1738.05	257.815	1792.80	273.936	1847.84	290.604	45
46	1684.50	242.489	1738.96	258.079	1793.71	274.209	1848.76	290.886	46
47	1685.40	242.745	1739.87	258.344	1794.63	274.483	1849.68	291.169	47
48	1686.31	243.000	1740.78	258.608	1795.54	274.756	1850.60	291.452	48
49	1687.21	243.256	1741.69	258.873	1796.46	275.030	1851.52	291.735	49
50	1688.12	243.512	1742.60	259.138	1797.37	275.304	1852.44	292.018	50
51	1689.02	243.768	1743.51	259.403	1798.29	275.578	1853.36	292.301	51
52	1689.93	244.024	1744.42	259.668	1799.20	275.852	1854.28	292.584	52
53	1690.83	244.281	1745.33	259.933	1800.12	276.127	1855.21	292.868	53
54	1691.74	244.537	1746.24	260.199	1801.03	276.401	1856.13	293.151	54
55	1692.65	244.794	1747.15	260.464	1801.95	276.676	1857.05	293.435	55
56	1693.55	245.051	1748.06	260.730	1802.86	276.951	1857.97	293.719	56
57	1694.46	245.308	1748.97	260.996	1803.78	277.226	1858.89	294.004	57
58	1695.37	245.565	1749.89	261.262	1804.70	277.501	1859.81	294.288	58
59	1696.27	245.822	1750.80	261.528	1805.61	277.776	1860.73	294.572	59

Table A-4. (*Continued*)

′	Δ = 36° T	Δ = 36° E	Δ = 37° T	Δ = 37° E	Δ = 38° T	Δ = 38° E	Δ = 39° T	Δ = 39° E	′
0	1861.65	294.857	1917.09	312.218	1972.85	330.142	2028.95	348.637	0
1	1862.57	295.142	1918.02	312.512	1973.78	330.446	2029.89	348.950	1
2	1863.50	295.427	1918.94	312.806	1974.72	330.749	2030.83	349.264	2
3	1864.42	295.712	1919.87	313.101	1975.65	331.053	2031.76	349.577	3
4	1865.34	295.997	1920.80	313.395	1976.58	331.357	2032.70	349.891	4
5	1866.26	296.282	1921.72	313.690	1977.51	331.662	2033.64	350.205	5
6	1867.18	296.568	1922.65	313.985	1978.45	331.966	2034.58	350.519	6
7	1868.10	296.853	1923.58	314.280	1979.38	332.270	2035.52	350.833	7
8	1869.03	297.139	1924.51	314.575	1980.31	332.575	2036.46	351.147	8
9	1869.95	297.426	1925.43	314.870	1981.24	332.880	2037.39	351.461	9
10	1870.87	297.712	1926.36	315.166	1982.18	333.185	2038.33	351.776	10
11	1871.79	297.998	1927.29	315.462	1983.11	333.490	2039.27	352.091	11
12	1872.71	298.284	1928.22	315.758	1984.04	333.795	2040.21	352.406	12
13	1873.64	298.571	1929.14	316.054	1984.98	334.101	2041.15	352.721	13
14	1874.56	298.858	1930.07	316.350	1985.91	334.406	2042.09	353.036	14
15	1875.48	299.145	1931.00	316.646	1986.84	334.712	2043.03	353.351	15
16	1876.40	299.432	1931.93	316.942	1987.78	335.018	2043.97	353.667	16
17	1877.33	299.719	1932.86	317.239	1988.71	335.324	2044.91	353.983	17
18	1878.25	300.006	1933.78	317.536	1989.65	335.630	2045.85	354.299	18
19	1879.17	300.294	1934.71	317.833	1990.58	335.937	2046.79	354.614	19
20	1880.10	300.582	1935.64	318.130	1991.51	336.243	2047.73	354.931	20
21	1881.02	300.870	1936.57	318.427	1992.45	336.550	2048.67	355.247	21
22	1881.94	301.158	1937.50	318.724	1993.38	336.857	2049.61	355.564	22
23	1882.87	301.446	1938.43	319.022	1994.32	337.164	2050.55	355.880	23
24	1883.79	301.734	1939.36	319.320	1995.25	337.471	2051.49	356.197	24
25	1884.71	302.023	1940.28	319.618	1996.18	337.779	2052.43	356.514	25
26	1885.64	302.311	1941.21	319.916	1997.12	338.086	2053.37	356.831	26
27	1886.56	302.600	1942.14	320.214	1998.05	338.394	2054.31	357.149	27
28	1887.48	302.889	1943.07	320.512	1998.99	338.702	2055.25	357.466	28
29	1888.41	303.178	1944.00	320.811	1999.92	339.010	2056.19	357.784	29
30	1889.33	303.467	1944.93	321.109	2000.86	339.318	2057.13	358.102	30
31	1890.26	303.757	1945.86	321.408	2001.79	339.626	2058.07	358.420	31
32	1891.18	304.047	1946.79	321.707	2002.73	339.934	2059.01	358.738	32
33	1892.10	304.336	1947.72	322.006	2003.66	340.243	2059.95	359.056	33
34	1893.03	304.626	1948.65	322.305	2004.60	340.552	2060.89	359.374	34
35	1893.95	304.916	1949.58	322.605	2005.53	340.861	2061.83	359.693	35
36	1894.88	305.206	1950.51	322.904	2006.47	341.170	2062.77	360.012	36
37	1895.80	305.497	1951.44	323.204	2007.41	341.480	2063.72	360.331	37
38	1896.73	305.787	1952.37	323.504	2008.34	341.789	2064.66	360.650	38
39	1897.65	306.078	1953.30	323.804	2009.28	342.099	2065.60	360.969	39
40	1898.58	306.369	1954.23	324.104	2010.21	342.408	2066.54	361.289	40
41	1899.50	306.660	1955.16	324.405	2011.15	342.718	2067.48	361.608	41
42	1900.43	306.951	1956.09	324.705	2012.08	343.029	2068.42	361.928	42
43	1901.35	307.242	1957.02	325.006	2013.02	343.339	2069.37	362.248	43
44	1902.28	307.534	1957.95	325.307	2013.96	343.649	2070.31	362.568	44
45	1903.20	307.825	1958.88	325.608	2014.89	343.960	2071.25	362.888	45
46	1904.13	308.117	1959.81	325.909	2015.83	344.270	2072.19	363.209	46
47	1905.05	308.409	1960.74	326.210	2016.77	344.581	2073.14	363.529	47
48	1905.98	308.701	1961.67	326.512	2017.70	344.892	2074.08	363.850	48
49	1906.90	308.993	1962.60	326.814	2018.64	345.203	2075.02	364.171	49
50	1907.83	309.286	1963.54	327.115	2019.58	345.515	2075.96	364.492	50
51	1908.75	309.578	1964.47	327.417	2020.51	345.826	2076.91	364.813	51
52	1909.68	309.871	1965.40	327.720	2021.45	346.138	2077.85	365.135	52
53	1910.61	310.164	1966.33	328.022	2022.39	346.450	2078.79	365.456	53
54	1911.53	310.457	1967.26	328.324	2023.32	346.762	2079.74	365.778	54
55	1912.46	310.750	1968.19	328.627	2024.26	347.074	2080.68	366.100	55
56	1913.38	311.043	1969.12	328.930	2025.20	347.387	2081.62	366.422	56
57	1914.31	311.336	1970.06	329.232	2026.14	347.699	2082.56	366.744	57
58	1915.24	311.630	1970.99	329.536	2027.07	348.012	2083.51	367.066	58
59	1916.16	311.924	1971.92	329.839	2028.01	348.324	2084.45	367.389	59

Table A-4. (*Continued*)

′	Δ = 40°		Δ = 41°		Δ = 42°		Δ = 43°		′
	T	E	T	E	T	E	T	E	
0	2085.40	367.71	2142.20	387.37	2199.38	407.63	2256.94	428.49	0
1	2086.34	368.03	2143.15	387.71	2200.33	407.97	2257.90	428.85	1
2	2087.28	368.36	2144.10	388.04	2201.30	408.32	2258.86	429.20	2
3	2088.23	368.68	2145.05	388.37	2202.25	408.66	2259.83	429.55	3
4	2089.17	369.00	2146.00	388.70	2203.20	409.00	2260.79	429.91	4
5	2090.12	369.33	2146.95	389.04	2204.16	409.35	2261.76	430.26	5
6	2091.06	369.65	2147.90	389.37	2205.12	409.69	2262.72	430.61	6
7	2092.00	369.97	2148.85	389.70	2206.07	410.03	2263.68	430.97	7
8	2092.95	370.30	2149.80	390.04	2207.03	410.38	2264.65	431.32	8
9	2093.89	370.62	2150.75	390.37	2207.99	410.72	2265.61	431.68	9
10	2094.84	370.95	2151.70	390.71	2208.95	411.07	2266.57	432.03	10
11	2095.78	371.27	2152.66	391.04	2209.90	411.41	2267.54	432.38	11
12	2096.73	371.60	2153.61	391.38	2210.86	411.75	2268.50	432.74	12
13	2097.67	371.92	2154.56	391.71	2211.82	412.10	2269.46	433.09	13
14	2098.62	372.25	2155.51	392.05	2212.78	412.44	2270.43	433.45	14
15	2099.56	372.57	2156.46	392.38	2213.73	412.79	2271.39	433.80	15
16	2100.51	372.90	2157.41	392.72	2214.69	413.13	2272.36	434.16	16
17	2101.45	373.22	2158.36	393.05	2215.65	413.48	2273.32	434.52	17
18	2102.40	373.55	2159.31	393.39	2216.61	413.83	2274.29	434.87	18
19	2103.34	373.87	2160.27	393.72	2217.56	414.17	2275.25	435.23	19
20	2104.29	374.20	2161.22	394.06	2218.52	414.52	2276.22	435.58	20
21	2105.24	374.53	2162.17	394.40	2219.48	414.86	2277.18	435.94	21
22	2106.18	374.85	2163.12	394.73	2220.44	415.21	2278.15	436.30	22
23	2107.13	375.18	2164.07	395.07	2221.40	415.56	2279.11	436.65	23
24	2108.07	375.51	2165.03	395.40	2222.36	415.90	2280.08	437.01	24
25	2109.02	375.83	2165.98	395.74	2223.32	416.25	2281.04	437.37	25
26	2109.97	376.16	2166.93	396.08	2224.27	416.60	2282.01	437.72	26
27	2110.91	376.49	2167.88	396.42	2225.23	416.94	2282.97	438.08	27
28	2111.86	376.81	2168.84	396.75	2226.19	417.29	2283.94	438.44	28
29	2112.81	377.14	2169.79	397.09	2227.15	417.64	2284.90	438.80	29
30	2113.75	377.47	2170.74	397.43	2228.11	417.99	2285.87	439.15	30
31	2114.70	377.80	2171.70	397.76	2229.07	418.33	2286.84	439.51	31
32	2115.65	378.12	2172.65	398.10	2230.03	418.68	2287.80	439.87	32
33	2116.59	378.45	2173.60	398.44	2230.99	419.03	2288.77	440.23	33
34	2117.54	378.78	2174.56	398.78	2231.95	419.38	2289.74	440.59	34
35	2118.49	379.11	2175.51	399.12	2232.91	419.73	2290.70	440.95	35
36	2119.44	379.44	2176.46	399.46	2233.87	420.07	2291.67	441.31	36
37	2120.38	379.77	2177.42	399.79	2234.83	420.42	2292.63	441.66	37
38	2121.33	380.10	2178.37	400.13	2235.79	420.77	2293.60	442.02	38
39	2122.28	380.42	2179.32	400.47	2236.75	421.12	2294.57	442.38	39
40	2123.23	380.75	2180.28	400.81	2237.71	421.47	2295.54	442.74	40
41	2124.17	381.08	2181.23	401.15	2238.67	421.82	2296.50	443.10	41
42	2125.12	381.41	2182.19	401.49	2239.63	422.17	2297.47	443.46	42
43	2126.07	381.74	2183.14	401.83	2240.59	422.52	2298.44	443.82	43
44	2127.02	382.07	2184.09	402.17	2241.55	422.87	2299.40	444.18	44
45	2127.97	382.40	2185.05	402.51	2242.51	423.22	2300.37	444.54	45
46	2128.91	382.73	2186.00	402.85	2243.47	423.57	2301.34	444.90	46
47	2129.86	383.06	2186.96	403.19	2244.44	423.92	2302.31	445.26	47
48	2130.81	383.39	2187.91	403.53	2245.40	424.27	2303.28	445.63	48
49	2131.76	383.72	2188.87	403.87	2246.36	424.62	2304.24	445.99	49
50	2132.71	384.05	2189.82	404.21	2247.32	424.97	2305.21	446.35	50
51	2133.66	384.39	2190.78	404.55	2248.28	425.32	2306.18	446.71	51
52	2134.61	384.72	2191.73	404.89	2249.24	425.68	2307.15	447.07	52
53	2135.56	385.05	2192.69	405.24	2250.20	426.03	2308.12	447.43	53
54	2136.50	385.38	2193.64	405.58	2251.17	426.38	2309.09	447.79	54
55	2137.45	385.71	2194.60	405.92	2252.13	426.73	2310.05	448.16	55
56	2138.40	386.04	2195.56	406.26	2253.09	427.08	2311.02	448.52	56
57	2139.35	386.38	2196.51	406.60	2254.05	427.44	2311.99	448.88	57
58	2140.30	386.71	2197.47	406.95	2255.02	427.79	2312.96	449.24	58
59	2141.25	387.04	2198.42	407.29	2255.98	428.14	2313.93	449.61	59

Table A-4. (*Continued*)

′	$\Delta = 44°$		$\Delta = 45°$		$\Delta = 46°$		$\Delta = 47°$		′
	T	E	T	E	T	E	T	E	
0	2314.90	449.97	2373.27	472.07	2432.06	494.81	2491.29	518.19	0
1	2315.87	450.33	2374.25	472.45	2433.05	495.19	2492.28	518.59	1
2	2316.84	450.70	2375.22	472.82	2434.03	495.58	2493.27	518.98	2
3	2317.81	451.06	2376.20	473.19	2435.01	495.96	2494.26	519.38	3
4	2318.78	451.42	2377.17	473.57	2436.00	496.35	2495.26	519.77	4
5	2319.75	451.79	2378.15	473.94	2436.98	496.73	2496.25	520.17	5
6	2320.72	452.15	2379.13	474.32	2437.96	497.12	2497.24	520.56	6
7	2321.69	452.51	2380.11	474.69	2438.95	497.50	2498.23	520.96	7
8	2322.66	452.88	2381.08	475.07	2439.93	497.89	2499.22	521.36	8
9	2323.63	453.25	2382.06	475.44	2440.92	498.27	2500.21	521.75	9
10	2324.60	453.61	2383.04	475.82	2441.90	498.66	2501.21	522.15	10
11	2325.57	453.98	2384.02	476.19	2442.89	499.05	2502.20	522.55	11
12	2326.54	454.34	2384.99	476.57	2443.87	499.43	2503.19	522.94	12
13	2327.51	454.71	2385.97	476.94	2444.86	499.82	2504.18	523.34	13
14	2328.48	455.07	2386.95	477.32	2445.84	500.21	2505.18	523.74	14
15	2329.45	455.44	2387.93	477.70	2446.83	500.59	2506.17	524.14	15
16	2330.42	455.80	2388.90	478.07	2447.81	500.98	2507.16	524.53	16
17	2331.40	456.17	2389.88	478.45	2448.80	501.37	2508.15	524.93	17
18	2332.37	456.54	2390.86	478.83	2449.78	501.76	2509.15	525.33	18
19	2333.34	456.90	2391.84	479.20	2450.77	502.14	2510.14	525.73	19
20	2334.31	457.27	2392.82	479.58	2451.76	502.53	2511.13	526.13	20
21	2335.28	457.63	2393.80	479.96	2452.74	502.92	2512.13	526.53	21
22	2336.25	458.00	2394.78	480.34	2453.73	503.31	2513.12	526.93	22
23	2337.23	458.37	2395.76	480.71	2454.71	503.69	2514.11	527.33	23
24	2338.20	458.74	2396.73	481.09	2455.70	504.08	2515.11	527.72	24
25	2339.17	459.10	2397.71	481.47	2456.69	504.47	2516.10	528.12	25
26	2340.14	459.47	2398.69	481.85	2457.67	504.86	2517.10	528.52	26
27	2341.11	459.84	2399.67	482.23	2458.66	505.25	2518.09	528.92	27
28	2342.09	460.21	2400.65	482.60	2459.65	505.64	2519.09	529.32	28
29	2343.06	460.57	2401.63	482.98	2460.63	506.03	2520.08	529.72	29
30	2344.03	460.94	2402.61	483.36	2461.62	506.42	2521.07	530.13	30
31	2345.01	461.31	2403.59	483.74	2462.61	506.81	2522.07	530.53	31
32	2345.98	461.68	2404.57	484.12	2463.60	507.20	2523.06	530.93	32
33	2346.95	462.05	2405.55	484.50	2464.58	507.59	2524.06	531.33	33
34	2347.92	462.42	2406.53	484.88	2465.57	507.98	2525.05	531.73	34
35	2348.90	462.79	2407.51	485.26	2466.56	508.37	2526.05	532.13	35
36	2349.87	463.16	2408.49	485.64	2467.55	508.76	2527.04	532.53	36
37	2350.84	463.53	2409.47	486.02	2468.53	509.15	2528.04	532.93	37
38	2351.82	463.90	2410.45	486.40	2469.52	509.54	2529.04	533.34	38
39	2352.79	464.27	2411.44	486.78	2470.51	509.93	2530.03	533.74	39
40	2353.77	464.64	2412.42	487.16	2471.50	510.32	2531.03	534.14	40
41	2354.74	465.01	2413.40	487.54	2472.49	510.71	2532.02	534.54	41
42	2355.71	465.38	2414.38	487.92	2473.48	511.11	2533.02	534.94	42
43	2356.69	465.75	2415.36	488.30	2474.46	511.50	2534.02	535.35	43
44	2357.66	466.12	2416.34	488.68	2475.45	511.89	2535.01	535.75	44
45	2358.64	466.49	2417.32	489.06	2476.44	512.28	2536.01	536.15	45
46	2359.61	466.86	2418.30	489.45	2477.43	512.68	2537.01	536.56	46
47	2360.59	467.23	2419.29	489.83	2478.42	513.07	2538.00	536.96	47
48	2361.56	467.60	2420.27	490.21	2479.41	513.46	2539 00	537.37	48
49	2362.54	467.97	2421.25	490.59	2480.40	513.85	2540.00	537.77	49
50	2363.51	468.35	2422.23	490.97	2481.39	514.25	2540.99	538.17	50
51	2364.49	468.72	2423.22	491.36	2482.38	514.64	2541.99	538.58	51
52	2365.46	469.09	2424.20	491.74	2483.37	515.03	2542.99	538.98	52
53	2366.44	469.46	2425.18	492.12	2484.36	515.43	2543.99	539.39	53
54	2367.41	469.83	2426.16	492.51	2485.35	515.82	2544.98	539.79	54
55	2368.39	470.21	2427.15	492.89	2486.34	516.22	2545.98	540.20	55
56	2369.36	470.58	2428.13	493.27	2487.33	516.61	2546.98	540.60	56
57	2370.34	470.95	2429.11	493.66	2488.32	517.00	2547.98	541.01	57
58	2371.32	471.33	2430.09	494.04	2489.31	517.40	2548.98	541.41	58
59	2372.29	471.70	2431.08	494.42	2490.30	517.79	2549.97	541.82	59

Table A-4. (*Continued*)

′	Δ = 48° T	Δ = 48° E	Δ = 49° T	Δ = 49° E	Δ = 50° T	Δ = 50° E	Δ = 51° T	Δ = 51° E	′
0	2550.97	542.23	2611.12	566.93	2671.75	592.31	2732.87	618.32	0
1	2551.97	542.63	2612.13	567.35	2672.76	592.74	2733.89	618.82	1
2	2552.97	543.04	2613.13	567.77	2673.78	593.17	2734.91	619.27	2
3	2553.97	543.45	2614.14	568.18	2674.79	593.60	2735.94	619.71	3
4	2554.97	543.85	2615.15	568.60	2675.81	594.03	2736.96	620.15	4
5	2555.97	544.26	2616.15	569.02	2676.82	594.46	2737.98	620.59	5
6	2556.97	544.67	2617.16	569.44	2677.84	594.90	2739.01	621.03	6
7	2557.97	545.07	2618.17	569.86	2678.85	595.32	2740.03	621.47	7
8	2558.96	545.48	2619.17	570.27	2679.87	595.75	2741.06	621.91	8
9	2559.96	545.89	2620.18	570.69	2680.88	596.18	2742.08	622.36	9
10	2560.96	546.30	2621.19	571.11	2681.90	596.61	2743.10	622.80	10
11	2561.96	546.71	2622.20	571.53	2682.91	597.04	2744.13	623.24	11
12	2562.96	547.11	2623.21	571.95	2683.93	597.47	2745.15	623.68	12
13	2563.96	547.52	2624.21	572.37	2684.95	597.90	2746.18	624.13	13
14	2564.96	547.93	2625.22	572.79	2685.96	598.33	2747.20	624.57	14
15	2565.96	548.34	2626.23	573.21	2686.98	598.76	2748.23	625.01	15
16	2566.97	548.75	2627.24	573.63	2688.00	599.20	2749.25	625.46	16
17	2567.97	549.16	2628.25	574.05	2689.01	599.63	2750.28	625.90	17
18	2568.97	549.57	2629.26	574.47	2690.03	600.06	2751.30	626.34	18
19	2569.97	549.98	2630.26	574.89	2691.05	600.50	2752.33	626.79	19
20	2570.97	550.39	2631.27	575.31	2692.06	600.93	2753.36	627.23	20
21	2571.97	550.80	2632.28	575.74	2693.08	601.36	2754.38	627.68	21
22	2572.97	551.21	2633.29	576.16	2694.10	601.79	2755.41	628.12	22
23	2573.97	551.62	2634.30	576.58	2695.12	602.22	2756.43	628.56	23
24	2574.97	552.03	2635.31	577.00	2696.13	602.66	2757.46	629.01	24
25	2575.98	552.44	2636.32	577.42	2697.15	603.09	2758.49	629.46	25
26	2576.98	552.85	2637.33	577.84	2698.17	603.52	2759.51	629.90	26
27	2577.98	553.26	2638.34	578.27	2699.19	603.96	2760.54	630.35	27
28	2578.98	553.67	2639.35	578.69	2700.21	604.39	2761.57	630.79	28
29	2579.98	554.08	2640.36	579.11	2701.23	604.83	2762.59	631.24	29
30	2580.99	554.49	2641.37	579.54	2702.24	605.26	2763.62	631.68	30
31	2581.99	554.91	2642.38	579.96	2703.26	605.70	2764.65	632.13	31
32	2582.99	555.32	2643.39	580.38	2704.28	606.13	2765.67	632.58	32
33	2583.99	555.73	2644.40	580.80	2705.30	606.57	2766.70	633.02	33
34	2585.00	556.14	2645.41	581.23	2706.32	607.00	2767.73	633.47	34
35	2586.00	556.55	2646.42	581.65	2707.34	607.44	2768.76	633.92	35
36	2587.00	556.97	2647.44	582.08	2708.36	607.87	2769.79	634.37	36
37	2588.01	557.38	2648.45	582.50	2709.38	608.31	2770.81	634.81	37
38	2589.01	557.79	2649.46	582.93	2710.40	608.74	2771.84	635.26	38
39	2590.01	558.21	2650.47	583.35	2711.42	609.18	2772.87	635.71	39
40	2591.02	558.62	2651.48	583.77	2712.44	609.62	2773.90	636.16	40
41	2592.02	559.03	2652.49	584.20	2713.46	610.05	2774.93	636.60	41
42	2593.03	559.45	2653.51	584.62	2714.48	610.49	2775.96	637.05	42
43	2594.03	559.86	2654.52	585.05	2715.50	610.93	2776.99	637.50	43
44	2595.03	560.28	2655.53	585.48	2716.52	611.36	2778.02	637.95	44
45	2596.04	560.69	2656.54	585.90	2717.54	611.80	2779.05	638.40	45
46	2597.04	561.10	2657.56	586.33	2718.56	612.24	2780.07	638.85	46
47	2598.05	561.52	2658.57	586.75	2719.58	612.68	2781.10	639.30	47
48	2599.05	561.93	2659.58	587.18	2720.60	613.12	2782.13	639.75	48
49	2600.06	562.35	2660.59	587.61	2721.62	613.55	2783.16	640.20	49
50	2601.06	562.77	2661.61	588.03	2722.65	613.99	2784.19	640.65	50
51	2602.07	563.18	2662.62	588.46	2723.67	614.43	2785.22	641.10	51
52	2603.07	563.60	2663.63	588.89	2724.69	614.87	2786.25	641.55	52
53	2604.08	564.01	2664.65	589.31	2725.71	615.31	2787.28	642.00	53
54	2605.08	564.43	2665.66	589.74	2726.73	615.75	2788.32	642.45	54
55	2606.09	564.85	2666.68	590.17	2727.76	616.19	2789.35	642.90	55
56	2607.09	565.26	2667.69	590.60	2728.78	616.63	2790.38	643.35	56
57	2608.10	565.68	2668.70	591.03	2729.80	617.07	2791.41	643.81	57
58	2609.11	566.09	2669.72	591.45	2730.82	617.51	2792.44	644.26	58
59	2610.11	566.51	2670.73	591.88	2731.85	617.95	2793.47	644.71	59

Table A-4. (*Continued*)

′	Δ = 52° T	Δ = 52° E	Δ = 53° T	Δ = 53° E	Δ = 54° T	Δ = 54° E	Δ = 55° T	Δ = 55° E	′
0	2794.50	645.16	2856.66	672.65	2919.37	700.88	2982.63	729.84	0
1	2795.53	645.61	2857.70	673.12	2920.42	701.36	2983.69	730.33	1
2	2796.57	646.07	2858.74	673.58	2921.47	701.83	2984.75	730.82	2
3	2797.60	646.52	2859.78	674.05	2922.52	702.31	2985.81	731.31	3
4	2798.63	646.97	2860.83	674.51	2923.57	702.79	2986.87	731.80	4
5	2799.66	647.42	2861.87	674.98	2924.62	703.26	2987.93	732.29	5
6	2800.69	647.88	2862.91	675.44	2925.67	703.74	2988.99	732.78	6
7	2801.73	648.33	2863.95	675.91	2926.72	704.22	2990.05	733.27	7
8	2802.76	648.79	2864.99	676.37	2927.77	704.70	2991.11	733.76	8
9	2803.79	649.24	2866.03	676.84	2928.82	705.18	2992.17	734.26	9
10	2804.82	649.69	2867.07	677.31	2929.87	705.65	2993.23	734.75	10
11	2805.86	650.15	2868.12	677.77	2930.92	706.13	2994.29	735.24	11
12	2806.89	650.60	2869.16	678.24	2931.97	706.61	2995.35	735.73	12
13	2807.92	651.06	2870.20	678.71	2933.02	707.09	2996.41	736.22	13
14	2808.96	651.51	2871.24	679.17	2934.08	707.57	2997.47	736.71	14
15	2809.99	651.97	2872.29	679.64	2935.13	708.05	2998.53	737.20	15
16	2811.03	652.42	2873.33	680.11	2936.18	708.53	2999.60	737.70	16
17	2812.06	652.88	2874.37	680.58	2937.23	709.01	3000.66	738.19	17
18	2813.09	653.33	2875.42	681.04	2938.29	709.49	3001.72	738.68	18
19	2814.13	653.79	2876.46	681.51	2939.34	709.97	3002.78	739.17	19
20	2815.16	654.25	2877.50	681.98	2940.39	710.45	3003.84	739.67	20
21	2816.20	654.70	2878.55	682.45	2941.44	710.93	3004.91	740.16	21
22	2817.23	655.16	2879.59	682.92	2942.50	711.41	3005.97	740.65	22
23	2818.27	655.62	2880.63	683.39	2943.55	711.89	3007.03	741.15	23
24	2819.30	656.07	2881.68	683.86	2944.60	712.37	3008.10	741.64	24
25	2820.34	656.53	2882.72	684.33	2945.66	712.86	3009.16	742.14	25
26	2821.37	656.99	2883.77	684.79	2946.71	713.34	3010.22	742.63	26
27	2822.41	657.44	2884.81	685.26	2947.76	713.82	3011.29	743.13	27
28	2823.44	657.90	2885.86	685.73	2948.82	714.30	3012.35	743.62	28
29	2824.48	658.36	2886.90	686.20	2949.87	714.78	3013.41	744.12	29
30	2825.52	658.82	2887.95	686.67	2950.93	715.27	3014.48	744.61	30
31	2826.55	659.28	2888.99	687.14	2951.98	715.75	3015.54	745.11	31
32	2827.59	659.74	2890.04	687.61	2953.04	716.23	3016.61	745.60	32
33	2828.62	660.19	2891.08	688.09	2954.09	716.72	3017.67	746.10	33
34	2829.66	660.65	2892.13	688.56	2955.15	717.20	3018.73	746.59	34
35	2830.70	661.11	2893.17	689.03	2956.20	717.68	3019.80	747.09	35
36	2831.73	661.57	2894.22	689.50	2957.26	718.17	3020.86	747.59	36
37	2832.77	662.03	2895.26	689.97	2958.31	718.65	3021.93	748.08	37
38	2833.81	662.49	2896.31	690.44	2959.37	719.14	3022.99	748.58	38
39	2834.84	662.95	2897.36	690.92	2960.42	719.62	3024.06	749.08	39
40	2835.88	663.41	2898.40	691.39	2961.48	720.11	3025.13	749.58	40
41	2836.92	663.87	2899.45	691.86	2962.53	720.59	3026.19	750.07	41
42	2837.96	664.33	2900.50	692.33	2963.59	721.08	3027.26	750.57	42
43	2839.00	664.79	2901.54	692.81	2964.65	721.56	3028.32	751.07	43
44	2840.03	665.25	2902.59	693.28	2965.70	722.05	3029.39	751.57	44
45	2841.07	665.71	2903.64	693.75	2966.76	722.53	3030.46	752.07	45
46	2842.11	666.17	2904.69	694.23	2967.82	723.02	3031.52	752.57	46
47	2843.15	666.64	2905.73	694.70	2968.87	723.51	3032.59	753.07	47
48	2844.19	667.10	2906.78	695.17	2969.93	723.99	3033.66	753.56	48
49	2845.23	667.56	2907.83	695.65	2970.99	724.48	3034.72	754.06	49
50	2846.26	668.02	2908.88	696.12	2972.05	724.96	3035.79	754.56	50
51	2847.30	668.48	2909.92	696.60	2973.10	725.45	3036.86	755.06	51
52	2848.34	668.95	2910.97	697.07	2974.16	725.94	3037.92	755.56	52
53	2849.38	669.41	2912.02	697.55	2975.22	726.43	3038.99	756.06	53
54	2850.42	669.87	2913.07	698.02	2976.28	726.91	3040.06	756.56	54
55	2851.46	670.34	2914.12	698.50	2977.34	727.40	3041.13	757.06	55
56	2852.50	670.80	2915.17	698.97	2978.39	727.89	3042.20	757.56	56
57	2853.54	671.26	2916.22	699.45	2979.45	728.38	3043.26	758.07	57
58	2854.58	671.73	2917.27	699.93	2980.51	728.87	3044.33	758.57	58
59	2855.62	672.19	2918.32	700.40	2981.57	729.36	3045.40	759.07	59

Table A-4. (*Continued*)

′	Δ = 56°		Δ = 57°		Δ = 58°		Δ = 59°		′
	T	E	T	E	T	E	T	E	
0	3046.47	759.57	3110.91	790.07	3175.96	821.36	3241.64	853.45	0
1	3047.54	760.07	3111.99	790.58	3177.05	821.89	3242.74	853.99	1
2	3048.61	760.57	3113.07	791.10	3178.14	822.41	3243.84	854.54	2
3	3049.68	761.08	3114.14	791.61	3179.23	822.94	3244.94	855.08	3
4	3050.75	761.58	3115.22	792.13	3180.32	823.47	3246.04	855.62	4
5	3051.82	762.08	3116.30	792.65	3181.41	824.00	3247.14	856.16	5
6	3052.89	762.58	3117.38	793.16	3182.50	824.53	3248.24	856.71	6
7	3053.96	763.09	3118.46	793.68	3183.59	825.06	3249.34	857.25	7
8	3055.03	763.59	3119.54	794.20	3184.68	825.59	3250.45	857.79	8
9	3056.10	764.10	3120.62	794.71	3185.77	826.12	3251.55	858.34	9
10	3057.17	764.60	3121.71	795.23	3186.86	826.65	3252.65	858.88	10
11	3058.24	765.10	3122.79	795.75	3187.95	827.18	3253.75	859.42	11
12	3059.31	765.61	3123.87	796.26	3189.04	827.71	3254.85	859.97	12
13	3060.38	766.11	3124.95	796.78	3190.13	828.24	3255.96	860.51	13
14	3061.45	766.62	3126.03	797.30	3191.23	828.77	3257.06	861.06	14
15	3062.52	767.12	3127.11	797.82	3192.32	829.31	3258.16	861.60	15
16	3063.59	767.63	3128.19	798.33	3193.41	829 84	3259.26	862.15	16
17	3064.67	768.13	3129.27	798.85	3194.50	830.37	3260.37	862.70	17
18	3065.74	768.64	3130.36	799.37	3195.59	830.90	3261.47	863.24	18
19	3066.81	769.14	3131.44	799.89	3196.69	831.43	3262.57	863.79	19
20	3067.88	769.65	3132.52	800.41	3197.78	831.97	3263.68	864.33	20
21	3068.95	770.16	3133.60	800.93	3198.87	832.50	3264.78	864.88	21
22	3070.03	770.66	3134.69	801.45	3199.97	833.03	3265.88	865.42	22
23	3071.10	771.17	3135.77	801.97	3201.06	833.56	3266.99	865.97	23
24	3072.17	771.68	3136.85	802.49	3202.15	834.10	3268.09	866.52	24
25	3073.25	772.18	3137.94	803.01	3203.25	834.63	3269.20	867.07	25
26	3074.32	772.69	3139.02	803.53	3204.34	835.17	3270.30	867.61	26
27	3075.39	773.20	3140.10	804.05	3205.43	835.70	3271.41	868.16	27
28	3076.47	773.71	3141.19	804.57	3206.53	836.23	3272.51	868.71	28
29	3077.54	774.21	3142.27	805.09	3207.62	836.77	3273.62	869.26	29
30	3078.61	774.72	3143.35	805.61	3208.72	837.30	3274.72	869.81	30
31	3079.69	775.23	3144.44	806.14	3209.81	837.84	3275.83	870.36	31
32	3080.76	775.74	3145.52	806.66	3210.91	838.37	3276.93	870.90	32
33	3081.84	776.25	3146.61	807.18	3212.00	838.91	3278.04	871.45	33
34	3082.91	776.76	3147.69	807.70	3213.10	839.45	3279.15	872.00	34
35	3083.98	777.27	3148.78	808.22	3214.19	839.98	3280.25	872.55	35
36	3085.06	777.78	3149.86	808.75	3215.29	840.52	3281.36	873.10	36
37	3086.13	778.29	3150.95	809.27	3216.38	841.05	3282.47	873.65	37
38	3087.21	778.80	3152.03	809.79	3217.48	841.59	3283.57	874.20	38
39	3088.29	779.31	3153.12	810.32	3218.58	842.13	3284.68	874.75	39
40	3089.36	779.82	3154.20	810.84	3219.67	842.66	3285.79	875.30	40
41	3090.44	780.33	3155.29	811.36	3220.77	843.20	3286.90	875.85	41
42	3091.51	780.84	3156.38	811.89	3221.87	843.74	3288.00	876.41	42
43	3092.59	781.35	3157.46	812.41	3222.96	844.28	3289.11	876.96	43
44	3093.66	781.86	3158.55	812.94	3224.06	844.81	3290.22	877.51	44
45	3094.74	782.37	3159.64	813.46	3225.16	845.35	3291.33	878.06	45
46	3095.82	782.88	3160.72	813.99	3226.26	845.89	3292.44	878.61	46
47	3096.89	783.39	3161.81	814.51	3227.35	846.43	3293.54	879.17	47
48	3097.97	783.91	3162.90	815.04	3228.45	846.97	3294.65	879.72	48
49	3099.05	784.42	3163.98	815.56	3229.55	847.51	3295.76	880.27	49
50	3100.13	784.93	3165.07	816.09	3230.65	848.05	3296.87	880.82	50
51	3101.20	785.44	3166.16	816.61	3231.75	848.59	3297.98	881.38	51
52	3102.28	785.96	3167.25	817.14	3232.84	849.13	3299.09	881.93	52
53	3103.36	786.47	3168.34	817.67	3233.94	849.67	3300.20	882.49	53
54	3104.44	786.98	3169.42	818.19	3235.04	850.21	3301.31	883.04	54
55	3105.51	787.50	3170.51	818.72	3236.14	850.75	3302.42	883.59	55
56	3106.59	788.01	3171.60	819.25	3237.24	851.29	3303.53	884.15	56
57	3107.67	788.53	3172.69	819.77	3238.34	851.83	3304.64	884.70	57
58	3108.75	789.04	3173.78	820.30	3239.44	852.37	3305.75	885.26	58
59	3109.83	789.55	3174.87	820.83	3240.54	852.91	3306.86	885.81	59

Table A-4. (*Continued*)

′	Δ = 60° T	Δ = 60° E	Δ = 61° T	Δ = 61° E	Δ = 62° T	Δ = 62° E	Δ = 63° T	Δ = 63° E	′
0	3308.0	886.4	3375.0	920.1	3442.7	954.7	3511.1	990.2	0
1	3309.1	886.9	3376.1	920.7	3443.8	955.3	3512.2	990.8	1
2	3310.2	887.5	3377.3	921.3	3445.0	955.9	3513.4	991.4	2
3	3311.3	888.0	3378.4	921.8	3446.1	956.5	3514.5	992.0	3
4	3312.5	888.6	3379.5	922.4	3447.2	957.1	3515.7	992.6	4
5	3313.6	889.2	3380.6	923.0	3448.4	957.7	3516.8	993.2	5
6	3314.7	889.7	3381.7	923.6	3449.5	958.3	3518.0	993.8	6
7	3315.8	890.3	3382.9	924.1	3450.6	958.8	3519.1	994.4	7
8	3316.9	890.8	3384.0	924.7	3451.8	959.4	3520.3	995.0	8
9	3318.0	891.4	3385.1	925.3	3452.9	960.0	3521.4	995.6	9
10	3319.1	891.9	3386.2	925.8	3454.0	960.6	3522.6	996.2	10
11	3320.2	892.5	3387.4	926.4	3455.2	961.2	3523.7	996.8	11
12	3321.4	893.1	3388.5	927.0	3456.3	961.8	3524.9	997.4	12
13	3322.5	893.6	3389.6	927.6	3457.5	962.4	3526.0	998.0	13
14	3323.6	894.2	3390.7	928.1	3458.6	963.0	3527.2	998.6	14
15	3324.7	894.7	3391.9	928.7	3459.7	963.5	3528.3	999.3	15
16	3325.8	895.3	3393.0	929.3	3460.9	964.1	3529.5	999.9	16
17	3326.9	895.9	3394.1	929.9	3462.0	964.7	3530.6	1000.5	17
18	3328.0	896.4	3395.2	930.4	3463.1	965.3	3531.8	1001.1	18
19	3329.1	897.0	3396.4	931.0	3464.3	965.9	3532.9	1001.7	19
20	3330.3	897.5	3397.5	931.6	3465.4	966.5	3534.1	1002.3	20
21	3331.4	898.1	3398.6	932.2	3466.6	967.1	3535.2	1002.9	21
22	3332.5	898.7	3399.7	932.7	3467.7	967.7	3536.4	1003.5	22
23	3333.6	899.2	3400.9	933.3	3468.8	968.2	3537.5	1004.1	23
24	3334.7	899.8	3402.0	933.9	3470.0	968.8	3538.7	1004.7	24
25	3335.8	900.3	3403.1	934.5	3471.1	969.4	3539.8	1005.3	25
26	3337.0	900.9	3404.2	935.0	3472.3	970.0	3541.0	1005.9	26
27	3338.1	901.5	3405.4	935.6	3473.4	970.6	3542.1	1006.5	27
28	3339.2	902.0	3406.5	936.2	3474.5	971.2	3543.3	1007.1	28
29	3340.3	902.6	3407.6	936.8	3475.7	971.8	3544.4	1007.7	29
30	3341.4	903.2	3408.8	937.3	3476.8	972.4	3545.6	1008.3	30
31	3342.5	903.7	3409.9	937.9	3478.0	973.0	3546.7	1008.9	31
32	3343.7	904.3	3411.0	938.5	3479.1	973.6	3547.9	1009.5	32
33	3344.8	904.8	3412.1	939.1	3480.2	974.2	3549.0	1010.1	33
34	3345.9	905.4	3413.3	939.6	3481.4	974.8	3550.2	1010.7	34
35	3347.0	906.0	3414.4	940.2	3482.5	975.3	3551.4	1011.3	35
36	3348.1	906.5	3415.5	940.8	3483.7	975.9	3552.5	1012.0	36
37	3349.2	907.1	3416.7	941.4	3484.8	976.5	3553.7	1012.6	37
38	3350.4	907.7	3417.8	942.0	3485.9	977.1	3554.8	1013.2	38
39	3351.5	908.2	3418.9	942.5	3487.1	977.7	3556.0	1013.8	39
40	3352.6	908.8	3420.0	943.1	3488.2	978.3	3557.1	1014.4	40
41	3353.7	909.4	3421.2	943.7	3489.4	978.9	3558.3	1015.0	41
42	3354.8	909.9	3422.3	944.3	3490.5	979.5	3559.4	1015.6	42
43	3356.0	910.5	3423.4	944.9	3491.7	980.1	3560.6	1016.2	43
44	3357.1	911.1	3424.6	945.4	3492.8	980.7	3561.7	1016.8	44
45	3358.2	911.6	3425.7	946.0	3493.9	981.3	3562.9	1017.4	45
46	3359.3	912.2	3426.8	946.6	3495.1	981.9	3564.1	1018.1	46
47	3360.4	912.7	3428.0	947.2	3496.2	982.5	3565.2	1018.7	47
48	3361.6	913.3	3429.1	947.8	3497.4	983.1	3566.4	1019.3	48
49	3362.7	913.9	3430.2	948.3	3498.5	983.7	3567.5	1019.9	49
50	3363.8	914.5	3431.4	948.9	3499.7	984.3	3568.7	1020.5	50
51	3364.9	915.0	3432.5	949.5	3500.8	984.9	3569.8	1021.1	51
52	3366.0	915.6	3433.6	950.1	3501.9	985.5	3571.0	1021.7	52
53	3367.2	916.2	3434.8	950.7	3503.1	986.1	3572.1	1022.3	53
54	3368.3	916.7	3435.9	951.2	3504.2	986.7	3573.3	1022.9	54
55	3369.4	917.3	3437.0	951.8	3505.4	987.2	3574.5	1023.5	55
56	3370.5	917.9	3438.2	952.4	3506.5	987.8	3575.6	1024.2	56
57	3371.6	918.4	3439.3	953.0	3507.7	988.4	3576.8	1024.8	57
58	3372.8	919.0	3440.4	953.6	3508.8	989.0	3577.9	1025.4	58
59	3373.9	919.6	3441.6	954.2	3509.9	989.6	3579.1	1026.0	59

Table A-4. (*Continued*)

′	Δ = 64°		Δ = 65°		Δ = 66°		Δ = 67°		′
	T	E	T	E	T	E	T	E	
0	3580.2	1026.6	3650.2	1063.9	3720.9	1102.2	3792.3	1141.4	0
1	3581.4	1027.2	3651.3	1064.5	3722.0	1102.8	3793.5	1142.0	1
2	3582.6	1027.8	3652.5	1065.2	3723.2	1103.5	3794.7	1142.7	2
3	3583.7	1028.5	3653.7	1065.8	3724.4	1104.1	3795.9	1143.4	3
4	3584.9	1029.1	3654.9	1066.4	3725.6	1104.8	3797.1	1144.0	4
5	3586.1	1029.7	3656.0	1067.1	3726.8	1105.4	3798.3	1144.7	5
6	3587.2	1030.3	3657.2	1067.7	3728.0	1106.0	3799.5	1145.3	6
7	3588.4	1030.9	3658.4	1068.3	3729.1	1106.7	3800.7	1146.0	7
8	3589.5	1031.6	3659.5	1069.0	3730.3	1107.3	3801.9	1146.7	8
9	3590.7	1032.2	3660.7	1069.6	3731.5	1108.0	3803.1	1147.3	9
10	3591.9	1032.8	3661.9	1070.2	3732.7	1108.6	3804.3	1148.0	10
11	3593.0	1033.4	3663.1	1070.9	3733.9	1109.3	3805.5	1148.7	11
12	3594.2	1034.0	3664.2	1071.5	3735.1	1109.9	3806.7	1149.3	12
13	3595.3	1034.6	3665.4	1072.1	3736.3	1110.6	3807.9	1150.0	13
14	3596.5	1035.2	3666.6	1072.8	3737.5	1111.2	3809.1	1150.7	14
15	3597.7	1035.9	3667.8	1073.4	3738.7	1111.9	3810.3	1151.3	15
16	3598.8	1036.5	3668.9	1074.0	3739.8	1112.5	3811.5	1152.0	16
17	3600.0	1037.1	3670.1	1074.7	3741.0	1113.2	3812.7	1152.7	17
18	3601.1	1037.7	3671.3	1075.3	3742.2	1113.8	3813.9	1153.3	18
19	3602.3	1038.3	3672.5	1075.9	3743.4	1114.5	3815.1	1154.0	19
20	3603.5	1039.0	3673.6	1076.6	3744.6	1115.1	3816.4	1154.7	20
21	3604.6	1039.6	3674.8	1077.2	3745.8	1115.8	3817.6	1155.3	21
22	3605.8	1040.2	3676.0	1077.8	3747.0	1116.4	3818.8	1156.0	22
23	3607.0	1040.8	3677.2	1078.5	3748.2	1117.1	3820.0	1156.7	23
24	3608.1	1041.4	3678.3	1079.1	3749.4	1117.7	3821.2	1157.3	24
25	3609.3	1042.1	3679.5	1079.7	3750.5	1118.4	3822.4	1158.0	25
26	3610.5	1042.7	3680.7	1080.4	3751.7	1119.0	3823.6	1158.7	26
27	3611.6	1043.3	3681.9	1081.0	3752.9	1119.7	3824.8	1159.3	27
28	3612.8	1043.9	3683.1	1081.6	3754.1	1120.3	3826.0	1160.0	28
29	3613.9	1044.5	3684.2	1082.3	3755.3	1121.0	3827.2	1160.7	29
30	3615.1	1045.2	3685.4	1082.9	3756.5	1121.6	3828.4	1161.3	30
31	3616.3	1045.8	3686.6	1083.5	3757.7	1122.3	3829.6	1162.0	31
32	3617.4	1046.4	3687.8	1084.2	3758.9	1123.0	3830.8	1162.7	32
33	3618.6	1047.0	3688.9	1084.8	3760.1	1123.6	3832.0	1163.3	33
34	3619.8	1047.6	3690.1	1085.5	3761.2	1124.3	3833.2	1164.0	34
35	3620.9	1048.3	3691.3	1086.1	3762.6	1124.9	3834.4	1164.7	35
36	3622.1	1048.9	3692.5	1086.7	3763.8	1125.6	3835.6	1165.4	36
37	3623.3	1049.5	3693.7	1087.4	3764.9	1126.2	3836.8	1166.0	37
38	3624.4	1050.1	3694.8	1088.0	3766.0	1126.9	3838.0	1166.7	38
39	3625.6	1050.8	3696.0	1088.7	3767.2	1127.5	3839.2	1167.4	39
40	3626.8	1051.4	3697.2	1089.3	3768.4	1128.2	3840.5	1168.0	40
41	3627.9	1052.0	3698.4	1089.9	3769.6	1128.9	3841.7	1168.7	41
42	3629.1	1052.6	3699.6	1090.6	3770.8	1129.5	3842.9	1169.4	42
43	3630.3	1053.3	3700.7	1091.2	3772.0	1130.2	3844.1	1170.1	43
44	3631.4	1053.9	3701.9	1091.9	3773.2	1130.8	3845.3	1170.7	44
45	3632.6	1054.5	3703.1	1092.5	3774.4	1131.5	3846.5	1171.4	45
46	3633.8	1055.1	3704.3	1093.1	3775.6	1132.1	3847.7	1172.1	46
47	3635.0	1055.8	3705.5	1093.8	3776.8	1132.8	3848.9	1172.8	47
48	3636.1	1056.4	3706.7	1094.4	3778.0	1133.5	3850.1	1173.4	48
49	3637.3	1057.0	3707.8	1095.1	3779.2	1134.1	3851.3	1174.1	49
50	3638.5	1057.6	3709.0	1095.7	3780.4	1134.8	3852.5	1174.8	50
51	3639.6	1058.3	3710.2	1096.4	3781.6	1135.4	3853.7	1175.5	51
52	3640.8	1058.9	3711.4	1097.0	3782.8	1136.1	3855.0	1176.1	52
53	3642.0	1059.5	3712.6	1097.6	3784.0	1136.7	3856.2	1176.8	53
54	3643.1	1060.2	3713.7	1098.3	3785.2	1137.4	3857.4	1177.5	54
55	3644.3	1060.8	3714.9	1098.9	3786.4	1138.1	3858.6	1178.2	55
56	3645.5	1061.4	3716.1	1099.6	3787.6	1138.7	3859.8	1178.8	56
57	3646.7	1062.0	3717.3	1100.2	3788.8	1139.4	3861.0	1179.5	57
58	3647.8	1062.7	3718.5	1100.9	3789.9	1140.0	3862.2	1180.2	58
59	3649.0	1063.3	3719.7	1101.5	3791.1	1140.7	3863.4	1180.9	59

Table A-4. (*Continued*)

′	Δ = 68°		Δ = 69°		Δ = 70°		Δ = 71°		′
	T	E	T	E	T	E	T	E	
0	3864.7	1181.6	3937.9	1222.7	4011.9	1265.0	4086.9	1308.2	0
1	3865.9	1182.2	3939.1	1223.4	4013.2	1265.7	4088.1	1308.9	1
2	3867.1	1182.9	3940.3	1224.1	4014.4	1266.4	4089.4	1309.7	2
3	3868.3	1183.6	3941.5	1224.8	4015.6	1267.1	4090.7	1310.4	3
4	3869.5	1184.3	3942.7	1225.5	4016.9	1267.8	4091.9	1311.2	4
5	3870.7	1184.9	3944.0	1226.2	4018.1	1268.5	4093.2	1311.9	5
6	3871.9	1185.6	3945.2	1226.9	4019.4	1269.2	4094.4	1312.6	6
7	3873.2	1186.3	3946.4	1227.6	4020.6	1269.9	4095.7	1313.3	7
8	3874.4	1187.0	3947.7	1228.3	4021.9	1270.7	4097.0	1314.1	8
9	3875.6	1187.7	3948.9	1229.0	4023.1	1271.4	4098.2	1314.8	9
10	3876.8	1188.3	3950.1	1229.7	4024.3	1272.1	4099.5	1315.5	10
11	3878.0	1189.0	3951.4	1230.4	4025.6	1272.8	4100.7	1316.3	11
12	3879.2	1189.7	3952.6	1231.1	4026.8	1273.5	4102.0	1317.0	12
13	3880.5	1190.4	3953.8	1231.8	4028.1	1274.2	4103.3	1317.7	13
14	3881.7	1191.1	3955.0	1232.5	4029.3	1275.0	4104.5	1318.5	14
15	3882.9	1191.8	3956.3	1233.2	4030.6	1275.7	4105.8	1319.2	15
16	3884.1	1192.4	3957.5	1233.9	4031.8	1276.4	4107.0	1319.9	16
17	3885.3	1193.1	3958.7	1234.6	4033.1	1277.1	4108.3	1320.7	17
18	3886.5	1193.8	3960.0	1235.3	4034.3	1277.8	4109.6	1321.4	18
19	3887.7	1194.5	3961.2	1236.0	4035.6	1278.5	4110.8	1322.2	19
20	3889.0	1195.2	3962.4	1236.7	4036.8	1279.3	4112.1	1322.9	20
21	3890.2	1195.9	3963.7	1237.4	4038.1	1280.0	4113.4	1323.6	21
22	3891.4	1196.5	3964.9	1238.1	4039.3	1280.7	4114.6	1324.4	22
23	3892.6	1197.2	3966.1	1238.8	4040.5	1281.4	4115.9	1325.1	23
24	3893.8	1197.9	3967.4	1239.5	4041.8	1282.1	4117.1	1325.8	24
25	3895.0	1198.6	3968.6	1240.2	4043.0	1282.9	4118.4	1326.6	25
26	3896.3	1199.3	3969.8	1240.9	4044.3	1283.6	4119.7	1327.3	26
27	3897.5	1200.0	3971.1	1241.6	4045.5	1284.3	4120.9	1328.0	27
28	3898.7	1200.6	3972.3	1242.3	4046.8	1285.0	4122.2	1328.8	28
29	3899.9	1201.3	3973.5	1243.0	4048.0	1285.7	4123.5	1329.5	29
30	3901.1	1202.0	3974.8	1243.7	4049.3	1286.5	4124.7	1330.3	30
31	3902.4	1202.7	3976.0	1244.4	4050.5	1287.2	4126.0	1331.0	31
32	3903.6	1203.4	3977.2	1245.1	4051.8	1287.9	4127.2	1331.7	32
33	3904.8	1204.1	3978.5	1245.8	4053.0	1288.6	4128.5	1332.5	33
34	3906.0	1204.8	3979.7	1246.5	4054.3	1289.3	4129.8	1333.2	34
35	3907.2	1205.4	3980.9	1247.2	4055.5	1290.0	4131.0	1334.0	35
36	3908.5	1206.1	3982.2	1247.9	4056.8	1290.8	4132.3	1334.7	36
37	3909.7	1206.8	3983.4	1248.6	4058.0	1291.5	4133.6	1335.4	37
38	3910.9	1207.5	3984.7	1249.4	4059.3	1292.2	4134.8	1336.2	38
39	3912.1	1208.2	3985.9	1250.1	4060.5	1292.9	4136.1	1336.9	39
40	3913.4	1208.9	3987.1	1250.8	4061.8	1293.7	4137.4	1337.7	40
41	3914.6	1209.6	3988.4	1251.5	4063.0	1294.4	4138.6	1338.4	41
42	3915.8	1210.3	3989.6	1252.2	4064.3	1295.1	4139.9	1339.2	42
43	3917.0	1211.0	3990.8	1252.9	4065.6	1295.8	4141.2	1339.9	43
44	3918.2	1211.7	3992.1	1253.6	4066.8	1296.6	4142.4	1340.6	44
45	3919.5	1212.3	3993.3	1254.3	4068.1	1297.3	4143.7	1341.4	45
46	3920.7	1213.0	3994.6	1255.0	4069.3	1298.0	4145.0	1342.1	46
47	3921.9	1213.7	3995.8	1255.7	4070.6	1298.7	4146.3	1342.9	47
48	3923.1	1214.4	3997.0	1256.4	4071.9	1299.5	4147.5	1343.6	48
49	3924.4	1215.1	3998.3	1257.1	4073.1	1300.2	4148.8	1344.4	49
50	3925.6	1215.8	3999.5	1257.8	4074.3	1300.9	4150.1	1345.1	50
51	3926.8	1216.5	4000.7	1258.6	4075.6	1301.6	4151.3	1345.8	51
52	3928.0	1217.2	4002.0	1259.3	4076.8	1302.4	4152.6	1346.6	52
53	3929.3	1217.9	4003.2	1260.0	4078.1	1303.1	4153.9	1347.3	53
54	3930.5	1218.6	4004.5	1260.7	4079.3	1303.8	4155.1	1348.1	54
55	3931.7	1219.2	4005.7	1261.4	4080.6	1304.6	4156.4	1348.8	55
56	3932.9	1219.9	4006.9	1262.1	4081.9	1305.3	4157.7	1349.6	56
57	3934.2	1220.6	4008.2	1262.8	4083.1	1306.0	4159.0	1350.3	57
58	3935.4	1221.3	4009.4	1263.5	4084.4	1306.7	4160.2	1351.1	58
59	3936.6	1222.0	4010.7	1264.2	4085.6	1307.5	4161.5	1351.8	59

Table A-4. (*Continued*)

′	Δ = 72° T	Δ = 72° E	Δ = 73° T	Δ = 73° E	Δ = 74° T	Δ = 74° E	Δ = 75° T	Δ = 75° E	′
0	4162.8	1352.6	4239.7	1398.0	4317.6	1444.6	4396.5	1492.4	0
1	4164.1	1353.3	4241.0	1398.8	4318.9	1445.4	4397.8	1493.2	1
2	4165.3	1354.1	4242.2	1399.6	4320.2	1446.2	4399.1	1494.0	2
3	4166.6	1354.8	4243.5	1400.3	4321.5	1447.0	4400.4	1494.8	3
4	4167.9	1355.6	4244.8	1401.1	4322.8	1447.8	4401.8	1495.6	4
5	4169.2	1356.3	4246.1	1401.9	4324.1	1448.6	4403.1	1496.4	5
6	4170.4	1357.1	4247.4	1402.6	4325.4	1449.3	4404.4	1497.2	6
7	4171.7	1357.8	4248.7	1403.4	4326.7	1450.1	4405.7	1498.1	7
8	4173.0	1358.6	4250.0	1404.2	4328.0	1450.9	4407.1	1498.9	8
9	4174.3	1359.3	4251.3	1404.9	4329.3	1451.7	4408.4	1499.7	9
10	4175.5	1360.1	4252.6	1405.7	4330.6	1452.5	4409.7	1500.5	10
11	4176.8	1360.8	4253.9	1406.5	4331.9	1453.3	4411.0	1501.3	11
12	4178.1	1361.6	4255.2	1407.3	4333.2	1454.1	4412.4	1502.1	12
13	4179.4	1362.3	4256.5	1408.0	4334.6	1454.9	4413.7	1502.9	13
14	4180.6	1363.1	4257.8	1408.8	4335.9	1455.7	4415.0	1503.7	14
15	4181.9	1363.8	4259.0	1409.6	4337.2	1456.5	4416.4	1504.5	15
16	4183.2	1364.6	4260.3	1410.4	4338.5	1457.2	4417.7	1505.3	16
17	4184.5	1365.3	4261.6	1411.1	4339.8	1458.0	4419.0	1506.2	17
18	4185.7	1366.1	4262.9	1411.9	4341.1	1458.8	4420.3	1507.0	18
19	4187.0	1366.8	4264.2	1412.7	4342.4	1459.6	4421.7	1507.8	19
20	4188.3	1367.6	4265.5	1413.4	4343.7	1460.4	4423.0	1508.6	20
21	4189.6	1368.4	4266.8	1414.2	4345.0	1461.2	4424.3	1509.4	21
22	4190.9	1369.1	4268.1	1415.0	4346.4	1462.0	4425.7	1510.2	22
23	4192.1	1369.9	4269.4	1415.8	4347.7	1462.8	4427.0	1511.0	23
24	4193.4	1370.6	4270.7	1416.5	4349.0	1463.6	4428.3	1511.8	24
25	4194.7	1371.4	4272.0	1417.3	4350.3	1464.4	4429.7	1512.6	25
26	4196.0	1372.1	4273.3	1418.1	4351.6	1465.2	4431.0	1513.5	26
27	4197.3	1372.9	4274.6	1418.9	4352.9	1466.0	4432.3	1514.3	27
28	4198.5	1373.7	4275.9	1419.6	4354.2	1466.8	4433.7	1515.1	28
29	4199.8	1374.4	4277.2	1420.4	4355.6	1467.6	4435.0	1515.9	29
30	4201.1	1375.2	4278.5	1421.2	4356.9	1468.4	4436.3	1516.7	30
31	4202.4	1375.9	4279.8	1422.0	4358.2	1469.2	4437.6	1517.5	31
32	4203.7	1376.7	4281.1	1422.8	4359.5	1470.0	4439.0	1518.4	32
33	4204.9	1377.4	4282.4	1423.5	4360.8	1470.8	4440.3	1519.2	33
34	4206.2	1378.2	4283.7	1424.3	4362.1	1471.6	4441.7	1520.0	34
35	4207.5	1379.0	4285.0	1425.1	4363.5	1472.4	4443.0	1520.8	35
36	4208.8	1379.7	4286.3	1425.9	4364.8	1473.2	4444.3	1521.6	36
37	4210.1	1380.5	4287.6	1426.6	4366.1	1474.0	4445.7	1522.4	37
38	4211.4	1381.2	4288.9	1427.4	4367.4	1474.8	4447.0	1523.3	38
39	4212.6	1382.0	4290.2	1428.2	4368.7	1475.6	4448.3	1524.1	39
40	4213.9	1382.8	4291.5	1429.0	4370.0	1476.4	4449.7	1524.9	40
41	4215.2	1383.5	4292.8	1429.8	4371.4	1477.2	4451.0	1525.7	41
42	4216.5	1384.3	4294.1	1430.5	4372.7	1478.0	4452.3	1526.5	42
43	4217.8	1385.0	4295.4	1431.3	4374.0	1478.8	4453.7	1527.4	43
44	4219.1	1385.8	4296.7	1432.1	4375.3	1479.6	4455.0	1528.2	44
45	4220.4	1386.6	4298.0	1432.9	4376.6	1480.4	4456.3	1529.0	45
46	4221.6	1387.3	4299.3	1433.7	4378.0	1481.2	4457.7	1529.8	46
47	4222.9	1388.1	4300.6	1434.4	4379.3	1482.0	4459.0	1530.6	47
48	4224.2	1388.9	4301.9	1435.2	4380.6	1482.8	4460.4	1531.5	48
49	4225.5	1389.6	4303.2	1436.0	4381.9	1483.6	4461.7	1532.3	49
50	4226.8	1390.4	4304.5	1436.8	4383.2	1484.4	4463.0	1533.1	50
51	4228.1	1391.2	4305.8	1437.6	4384.6	1485.2	4464.4	1533.9	51
52	4229.4	1391.9	4307.1	1438.4	4385.9	1486.0	4465.7	1534.8	52
53	4230.6	1392.7	4308.4	1439.1	4387.2	1486.8	4467.1	1535.6	53
54	4231.9	1393.4	4309.7	1439.9	4388.5	1487.6	4468.4	1536.4	54
55	4233.2	1394.2	4311.0	1440.7	4389.8	1488.4	4469.7	1537.2	55
56	4234.5	1395.0	4312.3	1441.5	4391.2	1489.2	4471.1	1538.1	56
57	4235.8	1395.7	4313.6	1442.3	4392.5	1490.0	4472.4	1538.9	57
58	4237.1	1396.5	4314.9	1443.0	4393.8	1490.8	4473.8	1539.7	58
59	4238.4	1397.3	4316.2	1443.8	4395.1	1491.6	4475.1	1540.5	59

Table A-4. (*Continued*)

′	Δ = 76°		Δ = 77°		Δ = 78°		Δ = 79°		′
	T	E	T	E	T	E	T	E	
0	4476.4	1541.4	4557.5	1591.6	4639.7	1643.0	4723.1	1695.8	0
1	4477.8	1542.2	4558.9	1592.4	4641.1	1643.9	4724.5	1696.7	1
2	4479.1	1543.0	4560.2	1593.3	4642.5	1644.7	4725.9	1697.6	2
3	4480.5	1543.9	4561.6	1594.1	4643.9	1645.6	4727.3	1698.4	3
4	4481.8	1544.7	4563.0	1595.0	4645.2	1646.5	4728.7	1699.3	4
5	4483.2	1545.5	4564.3	1595.8	4646.6	1647.3	4730.1	1700.2	5
6	4484.5	1546.3	4565.7	1596.7	4648.0	1648.2	4731.5	1701.1	6
7	4485.8	1547.2	4567.0	1597.5	4649.4	1649.1	4732.9	1702.0	7
8	4487.2	1548.0	4568.4	1598.4	4650.8	1650.0	4734.3	1702.9	8
9	4488.5	1548.9	4569.8	1599.2	4652.2	1650.8	4735.7	1703.8	9
10	4489.9	1549.6	4571.1	1600.1	4653.5	1651.7	4737.1	1704.7	10
11	4491.2	1550.5	4572.5	1600.9	4654.9	1652.6	4738.5	1705.6	11
12	4492.6	1551.3	4573.9	1601.8	4656.3	1653.5	4739.9	1706.5	12
13	4493.9	1552.1	4575.2	1602.6	4657.7	1654.3	4741.3	1707.4	13
14	4495.3	1553.0	4576.6	1603.5	4659.1	1655.2	4742.7	1708.3	14
15	4496.6	1553.6	4578.0	1604.3	4660.5	1656.1	4744.1	1709.2	15
16	4498.0	1554.6	4579.3	1605.2	4661.8	1657.0	4745.5	1710.1	16
17	4499.3	1555.5	4580.7	1606.0	4663.2	1657.8	4746.9	1711.0	17
18	4500.6	1556.3	4582.1	1606.9	4664.6	1658.7	4748.4	1711.9	18
19	4502.0	1557.2	4583.4	1607.7	4666.0	1659.6	4749.8	1712.8	19
20	4503.3	1558.0	4584.8	1608.6	4667.4	1660.5	4751.2	1713.7	20
21	4504.7	1558.8	4586.2	1609.4	4668.8	1661.3	4752.6	1714.6	21
22	4506.0	1559.6	4587.5	1610.3	4670.2	1662.2	4754.0	1715.5	22
23	4507.4	1560.5	4588.9	1611.1	4671.5	1663.1	4755.4	1716.3	23
24	4508.7	1561.3	4590.3	1612.0	4672.9	1664.0	4756.8	1717.2	24
25	4510.1	1562.1	4591.6	1612.8	4674.3	1664.8	4758.2	1718.2	25
26	4511.4	1563.0	4593.0	1613.7	4675.7	1665.7	4759.6	1719.1	26
27	4512.8	1563.8	4594.4	1614.5	4677.1	1666.6	4761.0	1719.9	27
28	4514.1	1564.6	4595.7	1615.4	4678.5	1667.5	4762.4	1720.9	28
29	4515.5	1565.5	4597.1	1616.3	4679.9	1668.3	4763.8	1721.8	29
30	4516.8	1566.3	4598.5	1617.1	4681.3	1669.2	4765.2	1722.7	30
31	4518.2	1567.1	4599.8	1618.0	4682.7	1670.1	4766.7	1723.6	31
32	4519.5	1568.0	4601.2	1618.8	4684.0	1671.0	4768.1	1724.5	32
33	4520.9	1568.8	4602.6	1619.7	4685.4	1671.9	4769.5	1725.4	33
34	4522.2	1569.7	4604.0	1620.5	4686.8	1672.8	4770.9	1726.3	34
35	4523.6	1570.5	4605.3	1621.4	4688.2	1673.6	4772.3	1727.2	35
36	4525.0	1571.3	4606.7	1622.3	4689.6	1674.5	4773.7	1728.1	36
37	4526.3	1572.2	4608.1	1623.1	4691.0	1675.4	4775.1	1729.0	37
38	4527.7	1573.0	4609.4	1624.0	4692.4	1676.5	4776.5	1729.9	38
39	4529.0	1573.9	4610.8	1624.8	4693.8	1677.2	4777.9	1730.8	39
40	4530.4	1574.7	4612.2	1625.7	4695.2	1678.1	4779.4	1731.7	40
41	4531.7	1575.5	4613.6	1626.6	4696.6	1678.9	4780.8	1732.6	41
42	4533.1	1576.4	4614.9	1627.4	4698.0	1679.8	4782.2	1733.5	42
43	4534.4	1577.2	4616.3	1628.3	4699.4	1680.7	4783.6	1734.4	43
44	4535.8	1578.1	4617.7	1629.2	4700.8	1681.6	4785.0	1735.3	44
45	4537.1	1578.9	4619.1	1630.0	4702.1	1682.5	4786.4	1736.2	45
46	4538.5	1579.7	4620.4	1630.9	4703.5	1683.3	4787.8	1737.1	46
47	4539.9	1580.6	4621.8	1631.8	4704.9	1684.3	4789.3	1738.1	47
48	4541.2	1581.4	4623.2	1632.6	4706.3	1685.1	4790.7	1738.9	48
49	4542.6	1582.3	4624.6	1633.5	4707.7	1686.0	4792.1	1739.8	48
50	4543.9	1583.1	4625.9	1634.4	4709.1	1686.9	4793.5	1740.8	50
51	4545.3	1583.9	4627.3	1635.2	4710.5	1687.8	4794.9	1741.7	51
52	4546.6	1584.8	4628.7	1636.1	4711.9	1688.7	4796.3	1742.6	52
53	4548.0	1585.6	4630.1	1637.0	4713.3	1689.6	4797.8	1743.5	53
54	4549.4	1586.5	4631.5	1637.8	4714.7	1690.4	4799.2	1744.4	54
55	4550.7	1587.3	4632.8	1638.7	4716.1	1691.3	4800.6	1745.3	55
56	4552.1	1588.2	4634.2	1639.6	4717.5	1692.2	4802.0	1746.2	56
57	4553.4	1589.0	4635.6	1640.4	4718.9	1693.1	4803.4	1747.1	57
58	4554.8	1589.9	4637.0	1641.3	4720.3	1694.0	4804.9	1748.0	58
59	4556.2	1590.7	4638.3	1642.1	4721.7	1694.9	4806.3	1749.0	59

Table A-4. (*Continued*)

′	Δ = 80°		Δ = 81°		Δ = 82°		Δ = 83°		′
	T	E	T	E	T	E	T	E	
0	4807.7	1749.9	4893.5	1805.3	4980.6	1862.2	5069.1	1920.5	0
1	4809.1	1750.8	4895.0	1806.2	4982.1	1863.2	5070.6	1921.5	1
2	4810.5	1751.7	4896.4	1807.2	4983.6	1864.1	5072.1	1922.5	2
3	4812.0	1752.6	4897.9	1808.1	4985.0	1865.1	5073.6	1923.5	3
4	4813.4	1753.5	4899.3	1809.1	4986.5	1866.0	5075.1	1924.5	4
5	4814.8	1754.4	4900.7	1810.0	4988.0	1867.0	5076.5	1925.4	5
6	4816.2	1755.3	4902.2	1810.9	4989.4	1868.0	5078.0	1926.4	6
7	4817.6	1756.3	4903.6	1811.9	4990.9	1868.9	5079.5	1927.4	7
8	4819.1	1757.2	4905.1	1812.8	4992.4	1869.9	5081.0	1928.4	8
9	4820.5	1758.1	4906.5	1813.8	4993.8	1870.8	5082.5	1929.4	9
10	4821.9	1759.0	4908.0	1814.7	4995.3	1871.8	5084.0	1930.4	10
11	4823.3	1759.9	4909.4	1815.6	4996.8	1872.8	5085.5	1931.4	11
12	4824.8	1760.8	4910.8	1816.6	4998.2	1873.7	5087.0	1932.4	12
13	4826.2	1761.8	4912.3	1817.5	4999.7	1874.7	5088.5	1933.4	13
14	4827.6	1762.7	4913.7	1818.5	5001.2	1875.7	5089.9	1934.3	14
15	4829.0	1763.6	4915.2	1819.4	5002.6	1876.6	5091.4	1935.3	15
16	4830.5	1764.5	4916.6	1820.3	5004.1	1877.6	5092.9	1936.3	16
17	4831.9	1765.4	4918.1	1821.3	5005.6	1878.6	5094.4	1937.3	17
18	4833.3	1766.4	4919.5	1822.2	5007.0	1879.6	5095.9	1938.3	18
19	4834.7	1767.3	4921.0	1823.2	5008.5	1880.5	5097.4	1939.3	19
20	4836.2	1768.2	4922.4	1824.1	5010.0	1881.5	5098.9	1940.3	20
21	4837.6	1769.1	4923.9	1825.0	5011.5	1882.4	5100.4	1941.3	21
22	4839.0	1770.0	4925.3	1826.0	5012.9	1883.4	5101.9	1942.3	22
23	4840.4	1771.0	4926.8	1826.9	5014.4	1884.4	5103.4	1943.3	23
24	4841.9	1771.9	4928.2	1827.9	5015.9	1885.4	5104.9	1944.3	24
25	4843.3	1772.8	4929.7	1828.8	5017.3	1886.3	5106.4	1945.3	25
26	4844.7	1773.7	4931.1	1829.8	5018.8	1887.3	5107.9	1946.3	26
27	4846.2	1774.6	4932.6	1830.7	5020.3	1888.3	5109.4	1947.3	27
28	4847.6	1775.6	4934.0	1831.7	5021.8	1889.3	5110.9	1948.3	28
29	4849.0	1776.5	4935.5	1832.6	5023.2	1890.2	5112.3	1949.3	29
30	4850.4	1777.4	4936.9	1833.6	5024.7	1891.2	5113.8	1950.3	30
31	4851.9	1778.3	4938.4	1834.5	5026.2	1892.1	5115.3	1951.3	31
32	4853.3	1779.3	4939.8	1835.5	5027.7	1893.1	5116.8	1952.3	32
33	4854.7	1780.2	4941.3	1836.4	5029.1	1894.1	5118.3	1953.3	33
34	4856.2	1781.1	4942.7	1837.4	5030.6	1895.1	5119.8	1954.3	34
35	4857.6	1782.0	4944.2	1838.3	5032.1	1896.0	5121.3	1955.2	35
36	4859.0	1783.0	4945.6	1839.3	5033.6	1897.0	5122.8	1956.2	36
37	4860.5	1783.9	4947.1	1840.2	5035.0	1898.0	5124.3	1957.2	37
38	4861.9	1784.8	4948.6	1841.2	5036.5	1899.0	5125.8	1958.2	38
39	4863.3	1785.8	4950.0	1842.1	5038.0	1899.9	5127.3	1959.2	39
40	4864.8	1786.7	4951.5	1843.1	5039.5	1900.9	5128.8	1960.2	40
41	4866.2	1787.6	4952.9	1844.0	5040.9	1901.9	5130.3	1961.2	41
42	4867.6	1788.5	4954.4	1845.0	5042.4	1902.9	5131.8	1962.2	42
43	4869.1	1789.5	4955.8	1845.9	5043.9	1903.8	5133.3	1963.2	43
44	4870.5	1790.4	4957.3	1846.9	5045.4	1904.8	5134.8	1964.2	44
45	4871.9	1791.3	4958.7	1847.8	5046.9	1905.8	5136.4	1965.2	45
46	4873.4	1792.3	4960.2	1848.8	5048.3	1906.8	5137.9	1966.3	46
47	4874.8	1793.2	4961.7	1849.7	5049.8	1907.8	5139.4	1967.3	47
48	4876.3	1794.1	4963.1	1850.7	5051.3	1908.7	5140.9	1968.3	48
49	4877.7	1795.1	4964.6	1851.6	5052.8	1909.7	5142.4	1969.3	49
50	4879.1	1796.0	4966.0	1852.6	5054.3	1910.7	5143.9	1970.3	50
51	4880.6	1796.9	4967.5	1853.6	5055.7	1911.7	5145.4	1971.3	51
52	4882.0	1797.9	4969.0	1854.5	5057.2	1912.7	5146.9	1972.3	52
53	4883.4	1798.8	4970.4	1855.5	5058.7	1913.6	5148.4	1973.3	53
54	4884.9	1799.7	4971.9	1856.4	5060.2	1914.6	5149.9	1974.3	54
55	4886.3	1800.6	4973.3	1857.4	5061.7	1915.6	5151.4	1975.3	55
56	4887.8	1801.6	4974.8	1858.4	5063.2	1916.6	5152.9	1976.4	56
57	4889.2	1802.5	4976.3	1859.3	5064.7	1917.6	5154.4	1977.4	57
58	4890.6	1803.4	4977.7	1860.3	5066.1	1918.5	5155.9	1978.4	58
59	4892.1	1804.4	4979.2	1861.2	5067.6	1919.5	5157.4	1979.4	59

Table A-4. (*Continued*)

′	Δ = 92°		Δ = 93°		Δ = 94°		Δ = 95°		′
	T	E	T	E	T	E	T	E	
0	5933.1	2518.5	6037.7	2594.0	6144.2	2671.6	6252.7	2751.3	0
1	5934.9	2519.7	6039.5	2595.3	6146.0	2672.9	6254.6	2752.6	1
2	5936.6	2521.0	6041.2	2596.6	6147.8	2674.2	6256.4	2754.0	2
3	5938.3	2522.2	6043.0	2597.8	6149.6	2675.5	6258.2	2755.3	3
4	5940.1	2523.5	6044.8	2599.1	6151.4	2676.8	6260.1	2756.7	4
5	5941.8	2524.7	6046.5	2600.4	6153.2	2678.1	6261.9	2758.0	5
6	5943.5	2525.9	6048.3	2601.7	6155.0	2679.5	6263.7	2759.4	6
7	5945.3	2527.2	6050.0	2603.0	6156.8	2680.8	6265.5	2760.7	7
8	5947.0	2528.4	6051.8	2604.2	6158.6	2682.1	6267.4	2762.1	8
9	5948.7	2529.7	6053.6	2605.5	6160.4	2683.4	6269.2	2763.4	9
10	5950.4	2530.9	6055.3	2606.8	6162.2	2684.7	6271.0	2764.8	10
11	5952.2	2532.2	6057.1	2608.1	6164.0	2686.0	6272.9	2766.1	11
12	5953.9	2533.4	6058.9	2609.4	6165.8	2687.4	6274.7	2767.5	12
13	5955.7	2534.7	6060.6	2610.6	6167.6	2688.7	6276.5	2768.8	13
14	5957.4	2535.9	6062.4	2611.9	6169.4	2690.0	6278.4	2770.2	14
15	5959.1	2537.2	6064.2	2613.2	6171.2	2691.3	6280.2	2771.5	15
16	5960.9	2538.4	6065.9	2614.5	6173.0	2692.6	6282.0	2772.9	16
17	5962.6	2539.7	6067.7	2615.8	6174.8	2693.9	6283.9	2774.2	17
18	5964.3	2540.9	6069.5	2617.1	6176.6	2695.3	6285.7	2775.6	18
19	5966.1	2542.2	6071.2	2618.4	6178.4	2696.6	6287.5	2777.0	19
20	5967.8	2543.4	6073.0	2619.6	6180.2	2697.9	6289.4	2778.3	20
21	5969.5	2544.7	6074.8	2620.9	6182.0	2699.2	6291.2	2779.7	21
22	5971.3	2545.9	6076.5	2622.2	6183.8	2700.6	6293.0	2781.0	22
23	5973.0	2547.2	6078.3	2623.5	6185.6	2701.9	6294.9	2782.4	23
24	5974.8	2548.5	6080.1	2624.8	6187.4	2703.2	6296.7	2783.8	24
25	5976.5	2549.7	6081.9	2626.1	6189.2	2704.5	6298.6	2785.1	25
26	5978.2	2551.0	6083.6	2627.4	6191.0	2705.9	6300.4	2786.5	26
27	5980.0	2552.2	6085.4	2628.7	6192.8	2707.2	6302.3	2787.8	27
28	5981.7	2553.5	6087.2	2630.0	6194.6	2708.5	6304.1	2789.2	28
29	5983.5	2554.7	6088.9	2631.2	6196.4	2709.8	6305.9	2790.6	29
30	5985.2	2556.0	6090.7	2632.5	6198.2	2711.2	6307.8	2791.9	30
31	5986.9	2557.3	6092.5	2633.8	6200.0	2712.5	6309.6	2793.3	31
32	5988.7	2558.5	6094.3	2635.1	6201.8	2713.8	6311.5	2794.7	32
33	5990.4	2559.8	6096.1	2636.4	6203.7	2715.2	6313.3	2796.0	33
34	5992.2	2561.0	6097.8	2637.7	6205.5	2716.5	6315.2	2797.4	34
35	5993.9	2562.3	6099.6	2639.0	6207.3	2717.8	6317.0	2798.8	35
36	5995.7	2563.6	6101.4	2640.3	6209.1	2719.1	6318.9	2800.1	36
37	5997.4	2564.8	6103.2	2641.6	6210.9	2720.5	6320.7	2801.5	37
38	5999.2	2566.1	6104.9	2642.9	6212.7	2721.8	6322.5	2802.9	38
39	6000.9	2567.3	6106.7	2644.2	6214.5	2723.1	6324.4	2804.2	39
40	6002.6	2568.6	6108.5	2645.5	6216.3	2724.5	6326.2	2805.6	40
41	6004.4	2569.9	6110.3	2646.8	6218.2	2725.8	6328.1	2807.0	41
42	6006.2	2571.1	6112.1	2648.1	6220.0	2727.1	6329.9	2808.4	42
43	6007.9	2572.4	6113.8	2649.4	6221.8	2728.5	6331.8	2809.7	43
44	6009.7	2573.7	6115.6	2650.7	6223.6	2729.8	6333.6	2811.1	44
45	6011.4	2574.9	6117.4	2652.0	6225.4	2731.2	6335.5	2812.5	45
46	6013.2	2576.2	6119.2	2653.3	6227.2	2732.5	6337.3	2813.8	46
47	6014.9	2577.5	6121.0	2654.6	6229.1	2733.8	6339.2	2815.2	47
48	6016.7	2578.7	6122.8	2655.9	6230.9	2735.2	6341.1	2816.6	48
49	6018.4	2580.0	6124.5	2657.2	6232.7	2736.5	6342.9	2818.0	49
50	6020.2	2581.3	6126.3	2658.5	6234.5	2737.8	6344.8	2819.3	50
51	6021.9	2582.6	6128.1	2659.8	6236.3	2739.2	6346.6	2820.7	51
52	6023.7	2583.8	6129.9	2661.1	6238.2	2740.5	6348.5	2822.1	52
53	6025.4	2585.1	6131.7	2662.4	6240.0	2741.9	6350.3	2823.5	53
54	6027.2	2586.4	6133.5	2663.7	6241.8	2743.2	6352.2	2824.9	54
55	6028.9	2587.6	6135.3	2665.0	6243.6	2744.6	6354.0	2826.2	55
56	6030.7	2588.9	6137.1	2666.4	6245.4	2745.9	6355.9	2827.6	56
57	6032.4	2590.2	6138.9	2667.7	6247.3	2747.2	6357.8	2829.0	57
58	6034.2	2591.5	6140.6	2669.0	6249.1	2748.6	6359.6	2830.4	58
59	6036.0	2592.7	6142.4	2670.3	6250.9	2749.9	6361.5	2831.8	59

Table A-4. (*Continued*)

′	Δ = 96°		Δ = 97°		Δ = 98°		Δ = 99°		′
	T	E	T	E	T	E	T	E	
0	6363.3	2833.1	6476.1	2917.3	6591.1	3003.7	6708.5	3092.6	0
1	6365.2	2834.5	6478.0	2918.7	6593.1	3005.2	6710.5	3094.2	1
2	6367.1	2835.9	6479.9	2920.1	6595.0	3006.7	6712.4	3095.7	2
3	6368.9	2837.3	6481.8	2921.5	6596.9	3008.1	6714.4	3097.2	3
4	6370.8	2838.7	6483.7	2923.0	6598.9	3009.6	6716.4	3098.7	4
5	6372.7	2840.1	6485.6	2924.4	6600.8	3011.1	6718.4	3100.2	5
6	6374.5	2841.5	6487.5	2925.8	6602.8	3012.5	6720.3	3101.7	6
7	6376.4	2842.8	6489.4	2927.2	6604.7	3014.0	6722.3	3103.2	7
8	6378.3	2844.2	6491.3	2928.7	6606.6	3015.5	6724.3	3104.7	8
9	6380.1	2845.6	6493.2	2930.1	6608.6	3016.9	6726.3	3106.2	9
10	6382.0	2847.0	6495.1	2931.5	6610.5	3018.4	6728.3	3107.7	10
11	6383.9	2848.4	6497.0	2933.0	6612.5	3019.9	6730.3	3109.2	11
12	6385.7	2849.8	6498.9	2934.4	6614.4	3021.3	6732.2	3110.7	12
13	6387.6	2851.2	6500.8	2935.8	6616.4	3022.8	6734.2	3112.2	13
14	6389.5	2852.6	6502.8	2937.2	6618.3	3024.3	6736.2	3113.8	14
15	6391.3	2854.0	6504.7	2938.7	6620.2	3025.8	6738.2	3115.3	15
16	6393.2	2855.4	6506.6	2940.1	6622.2	3027.2	6740.2	3116.8	16
17	6395.1	2856.8	6508.5	2941.5	6624.1	3028.7	6742.2	3118.3	17
18	6396.9	2858.1	6510.4	2943.0	6626.1	3030.2	6744.2	3119.8	18
19	6398.8	2859.5	6512.3	2944.4	6628.0	3031.6	6746.1	3121.3	19
20	6400.7	2860.9	6514.2	2945.8	6630.0	3033.1	6748.1	3122.8	20
21	6402.6	2862.3	6516.1	2947.3	6631.9	3034.6	6750.1	3124.4	21
22	6404.4	2863.7	6518.0	2948.7	6633.9	3036.1	6752.1	3125.9	22
23	6406.3	2865.1	6519.9	2950.1	6635.8	3037.5	6754.1	3127.4	23
24	6408.2	2866.5	6521.8	2951.6	6637.8	3039.0	6756.1	3128.9	24
25	6410.1	2867.9	6523.8	2953.0	6639.7	3040.5	6758.1	3130.4	25
26	6411.9	2869.3	6525.7	2954.5	6641.7	3042.0	6760.1	3132.0	26
27	6413.8	2870.7	6527.6	2955.9	6643.6	3043.5	6762.1	3133.5	27
28	6415.7	2872.1	6529.5	2957.3	6645.6	3044.9	6764.1	3135.0	28
29	6417.6	2873.5	6531.4	2958.8	6647.5	3046.4	6766.1	3136.5	29
30	6419.5	2874.9	6533.3	2960.2	6649.5	3047.9	6768.1	3138.0	30
31	6421.3	2876.3	6535.3	2961.7	6651.5	3049.4	6770.1	3139.6	31
32	6423.2	2877.7	6537.2	2963.1	6653.4	3050.9	6772.0	3141.1	32
33	6425.1	2879.1	6539.1	2964.5	6655.4	3052.3	6774.0	3142.6	33
34	6427.0	2880.5	6541.0	2966.0	6657.3	3053.8	6776.0	3144.1	34
35	6428.9	2881.9	6542.9	2967.4	6659.3	3055.3	6778.0	3145.7	35
36	6430.7	2883.4	6544.9	2968.9	6661.3	3056.8	6780.0	3147.2	36
37	6432.6	2884.8	6546.8	2970.3	6663.2	3058.3	6782.0	3148.7	37
38	6434.5	2886.2	6548.7	2971.8	6665.2	3059.8	6784.1	3150.3	38
39	6436.4	2887.6	6550.6	2973.2	6667.1	3061.3	6786.0	3151.8	39
40	6438.3	2889.0	6552.5	2974.7	6669.1	3062.7	6788.0	3153.3	40
41	6440.2	2890.4	6554.5	2976.1	6671.1	3064.2	6790.1	3154.8	41
42	6442.1	2891.8	6556.4	2977.6	6673.0	3065.7	6792.1	3156.4	42
43	6443.9	2893.2	6558.3	2979.0	6675.0	3067.2	6794.1	3157.9	43
44	6445.8	2894.6	6560.2	2980.5	6677.0	3068.7	6796.1	3159.4	44
45	6447.7	2896.0	6562.2	2981.9	6678.9	3070.2	6798.1	3161.0	45
46	6449.6	2897.4	6564.1	2983.4	6680.9	3071.7	6800.1	3162.5	46
47	6451.5	2898.9	6566.0	2984.8	6682.9	3073.2	6802.1	3164.0	47
48	6453.4	2900.3	6567.9	2986.3	6684.8	3074.7	6804.1	3165.6	48
49	6455.3	2901.7	6569.9	2987.7	6686.8	3076.2	6806.1	3167.1	49
50	6457.2	2903.1	6571.8	2989.2	6688.8	3077.7	6808.1	3168.6	50
51	6459.1	2904.5	6573.7	2990.6	6690.7	3079.2	6810.1	3170.2	51
52	6461.0	2905.9	6575.7	2992.1	6692.7	3080.7	6812.1	3171.7	52
53	6462.8	2907.3	6577.6	2993.5	6694.7	3082.2	6814.2	3173.3	53
54	6464.7	2908.8	6579.5	2995.0	6696.6	3083.7	6816.2	3174.8	54
55	6466.6	2910.2	6581.5	2996.5	6698.6	3085.2	6818.2	3176.4	55
56	6468.5	2911.6	6583.4	2997.9	6700.6	3086.7	6820.2	3177.9	56
57	6470.4	2913.0	6585.3	2999.4	6702.6	3088.2	6822.2	3179.4	57
58	6472.3	2914.4	6587.3	3000.8	6704.5	3089.7	6824.2	3181.0	58
59	6474.2	2915.9	6589.2	3002.3	6706.5	3091.2	6826.2	3182.5	59

Table A-4. (*Continued*)

′	Δ = 100°		Δ = 101°		Δ = 102°		Δ = 103°		′
	T	E	T	E	T	E	T	E	
0	6828.2	3184.1	6950.5	3278.1	7075.4	3374.8	7203.1	3474.3	0
10	6848.4	3199.6	6971.2	3294.0	7096.5	3391.2	7224.6	3491.2	10
20	6868.7	3215.1	6991.9	3310.0	7117.7	3407.7	7246.2	3508.2	20
30	6889.1	3230.7	7012.7	3326.1	7138.9	3424.2	7267.9	3525.2	30
40	6909.5	3246.4	7033.5	3342.3	7160.2	3440.8	7289.7	3542.3	40
50	6930.0	3262.2	7054.4	3358.5	7181.6	3457.6	7311.6	3559.5	50
60	6950.5	3278.1	7075.4	3374.8	7203.1	3474.3	7333.5	3576.8	60

′	Δ = 104°		Δ = 105°		Δ = 106°		Δ = 107°		′
	T	E	T	E	T	E	T	E	
0	7333.5	3576.8	7466.9	3682.3	7603.4	3790.9	7743.1	3902.8	0
10	7355.6	3594.2	7489.5	3700.2	7626.5	3809.3	7766.7	3921.8	10
20	7377.7	3611.6	7512.1	3718.1	7649.6	3827.9	7790.4	3940.9	20
30	7399.9	3629.2	7534.8	3736.2	7672.8	3846.5	7814.2	3960.1	30
40	7422.1	3646.8	7557.6	3754.3	7696.2	3865.2	7838.0	3979.3	40
50	7444.5	3664.5	7580.4	3772.6	7719.6	3883.9	7862.0	3998.7	50
60	7466.9	3682.3	7603.4	3790.9	7743.1	3902.8	7886.1	4018.2	60

′	Δ = 108°		Δ = 109°		Δ = 110°		Δ = 111°		′
	T	E	T	E	T	E	T	E	
0	7886.1	4018.2	8032.6	4137.0	8182.7	4259.6	8336.6	4386.1	0
10	7910.3	4037.7	8057.3	4157.2	8208.1	4280.4	8362.6	4407.6	10
20	7934.5	4057.4	8082.2	4177.5	8233.6	4301.4	8388.8	4429.1	20
30	7958.9	4077.2	8107.2	4197.9	8259.2	4322.4	8415.0	4450.8	30
40	7983.3	4097.0	8132.2	4218.4	8284.9	4343.5	8441.4	4472.6	40
50	8007.9	4117.0	8157.4	4238.9	8310.7	4364.7	8467.9	4494.5	50
60	8032.6	4137.0	8182.7	4259.6	8336.6	4386.1	8494.4	4516.6	60

′	Δ = 112°		Δ = 113°		Δ = 114°		Δ = 115°		′
	T	E	T	E	T	E	T	E	
0	8494.4	4516.6	8656.4	4651.3	8822.8	4790.4	8993.6	4934.1	0
10	8521.2	4538.7	8683.9	4674.1	8850.9	4814.0	9022.6	4958.5	10
20	8548.0	4561.0	8711.4	4697.1	8879.2	4837.8	9051.6	4983.0	20
30	8574.9	4583.4	8739.1	4720.3	8907.6	4861.6	9080.8	5007.7	30
40	8602.0	4605.9	8766.8	4743.5	8936.2	4885.7	9110.2	5032.5	40
50	8629.2	4628.5	8794.7	4766.9	8964.8	4909.8	9139.6	5057.5	50
60	8656.4	4651.3	8822.8	4790.4	8993.6	4934.1	9169.2	5082.6	60

′	Δ = 116°		Δ = 117°		Δ = 118°		Δ = 119°		′
	T	E	T	E	T	E	T	E	
0	9169.2	5082.6	9349.8	5236.2	9535.6	5395.0	9726.9	5559.4	0
10	9199.0	5107.8	9380.4	5262.3	9567.1	5422.0	9759.3	5587.3	10
20	9228.9	5133.2	9411.2	5288.5	9598.8	5449.2	9791.9	5615.5	20
30	9258.9	5158.7	9442.1	5314.9	9630.6	5476.5	9824.7	5643.7	30
40	9289.1	5184.4	9473.1	5341.4	9662.5	5504.0	9857.6	5672.2	40
50	9319.4	5210.2	9504.3	5368.1	9694.6	5531.6	9890.7	5700.8	50
60	9349.8	5236.2	9535.6	5395.0	9726.9	5559.4	9923.9	5729.6	60

Table A-5. Direction of Angle α and Distance r from PI to First (or Third) Quarter Point of a Circular Curve

Δ	α	r (ft)	Δ	α	r (ft)	Δ	α	r (ft)
10°	1°14.6′	251.4	47°	5°10.8′	1330	84°	6°59.2′	3129
11	1 21.9	276.9	48	5 15.6	1366	85	6 59.9	3197
12	1 29.3	302.4	49	5 20.4	1402	86	7 00.5	3267
13	1 36.6	328.1	50	5 25.1	1438	87	7 01.0	3339
14	1 43.8	353.9	51	5 29.7	1475	88	7 01.4	3412
15	1 51.1	379.8	52	5 34.2	1513	89	7 01.6	3487
16	1 58.3	405.8	53	5 38.6	1551	90	7 01.7	3564
17	2 05.4	432.0	54	5 42.9	1590	91	7 01.8	3643
18	2 12.5	458.3	55	5 47.1	1629	92	7 01.7	3723
19	2 19.6	484.7	56	5 51.2	1669	93	7 01.5	3805
20	2 26.6	511.4	57	5 55.1	1710	94	7 01.2	3889
21	2 33.6	538.2	58	5 59.0	1751	95	7 00.8	3975
22	2 40.5	565.2	59	6 02.8	1793	96	7 00.2	4063
23	2 47.4	592.4	60	6 06.4	1836	97	6 59.5	4154
24	2 54.2	619.8	61	6 09.9	1879	98	6 58.7	4247
25	3 00.9	647.4	62	6 13.3	1923	99	6 57.8	4342
26	3 07.6	675.2	63	6 16.6	1968	100	6 56.7	4439
27	3 14.2	703.2	64	6 19.8	2013	101	6 55.5	4539
28	3 20.8	731.5	65	6 22.9	2060	102	6 54.3	4642
29	3 27.3	760.1	66	6 25.8	2107	103	6 52.9	4748
30	3 33.7	788.9	67	6 28.6	2155	104	6 51.4	4856
31	3 40.1	818.0	68	6 31.4	2204	105	6 49.8	4968
32	3 46.4	847.3	69	6 34.0	2253	106	6 48.1	5083
33	3 52.6	877.0	70	6 36.5	2304	107	6 46.3	5201
34	3 58.7	907.0	71	6 38.9	2356	108	6 44.4	5322
35	4 04.8	937.3	72	6 41.2	2409	109	6 42.3	5447
36	4 10.8	967.9	73	6 43.3	2462	110	6 40.1	5575
37	4 16.7	998.9	74	6 45.3	2517	111	6 37.8	5707
38	4 22.5	1030.2	75	6 47.2	2573	112	6 35.4	5843
39	4 28.2	1061.9	76	6 49.1	2630	113	6 32.9	5984
40	4 33.8	1094.0	77	6 50.8	2688	114	6 30.3	6129
41	4 39.3	1126.4	78	6 52.3	2747	115	6 27.6	6278
42	4 44.8	1159.2	79	6 53.7	2807	116	6 24.8	6432
43	4 50.2	1192.5	80	6 55.1	2869	117	6 21.9	6591
44	4 55.5	1226.2	81	6 56.3	2932	118	6 18.8	6755
45	5 00.7	1260 3	82	6 57 4	2996	119	6 15.6	6925
46	5 05.8	1294.9	83	6 58.3	3062	120	6 12.4	7101

Appendix B
Instructions for Using Computer Programs

INTRODUCTION

The software provided on the diskette* consists of a set of programs which simplify performing some of the calculations discussed in Chapters 1 and 2. The programs, collectively called the Computer Solutions Programs (CSP), were developed on an Apple IIc computer with 128K bytes of memory and the ProDOS operating system. CSP is written in Applesoft BASIC. A printer, although not required, is useful for obtaining a paper record of your computer session.

The speed and flexibility of the CSP allows you to easily perform calculations and solve problems that otherwise would require a tedious manual process. Solutions of triangles and area and volume calculations of the type shown in Chapters 1 and 2 can be computationally intensive and error-prone. Use of a computer-based system like CSP ensures that results have a high degree of accuracy.

These instructions assume that you are familiar with the basic operation of your computer; that is, you can power-up and power-down, can insert and remove diskettes, and know the functions of the keys on the keyboard. Refer to your computer manuals for this type of information.

PROGRAM DESIGN PHILOSOPHY

The CSP has been designed to ensure ease of operation for all users regardless of the level of their computer sophistication. The programs are completely menu-driven, so that someone with little or no previous experience with automated systems can successfully conduct a CSP session on their first attempt.

The main menu is displayed upon entry to the CSP. Major options are selected from this menu. Once a menu item is selected, either a submenu appears listing additional options or the user is prompted for more input. Often, choices are presented which require selecting one option from a list of two or more. For example, in finding the area of a regular polygon you must specify whether the length of the radius or apothem is given; the computer displays the following:

---

*Disk is available free of charge to instructors using the book. It can be obtained by writing to the Engineering Editor, Van Nostrand Reinhold, 115 Fifth Avenue, New York, NY 10003.

WHICH VALUE IS KNOWN:
1 RADIUS
2 APOTHEM
ENTER 1 or 2

In this way the various logical paths within the program are traversed.

If at any time you enter a set of values which would result in an indeterminate solution, then an appropriate error message is given. For example, suppose the sides of a triangle are entered as 10, 20, and 40. Clearly, this is an impossible situation (in any triangle the longest side must be less than the sum of the other two sides). The program will detect the error and display an informative message with instructions about what to do next.

PROGRAM OPTIONS

Figure B-1 shows the main menu as it appears upon entering the CSP. This section briefly indicates the function of each of the menu items.

1. Solution of right triangles
   Given 2 parts (one part must be a side) of a right triangle this option computes the remaining parts. Begin by entering a number (1 through 9) corresponding to

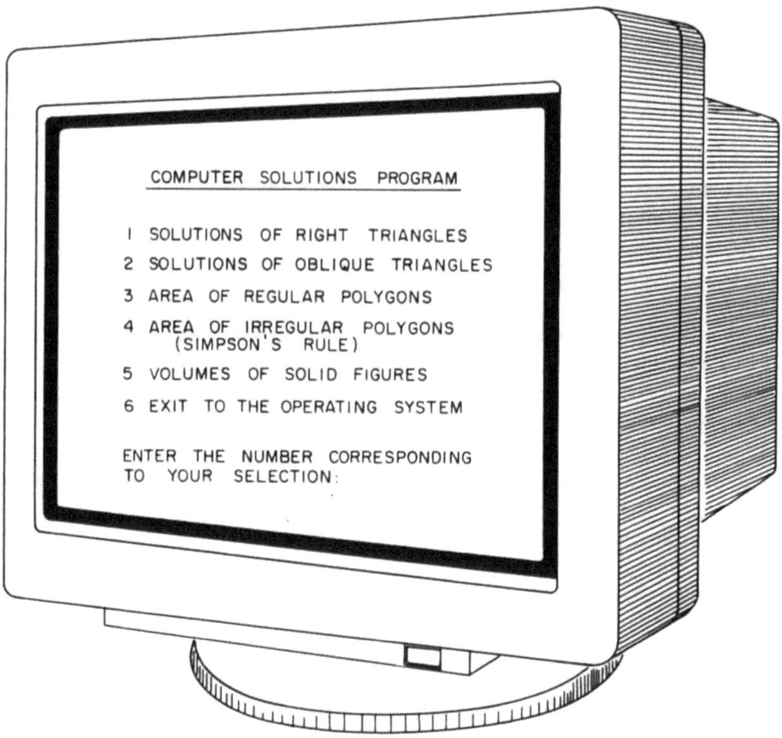

Fig. B-1. The main menu as it appears upon entering the CSP.

the known parts of the triangle shown in the GIVEN column. Then enter values when prompted.

2. Solution of oblique triangles
   This option provides capabilities for solving oblique triangles. Oblique triangles have six parts, three sides and three angles. If any three parts are known, at least one being a side, the remaining three can be determined. There are four possible arrangements of these parts:

CASE 1	Two angles and one side known
	Condition—The sum of the angles must be less than 180°
CASE 2	Two sides and the included angle known
	Condition—The angle must be less than 180°
CASE 3	Two sides and the angle opposite one of them known
CASE 4	Three sides known
	Condition—The longest side must be less than the sum of the other two sides

3. Area of regular polygons
   In this option enter the length of a side, the number of sides, and the length of the radius or apothem. The total area of the polygon is then computed.
4. Area of irregular polygons (Simpson's Rule)
   Follow the instructions given by the on-screen prompts to compute the area within a closed polygonal boundary. Remember that the number of offsets must be odd to produce an even number of equal strips.
5. Volumes of solid figures
   If this menu item is selected then another menu is displayed offering a choice of three types of volume calculations:
   1. Figures with constant cross-sectional area
   2. Figures with changing cross-sectional area
      a. Average the areas of the two end sections
      b. Prismoidal formula

   Select the desired item and enter data when prompted.
6. Exit to the operating system
   Select this option to terminate the CSP and return to the operating system.

BACKUP YOUR CSP DISKETTE

The diskette contains the CSP programs. To prevent loss of the CSP in the event the diskette becomes damaged you should make a backup copy before proceeding. Refer to your computer manuals for instructions on how to make a copy of a diskette. After you have created a backup copy, store the original in a safe place and use the copy for all operations. Be sure to place a write protect tab on both diskettes.

GETTING STARTED

To initiate a CSP session perform the following steps:

1. Bring up BASIC on your computer just as you normally do.

2. Insert the backup copy of the CSP diskette into the drive.
3. Type RUN CSP, then press the return key.

You are now in the program and are ready to begin. Simply select a menu item and follow the instructions as they appear on your screen.

When prompted for data, enter all lengths (use any linear measure) with a maximum of two digits to the right of the decimal. Do not enter the units. Similarly, enter all angles in degrees and, if necessary, in hundredths of degrees.

Examples

756.345 yards	enter 756.34
21.2 meters	enter 21.2
63 feet	enter 63
45°40′	enter 45.67

Computer Solutions Programs

1. Solutions of right triangles
2. Solutions of oblique triangles
3. Area of regular polygons
4. Area of irregular polygons (Simpson's Rule)
5. Volumes of solid figures
6. Exit to the operating system

Enter the number corresponding to your selection.

Appendix C
Tables

Table C-1. Trigonometric Formulas (Right Triangles).

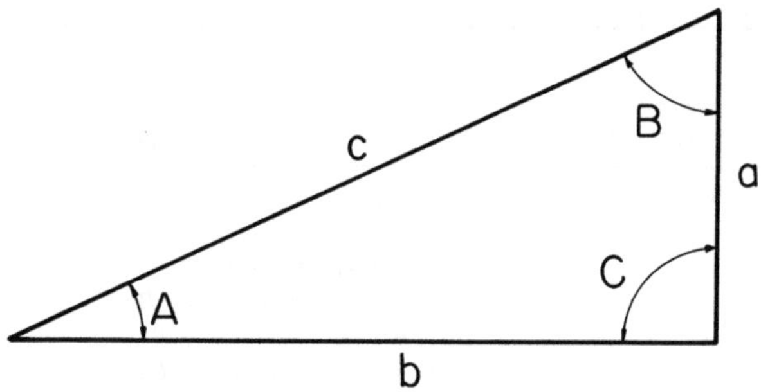

Given	To find	Equations		
ab	A	$\tan A = a/b$	or	$\cot A = b/a$
ab	B	$\cot B = a/b$	or	$\tan B = b/a$
ac	A	$\sin A = a/c$	or	$\operatorname{cosec} A = c/a$
ac	B	$\cos B = a/c$	or	$\sec B = c/a$
bc	A	$\sec A = c/b$	or	$\cos A = b/c$
bc	B	$\operatorname{cosec} B = c/b$	or	$\sin B = b/c$
Aa	b	$b = a \cot A$	or	$b = a/\tan A$
Aa	c	$c = a \operatorname{cosec} A$	or	$c = a/\sin A$
Ab	a	$a = b \tan A$	or	$a = b/\cot A$
Ab	c	$c = b \sec A$	or	$C = b/\cos A$
Ac	a	$a = c \sin A$	or	$a = c/\operatorname{cosec} A$
Ac	b	$b = c \cos A$	or	$b = c/\sec A$

Table C-2. Trigonometric Formulas (Oblique Triangles).

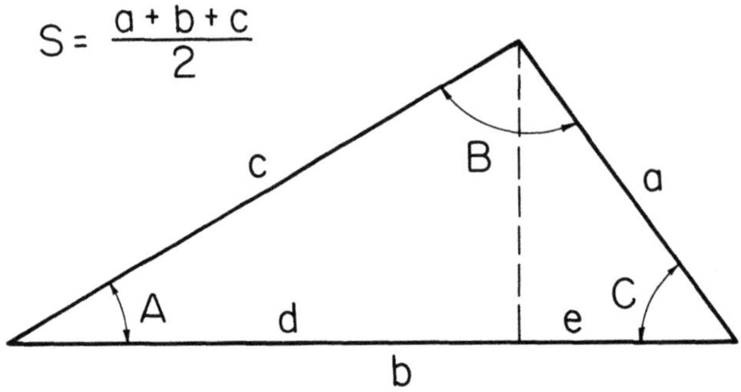

Given	To Find	Equations
ABa	*b*	$b = a \sin B/\sin A$
ABa	*c*	$c = a \sin (A + B)/\sin A$
ABb	*a*	$a = B \sin A/\sin B$
ABb	*c*	$c = b \sin (A + B)/\sin B$
ABc	*a*	$a = c \sin A/\sin (A + B)$
ABc	*b*	$b = c \sin B/\sin (A + B)$
ACa	*b*	$b = a \sin (A + C)/\sin A$
ACa	*c*	$c = a \sin C/\sin A$
ACb	*a*	$a = b \sin A/\sin (A + C)$
ACb	*c*	$c = b \sin C/\sin (A + C)$
ACc	*a*	$a = c \sin A/\sin C$
ACc	*b*	$b = c \sin (A + C)/\sin C$
BCa	*b*	$b = a \sin B/\sin (B + C)$
BCa	*c*	$c = a \sin C/\sin (B + C)$
BCb	*a*	$a = b \sin (B + C)/\sin B$
BCb	*c*	$c = b \sin C/\sin B$
BCc	*a*	$a = c \sin (B + C)/\sin C$
BCc	*b*	$b = c \sin B/\sin C$
abc	*S*	$S = (a + b + c)/2$
abcs	*A*	$\sin \frac{1}{2}A = \sqrt{(s - b)(s - c)/bc}$
abcs	*A*	$\cos \frac{1}{2}A = \sqrt{s(s - a)/bc}$
abcs	*A*	$\tan \frac{1}{2}A = \sqrt{(s - b)(s - c)/s(s - a)}$
abcs	*B*	$\sin \frac{1}{2}B = \sqrt{(s - a)(s - c)/ac}$
abcs	*B*	$\tan \frac{1}{2}B = \sqrt{s(s - b)/ac}$
abcs	*B*	$\cos \frac{1}{2}B = \sqrt{(s - a)(s - c)/s(s - b)}$
abcs	*C*	$\sin \frac{1}{2}C = \sqrt{(s - a)(s - b)/ab}$
abcs	*C*	$\cos \frac{1}{2}C = \sqrt{s(s - c)/ab}$
abcs	*C*	$\tan \frac{1}{2}C = \sqrt{(s - a)(s - b)/s(s - c)}$
abcs	*d*	$d = (b^2 + c^2 - a^2)/2b$
abcs	*e*	$e = (a^2 + b^2 - c^2)/2b$
Aab	*B*	$\sin B = b \sin A/a$
Aac	*c*	$\sin C = c \sin A/a$
Bab	*A*	$\sin A = a \sin B/b$

Table C-2. (*Continued*)

Given	To Find	Equations
Bbc	C	$\sin C = c \sin B/b$
Cac	A	$\sin A = a \sin C/c$
Cbc	B	$\sin B = b \sin C/c$
Abc	$\frac{1}{2}(B + C)$	$\frac{1}{2}(B + C) = 90° - \frac{1}{2}A$
Abc	$\frac{1}{2}(B - C)$	$\tan \frac{1}{2}(B - C) = [(b - c) \tan (90° - \frac{1}{2}A)]/(b + c)$
Abc	B	$B = \frac{1}{2}(B + C) + \frac{1}{2}(B - C)$
Abc	C	$C = \frac{1}{2}(B + C) - \frac{1}{2}(B - C)$
Abc	a	$a = \sqrt{b^2 + c^2 - 2bc \cos A}$
Bac	$\frac{1}{2}(A + C)$	$\frac{1}{2}(A + C) = 90° - \frac{1}{2}B$
Bac	$\frac{1}{2}(A - C)$	$\tan \frac{1}{2}(A - C) = [(a - c) \tan (90° - \frac{1}{2}B)]/(a + c)$
Bac	A	$A = \frac{1}{2}(A + C) + \frac{1}{2}(A - C)$
Bac	C	$C = \frac{1}{2}(A + C) - \frac{1}{2}(A - C)$
Bac	b	$b = \sqrt{a^2 + c^2 - 2ac \cos B}$
Cab	$\frac{1}{2}(A + B)$	$\frac{1}{2}(A + B) = 90° - \frac{1}{2}C$
Cab	$\frac{1}{2}(A - B)$	$\tan \frac{1}{2}(A - B) = [(a - b) \tan (90° - \frac{1}{2}C)]/(a + b)$
Cab	A	$A = \frac{1}{2}(A + B) + \frac{1}{2}(A - B)$
Cab	B	$B = \frac{1}{2}(A + B) - \frac{1}{2}(-B)$
Cab	c	$c = \sqrt{a^2 + b^2 - 2ab \cos C}$

Appendix D
Tables

Table D-1. Converting Inches into Decimals of a Foot

Inches		Decimal of a Ft.	Inches		Decimal of a Ft.	Inches		Decimal of a Ft.
	1/16	.005208		1/16	.338542		1/16	.671875
	1/8	.010416		1/8	.343750		1/8	.677083
	3/16	.015625		3/16	.348958		3/16	.682292
	1/4	.020833		1/4	.354166		1/4	.687500
	5/16	.026042		5/16	.359375		5/16	.692708
	3/8	.031250		3/8	.364583		3/8	.697916
	7/16	.036458	4"	7/16	.369792	8"	7/16	.703125
0"	1/2	:041666	.33333	1/2	.375000	.66666	1/2	.708333
	9/16	.046875		9/16	.380208		9/16	.713542
	5/8	.052083		5/8	.385416		5/8	.718750
	11/16	.057292		11/16	.390625		11/16	.723958
	3/4	.062500		3/4	.395833		3/4	.729166
	13/16	.067708		13/16	.401042		13/16	.734375
	7/8	.072916		7/8	.406250		7/8	.739583
	15/16	.078125		15/16	.411458		15/16	.744792
	1/16	.088542		1/16	.421875		1/16	.755208
	1/8	.093750		1/8	.427083		1/8	.760416
	3/16	.098958		3/16	.432292		3/16	.765625
	1/4	.104166		1/4	.437500		1/4	.770833
	5/16	.109375		5/16	.442708		5/16	.776042
	3/8	.114583		3/8	.447916		3/8	.781250
	7/16	.119792		7/16	.453125		7/16	.786458
1"	1/2	.125000	5"	1/2	.458333	9"	1/2	.791666
.083333	9/16	.130208	.416666	9/16	.463542	.750	9/16	.796875
	5/8	.135416		5/8	.468750		5/8	.802083
	11/16	.140625		11/16	.473958		11/16	.807292
	3/4	.145833		3/4	.479166		3/4	.812500
	13/16	.151042		13/16	.484375		13/16	.817708
	7/8	.156250		7/8	.489583		7/8	.822916
	15/16	.161458		15/16	.494792		15/16	.828125
	1/16	.171875		1/16	.505208		1/16	.838542
	1/8	.177083		1/8	.510416		1/8	.843750
	3/16	.182292		3/16	.515625		3/16	.848958
	1/4	.187500		1/4	.520833		1/4	.854166
	5/16	.192708		5/16	.526042		5/16	.859375
	3/8	.197916		3/8	.531250		3/8	.864583
	7/16	.203125		7/16	.536458		7/16	.869792
2"	1/2	.208333	6"	1/2	.541666	10"	1/2	.875000
.166666	9/16	.213542	.50	9/16	.546875	.833333	9/16	.880208
	5/8	.218750		5/8	.552083		5/8	.885416
	11/16	.223958		11/16	.557292		11/16	.890625
	3/4	.229166		3/4	.562500		3/4	.895833
	13/16	.234375		13/16	.567708		13/16	.901042
	7/8	.239583		7/8	.572916		7/8	.906250
	15/16	.244792		15/16	.578125		15/16	.911458

Table D-1. (*Continued*)

	1/16	.255208		1/16	.588542		1/16	.921875
	1/8	.260416		1/8	.593750		1/8	.927083
	3/16	.265625		3/16	.598958		3/16	.932292
	1/4	.270833		1/4	.604166		1/4	.937500
	5/16	.276042		5/16	.609375		5/16	.942708
	3/8	.281250		3/8	.614583		3/8	.947916
	7/16	.286458		7/16	.619792		7/16	.953125
3″	1/2	.291666	**7″**	1/2	.625000	**11″**	1/2	.958333
.250	9/16	.296875	**.583333**	9/16	.630208	**.916666**	9/16	.963542
	5/8	.302083		5/8	.635416		5/8	.968750
	11/16	.307292		11/16	.640625		11/16	.973958
	3/4	.312500		3/4	.645833		3/4	.979166
	13/16	.317708		13/16	.651042		13/16	.984375
	7/8	.322916		7/8	.656250		7/8	.989593
	15/16	.328125		15/16	.661458		15/16	.994792

Table D-2. Minutes of Arc in Decimals of a Degree

′	0″	10″	15″	20″	30″	40″	45″	50″	′
0	.00000	.00278	.00417	.00556	.00833	.01111	.01250	.01389	0
1	.01667	.01944	.02083	.02222	.02500	.02778	.02917	.03055	1
2	.03333	.03611	.03750	.03889	.04167	.04444	.04583	.04722	2
3	.05000	.05278	.05417	.05556	.05833	.06111	.06250	.06389	3
4	.06667	.06944	.07083	.07222	.07500	.07778	.07917	.08056	4
5	.08333	.08611	.08750	.08889	.09167	.09444	.09583	.09722	5
6	.10000	.10278	.10417	.10556	.10833	.11111	.11250	.11389	6
7	.11667	.11944	.12083	.12222	.12500	.12778	.12917	.13056	7
8	.13333	.13611	.13750	.13889	.14167	.14444	.14583	.14722	8
9	.15000	.15278	.15417	.15556	.15833	.16111	.16250	.16389	9
10	.16667	.16944	.17083	.17222	.17500	.17778	.17917	.18056	10
11	.18333	.18611	.18750	.18889	.19167	.19444	.19583	.19722	11
12	.20000	.20278	.20417	.20556	.20833	.21111	.21250	.21389	12
13	.21667	.21944	.22083	.22222	.22500	.22778	.22917	.23056	13
14	.23333	.23611	.23750	.23889	.24167	.24444	.24583	.24722	14
15	.25000	.25278	.25417	.25556	.25833	.26111	.26250	.26389	15
16	.26667	.26944	.27083	.27222	.27500	.27778	.27917	.28056	16
17	.28333	.28611	.28750	.28889	.29167	.29444	.29583	.29722	17
18	.30000	.30278	.30417	.30556	.30833	.31111	.31250	.31389	18
19	.31667	.31944	.32083	.32222	.32500	.32778	.32917	.33056	19
20	.33333	.33611	.33750	.33889	.34167	.34444	.34583	.34722	20
21	.35000	.35278	.35417	.35556	.35833	.36111	.36250	.36389	21
22	.36667	.36944	.37083	.37222	.37500	.37778	.37917	.38056	22
23	.38333	.38611	.38750	.38889	.39167	.39444	.39583	.39722	23
24	.40000	.40278	.40417	.40556	.40833	.41111	.41250	.41389	24
25	.41667	.41944	.42083	.42222	.42500	.42778	.42917	.43056	25
26	.43333	.43611	.43750	.43889	.44167	.44444	.44583	.44722	26
27	.45000	.45278	.45417	.45556	.45833	.46111	.46250	.46389	27
28	.46667	.46944	.47083	.47222	.47500	.47778	.47917	.48056	28
29	.48333	.48611	.48750	.48889	.49167	.49444	.49583	.49722	29
30	.50000	.50278	.50417	.50556	.50833	.51111	.51250	.51389	30
31	.51667	.51944	.52083	.52222	.52500	.52778	.52917	.53056	31
32	.53333	.53611	.53750	.53889	.54167	.54444	.54583	.54722	32
33	.55000	.55278	.55417	.55556	.55833	.56111	.56250	.56389	33
34	.56667	.56944	.57083	.57222	.57500	.57778	.57917	.58056	34
35	.58333	.58611	.58750	.58889	.59167	.59444	.59583	.59722	35
36	.60000	.60278	.60417	.60556	.60833	.61111	.61250	.61389	36
37	.61667	.61944	.62083	.62222	.62500	.62778	.62917	.63056	37
38	.63333	.63611	.63750	.63889	.64167	.64444	.64583	.64722	38
39	.65000	.65278	.65417	.65556	.65833	.66111	.66250	.66389	39
40	.66667	.66944	.67083	.67222	.67500	.67778	.67917	.68056	40
41	.68333	.68611	.68750	.68889	.69167	.69444	.69583	.69722	41
42	.70000	.70278	.70417	.70556	.70833	.71111	.71250	.71389	42
43	.71667	.71944	.72083	.72222	.72500	.72778	.72917	.73056	43
44	.73333	.73611	.73750	.73889	.74167	.74444	.74583	.74722	44
45	.75000	.75278	.75417	.75556	.75833	.76111	.76250	.76389	45
46	.76667	.76944	.77083	.77222	.77500	.77778	.77917	.78056	46
47	.78333	.78611	.78750	.78889	.79167	.79444	.79583	.79722	47
48	.80000	.80278	.80417	.80556	.80833	.81111	.81250	.81389	48
49	.81667	.81944	.82083	.82222	.82500	.82778	.82917	.83056	49
50	.83333	.83611	.83750	.83889	.84167	.84444	.84583	.84722	50
51	.85000	.85278	.85417	.85556	.85833	.86111	.86250	.86389	51
52	.86667	.86944	.87083	.87222	.87500	.87778	.87917	.88056	52
53	.88333	.88611	.88750	.88889	.89167	.89444	.89583	.89722	53
54	.90000	.90278	.90417	.90556	.90833	.91111	.91250	.91389	54
55	.91667	.91944	.92083	.92222	.92500	.92778	.92917	.93056	55
56	.93333	.93611	.93750	.93889	.94167	.94444	.94583	.94722	56
57	.95000	.95278	.95417	.95556	.95833	.96111	.96250	.96389	57
58	.96667	.96944	.97083	.97222	.97500	.97778	.97917	.98056	58
59	.98333	.98611	.98750	.98889	.99167	.99444	.99583	.99722	59
′	0″	10″	15″	20″	30″	40″	45″	50″	′

Table D-3. Decimal Equivalents of Inch Fractions.

Inches	Decimals	Inches	Decimals
1/16	0.0625	9/16	0.5625
1/8	0.1250	5/8	0.6250
3/16	0.1875	11/16	0.6875
1/4	0.2500	3/4	0.7500
5/16	0.3125	13/16	0.8125
3/8	0.3750	7/8	0.8750
7/16	0.4375	15/16	0.9375
1/2	0.5000	1	1.0000

Table D-4. Ground Distances at Various Map Scales.

Scale	1 inch	1 cm equals
1:500	500 inches	500 cm
	41.67 ft	16.40 ft
	12.70 meters	5.00 meters
1:1000	1000 inches	1000 cm
	83.33 ft	32.81 ft
	25.40 meters	10.00 meters
1:1250	1250 inches	1250 cm
	104.17 ft	41.01 ft
	31.70 meters	12.50 meters
1:2000	2000 inches	2000 cm
	166.67 ft	65.52 ft
	50.80 meters	20.00 meters
1:2500	2500 inches	2500 cm
	208.33 ft	82.02 ft
	63.50 meters	25.00 meters
1:5000	5000 inches	5000 cm
	416.67 ft	164.00 ft
	127.00 meters	50.00 meters
1:10,000	10,000 inches	10,000 cm
	833.33 ft	328.08 ft
	254.00 meters	100.00 meters
1:12,500	12,500 inches	12,500 cm
	1041.66 ft	410.10 ft
	317.00 meters	125.00 meters

Table D-4. (*Continued*)

Scale	1 inch	1 cm equals
1:20,000	20,000 inches 1666.67 ft 508.00 meters	20,000 cm 656.20 ft 200.00 meters
1:25,000	25,000 inches 2083.3 ft 635.0 meters	25,000 cm 820.2 ft 250.0 meters
1:50,000	50,000 inches 4166.7 ft 1270.0 meters	50,000 cm 1640.4 ft 500.0 meters
1:63,360	63,360 inches 5280.0 ft 1609.3 meters	63,360 cm 2078.7 ft 633.6 meters
1:100,000	100,000 inches 8333.3 ft 2540.0 meters	100,000 cm 3280.8 ft 1000.0 meters
1:250,000	250,000 inches 20833. ft 6350.0 meters	250,000 cm 8202.0 ft 2500 meters
1:500,000	500,000 inches 41667 ft 12700 ft	500,000 cm 16404 ft 5000 meters

Table D-5. Conversion Factors.

To convert	Into	Multiply by
acres	sq chain (gunters)	10
acres	rods	160
acres	sq links (gunters)	1×10^5
acres	hectare or sq hectometer	0.4047
acres	sq feet	43,560.0
acres	sq meters	4,047.0
acres	sq miles	1.562×10^{-3}
acres	sq yards	4,840.0
acre-feet	cubic feet	43,560.0
acre-feet	gallons	3.259×10^5
barrels (U.S. dry)	cubic inches	7,056.0
barrels (U.S. dry)	quarts (dry)	105.0
barrels (U.S. liq)	gallons	31.5
barrels (oil)	gallons (oil)	42.0

Table D-5. (*Continued*)

To convert	Into	Multiply by
centimeters	feet	3.281×10^{-2}
centimeters	inches	0.3937
centimeters	kilometers	1×10^{-5}
centimeters	meters	0.01
centimeters	miles	6.214×10^{-6}
centimeters	millimeters	10.0
centimeters	mils	393.7
centimeters	yards	$1{,}094 \times 10^{-2}$
chains	inches	792.00
chains	meters	20.12
chains (Gunters/surveyors)	yards	22.00
circumference	radians	6.283
circular mils	sq inches	7.854×10^{-7}
cubic centimeters	cubic feet	3.531×10^{-5}
cubic centimeters	cubic inches	0.06102
cubic centimeters	cubic meters	1×10^{-6}
cubic centimeters	cubic yards	1.308×10^{-6}
cubic centimeters	gallons (U.S. liquid)	2.642×10^{-4}
cubic centimeters	liters	0.001
cubic centimeters	pints (U.S. liquid)	2.113×10^{-3}
cubic centimeters	quarts (U.S. liquid)	1.057×10^{-3}
cubic feet	bushels (dry)	0.8036
cubic feet	cubic centimeters	28,320.0
cubic feet	cubic inches	1,728.0
cubic feet	cubic meters	0.02832
cubic feet	cubic yards	0.03704
cubic feet	gallons (U.S. liquid)	7.48052
cubic feet	quarts (U.S. liquid)	29.92
cubic feet	liters	28.32
cubic feet	pints (U.S. liquid)	59.84
cubic inches	cubic centimeters	16.39
cubic inches	cubic feet	5.787×10^{-4}
cubic inches	cubic meters	1.639×10^{-5}
cubic inches	cubic yards	2.143×10^{-5}
cubic inches	gallons	4.329×10^{-3}
cubic inches	liters	0.01639
cubic inches	mil-feet	1.061×10^{5}
cubic inches	pints (U.S. liquid)	0.03463
cubic inches	quarts (U.S. liquid)	0.01732
cubic meters	bushels (dry)	28.38
cubic meters	cubic centimeters	1×10^{6}
cubic meters	cubic feet	35.31
cubic meters	cubic inches	61,023.0
cubic meters	cubic yards	1.308
cubic meters	gallons (U.S. liquid)	264.2
cubic meters	liters	1,000.0
cubic meters	pints (U.S. liquid)	2,113.0

Table D-5. (*Continued*)

To convert	Into	Multiply by
cubic meters	quarts (U.S. liquid)	1,057.0
cubic yards	cubic centimeters	7.646×10^5
cubic yards	cubic feet	27.0
cubic yards	cubic inches	46,656.0
cubic yards	cubic meters	0.7646
cubic yards	gallons (U.S. liquid)	202.0
cubic yards	liters	764.6
cubic yards	pints (U.S. liquid)	1,615.9
cubic yards	quarts (U.S. liquid)	807.9
decimeters	meters	0.1
degrees (angle)	quadrants	0.01111
degrees (angle)	radians	0.01745
degrees (angle)	seconds	3,600.0
dekameters	meters	10.0
fathoms	meters	1.828804
fathoms	feet	6.0
feet	centimeters	30.48
feet	kilometers	3.048×10^{-4}
feet	meters	0.3048
feet	miles (naut.)	1.645×10^{-4}
feet	miles (stat.)	1.89×10^{-4}
feet	millimeters	304.8
feet	mils	1.2×10^4
feet of water	atmospheres	0.02950
feet of water	inches of mercury	0.8826
feet of water	kg/sq centimeter	0.03048
feet of water	kg/sq meter	304.8
feet of water	lb/sq foot	62.43
feet of water	lb/sq inches	0.4335
gallons	cubic centimeters	3,785.0
gallons	cubic feet	0.1337
gallons	cubic inches	231.0
gallons	cubic meters	3.785×10^{-3}
gallons	cubic yards	4.951×10^{-3}
gallons (liq. Br. imp.)	gallons (U.S. liquid)	1.20095
gallons (U.S.)	gallons (imp.)	0.83267
gallons of water	pounds of water	8.3453
gallons/min	cubic ft/sec	2.228×10^{-3}
gallons/min	liters/sec	0.06308
gallons/min	cubic ft/hour	8.0208
hand	centimeters	10.16
hectares	acres	2.471
hectares	square feet	1.076×10^5
hogsheads (British)	cubic feet	10.114

Table D-5. (*Continued*)

To convert	Into	Multiply by
hogsheads (U.S.)	cubic feet	8.42184
hogsheads (U.S.)	gallons (U.S.)	63
inches	centimeters	2.540
inches	meters	2.540×10^{-2}
inches	miles	1.578×10^{-5}
inches	millimeters	25.40
inches	mils	1,000.0
inches	yards	2.778×10^{-2}
kilometers	centimeters	1×10^{5}
kilometers	feet	3,281.0
kilometers	inches	8.937×10^{4}
kilometers	meters	1,000.0
kilometers	miles	0.6214
kilometers	millimeters	1×10^{6}
kilometers	yards	1,094.0
knots	feet/hr	6,080.0
knots	kilometers/hr	1.8532
knots	naut. miles/hr	1.0
knots	stat. miles/hr	1.151
knots	yards/hr	2,027.0
knots	feet/sec	1.689
liters	bushels (U.S. dry)	0.02838
liters	cubic centimeters	1,000.0
liters	cubic feet	0.03531
liters	cubic inches	61.02
liters	cubic meters	0.001
liters	cubic yards	1.308×10^{-3}
liters	gallons (U.S. liquid)	0.2642
liters	pints (U.S. liquid)	2.113
liters	quarts (U.S. liquid)	1.057
liters/min	cubic ft/sec	5.886×10^{-4}
liters/min	gallons/sec	4.403×10^{-3}
meters	centimeters	100.0
meters	feet	3.281
meters	inches	39.37
meters	kilometers	0.001
meters	miles (naut)	5.396×10^{-4}
meters	miles (stat)	6.214×10^{-4}
meters	millimeters	1,000.0
meters	yards	1.094
meters	yards	1.179
meters/min	centimeters/sec	1.667
meters/min	feet/min	3.281
meters/min	feet/sec	0.05468
meters/min	kilometers/hour	0.06
meters/min	knots	0.03238

Table D-5. (*Continued*)

To convert	Into	Multiply by
meters/min	miles/hour	0.03728
meters/sec	feet/min	196.8
meters/sec	feet/sec	3.281
meters/sec	kilometers/hour	3.6
meters/sec	kilometers/min	0.06
miles (naut)	meters	1.853.0
miles (naut)	miles (stat)	1.1516
miles (naut)	yards	2,027.0
miles (stat)	centimeters	1.609×10^5
miles (stat)	feet	5,280.0
miles (stat)	inches	6.336×10^4
miles (stat)	kilometers	1.609
miles (stat)	meters	1,609.0
miles (stat)	miles (naut.)	0.8684
miles (stat)	yards	1,760.0
miles/hr	centimeters/sec	44.70
miles/hr	feet/min	88.0
miles/hr	feet/sec	1.467
miles/hr	kilometers/hour	1.609
miles/hr	kilometers/min	0.02682
miles/hr	knots	0.8684
miles/hr	meters/min	26.82
miles/hr	miles/min	0.1667
miles/min	centimeters/sec	2,682.0
miles/min	feet/sec	88.0
miles/min	kilometers/min	1.609
miles/min	knots/min	0.8684
miles/min	miles/hour	60.0
million gallons/day	cubic ft/sec	1.54723
miner's inches	cubic ft/min	1.5
minutes (angles)	degrees	0.01667
minutes (angles)	quadrants	1.852×10^{-4}
minutes (angles)	radians	2.909×10^{-4}
minutes (angles)	seconds	60.0
pints (dry)	cubic inches	33.60
pints (liquid)	cubic centimeters	473.2
pints (liquid)	cubic feet	0.01671
pints (liquid)	cubic inches	28.87
pints (liquid)	cubic meters	4.732×10^{-4}
pints (liquid)	cubic yards	6.189×10^{-4}
pints (liquid)	gallons	0.125
pints (liquid)	liters	0.4732
pints (liquid)	quarts (liquid)	0.5
quadrants (angle)	degrees	90.0
quadrants (angle)	minutes	5,400.0
quadrants (angle)	radians	1.571
quadrants (angle)	seconds	3.24×10^5

Table D-5. (*Continued*)

To convert	Into	Multiply by
radians	degrees	57.30
radians	minutes	3,438.0
radians	quadrants	0.6366
radians	seconds	2.063×10^5
rods	chains (gunters)	0.25
rods	meters	5.029
rods (surveyor's chains)	yards	5.5
rods	feet	16.5
seconds (angle)	degrees	2.778×10^{-4}
seconds (angle)	minutes	0.01667
seconds (angle)	quadrants	3.087×10^{-6}
seconds (angle)	radians	4.848×10^{-6}
square centimeters	circular mils	1.973×10^5
square centimeters	sq feet	1.076×10^{-3}
square centimeters	sq inches	0.1550
square centimeters	sq meters	0.0001
square centimeters	sq miles	3.861×10^{-11}
square centimeters	sq millimeters	100.0
square centimeters	sq yards	1.196×10^{-4}
square feet	acres	2.296×10^{-5}
square feet	circular mils	1.833×10^8
square feet	sq centimeters	929.0
square feet	sq inches	144.0
square feet	sq meters	0.09290
square feet	sq miles	3.587×10^{-8}
square feet	sq millimeters	9.290×10^4
square feet	sq yards	0.111
square inches	circular mils	1.273×10^6
square inches	sq centimeters	6.452
square inches	sq feet	6.944×10^{-3}
square inches	sq millimeters	645.2
square inches	sq mils	1×10^6
square inches	sq yards	7.716×10^{-4}
square kilometers	acres	247.1
square kilometers	sq centimeters	1×10^6
square kilometers	sq feet	1.076×10^7
square kilometers	sq inches	1.550×10^9
square kilometers	sq meters	1×10^6
square kilometers	sq miles	0.3861
square kilometers	sq yards	1.196×10^6
square meters	acres	2.471×10^{-4}
square meters	sq centimeters	1×10^4
square meters	sq feet	10.76
square meters	sq inches	1,550.0
square meters	sq miles	3.861×10^{-7}
square meters	sq millimeters	1×10^6
square meters	sq yards	1.196

Table D-5. (*Continued*)

To convert	Into	Multiply by
square miles	acres	640.0
square miles	sq feet	2.788×10^{7}
square miles	sq kilometers	2.590
square miles	sq meters	2.590×10^{6}
square miles	sq yards	3.098×10^{6}
square millimeters	circular mils	1,973.0
square millimeters	sq centimeters	0.01
square millimeters	sq feet	1.076×10^{-5}
square millimeters	sq inches	1.550×10^{-3}
square mils	circular mils	1.273
square mils	sq centimeters	6.452×10^{-6}
square miles	sq inches	1×10^{-6}
square yards	acres	2.066×10^{-4}
square yards	sq centimeters	8,361.0
square yards	sq feet	9.0
square yards	sq inches	1,296.0
square yards	sq meters	0.8361
square yards	sq miles	3.228×10^{-7}
square yards	sq millimeters	8.361×10^{5}
temp. (°C) + 273	absolute temperature (°C)	1.0
temp. (°C) + 17.78	temperature (°F)	1.8
temp. (°F) + 460	absolute temperature (°F)	1.0
temp. (°F) − 32	temperature (°C)	5/9
tons (long)	kilograms	1,016.0
tons (long)	pounds	2,240.0
tons (long)	tons (short)	1.120
tons (metric)	kilograms	1,000.0
tons (metric)	pounds	2,205.0
tons (short)	kilograms	907.1848
tons (short)	ounces	32,000.0
tons (short)	ounces (troy)	29,166.72
tons (short)	pounds	2,000.0
tons (short)	pounds (troy)	2,430.56
tons (short)	tons (long)	0.89287
tons (short)	tons (metric)	0.9078
tons (short)/sq. ft	kilograms/sq meter	9,765.0
tons (short)/sq ft	pounds/sq inch	2,000.0
tons of water /24 hr	pounds of water/hr	83.333
tons of water /24 hr	gallons/min	0.16643
tons of water /24 hr	cubic ft/hr	1.3349
yards	centimeters	91.44
yards	kilometers	9.144×10^{-4}
yards	meters	0.9144
yards	miles (naut)	9.434×10^{-4}
yards	miles (stat)	5.682×10^{-4}
yards	millimeters	914.4

Appendix E
Geometric Field Constructions

The constructions described below can be used in a number of layout problems, and can be used as starting points to solve even more complex problems. Only a tape is required.

1. Angle Copy

Construct: An angle that is equal to angle *ABC* (see Fig. E-1).

Procedure:

a. Lay out arc *XY* (radius *r*) so that it intersects *AB* and *BC* at *D* and *E*, respectively.
b. Determine arc distance *DE* by swinging tape about point *E*.
c. Let *Q* be any point on line *MN*.
d. Swing an arc (radius *r*) from *Q* on *MN* to locate *R*.
e. From *R* swing an arc equal to *DE* in step *b* to locate *P*.
f. Angle *ABC* = *ZQN*.

2. Bisecting an Angle

Construct: Bisector *BP* for angle *ABC* (see Fig. E-2).

Procedure:

a. From *B* swing an arc (radius *r*) to locate *X* and *Y*.
b. From points *X* and *Y* (using any convenient radius) swing arcs to define point *P*.
c. Line *BP* bisects angle *ABC*.

3. Line Parallel to a Given Line through a Point on the Given Line

Construct: Line *XY* parallel to given *AB* through point *Q* (see Fig. E-3).

Procedure:

a. Draw any line *MN* through *Q* and intersecting *AB* at *C*.
b. Use angle copy procedure so as to make angle *NQY* = *QCB*.
c. Line *XY* is parallel to line *AB*.

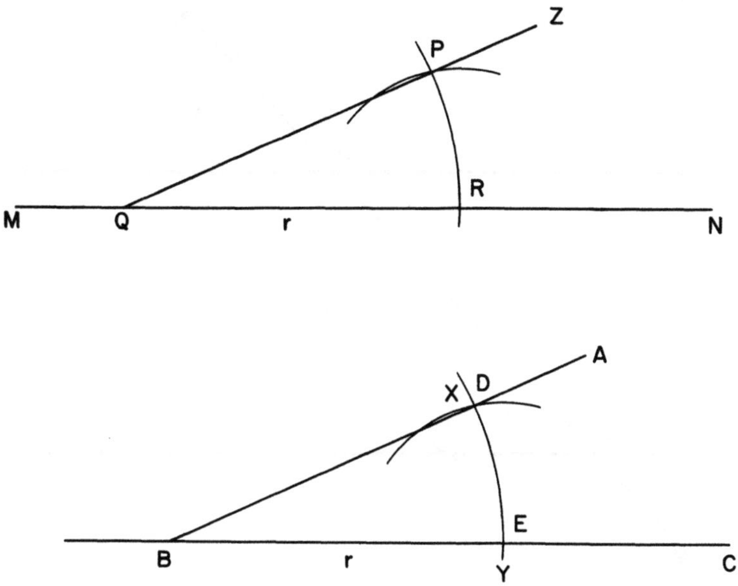

Fig. E-1. Layout procedure for copying an angle in the field.

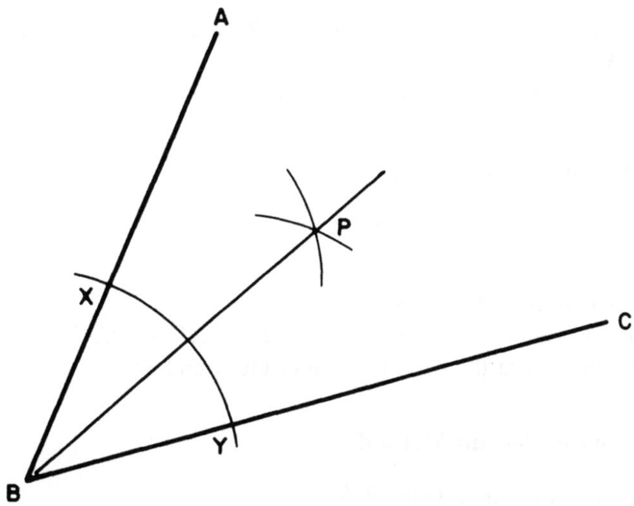

Fig. E-2. Bisecting angle *ABC*.

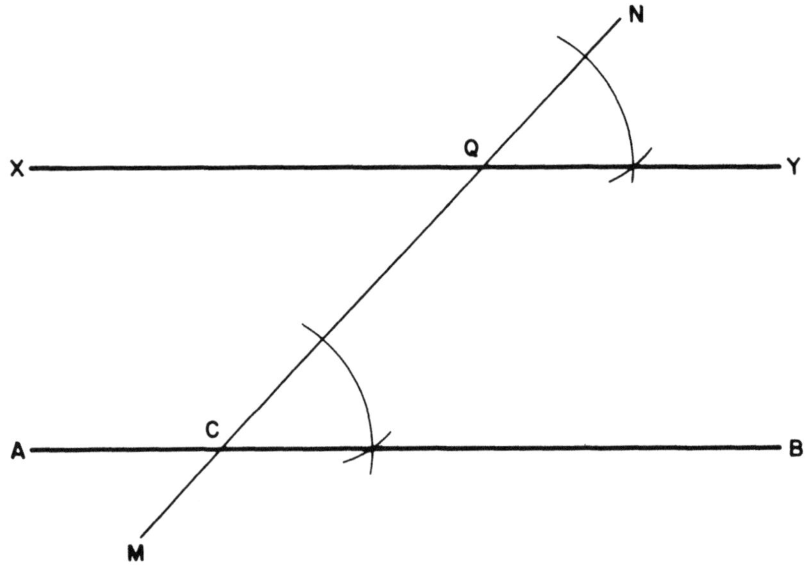

Fig. E-3. Using the angle copy procedure to lay out *XY* parallel to a given line *AB*.

4. Perpendicular Bisector of a Line

Construct: *XY* the perpendicular bisector of *AB* (see Fig. E-4).
Procedure:

a. Use any convenient radius (greater than $\frac{1}{2}$ *AB*) and swing arcs from *A* and *B* to define *X* and *Y*.
b. Line *PQ* is perpendicular to and bisects *AB*.

5. Center of a Circle (First Method)

Locate: Center of a given circle (Fig. E-5).
Procedure:

a. Draw any two chords *AB* and *BC*.
b. Construct perpendicular bisectors *PQ* of *AB* and *XY* of *BC*.
c. *PQ* and *XY* intersect at the center of the circle, point *O*.

6. Center of a Circle (Second Method)

Locate: Center of a given circle (Fig. E-6).
Procedure:

a. Select any three widely spaced points, *A*, *B*, and *C* on the circumference.

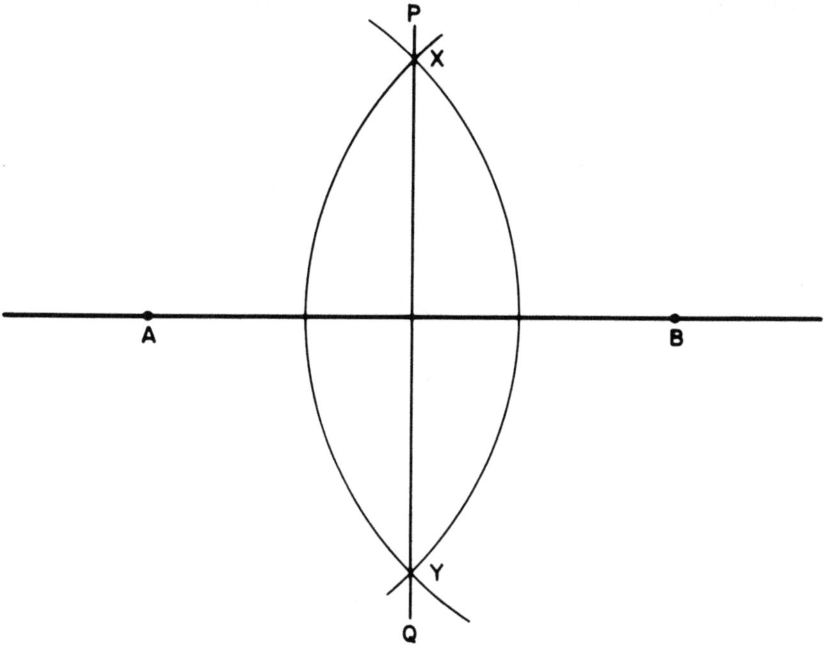

Fig. E-4. Layout procedure for determining the perpendicular bisector of line *AB*.

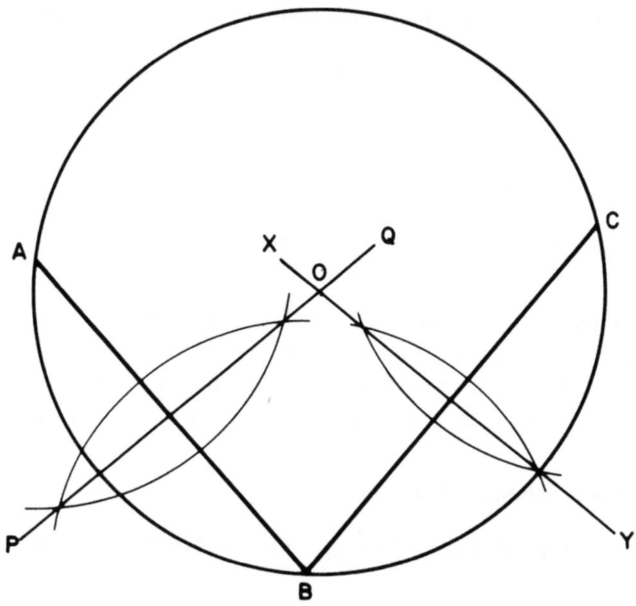

Fig. E-5. A method for locating the center of a circle using perpendicular bisectors of any two chords.

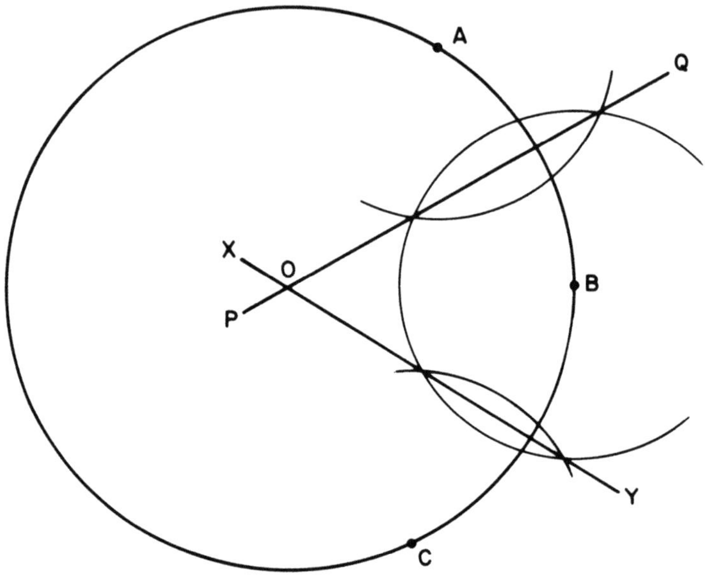

Fig. E-6. A method for locating the center of a circle using arcs drawn from three points on the circumference.

b. With three equal radii, swing arcs about these points so that they intersect each other.
c. Draw lines *XY* and *PQ* through these intersections.
d. *XY* and *PQ* intersect at the center of the circle, point *O*.

7. Lay Out a Hexagon

Construct: Hexagon on a given straight line *AB* (see Fig. E-7).
Procedure:

a. From *A* and *B* swing arcs equal to *AB* to define point *O*.
b. From *O*, using *OA* as radius, describe a circle.
c. With the same radius lay off arcs *AF*, *FE* and *BC*, *CD*.
d. The intersection of these arcs and the circle define the hexagon.

8. Lay Out Any Regular Polygon

Construct: Any regular polygon using a circumscribed circle as a guide (see Fig. E-8).
Procedure:

a. Draw diameter DH.

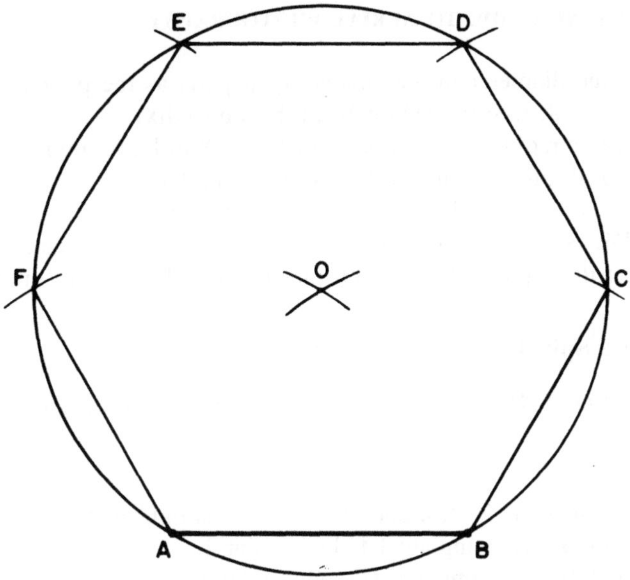

Fig. E-7. Procedure for laying out a hexagon starting with line *AB*.

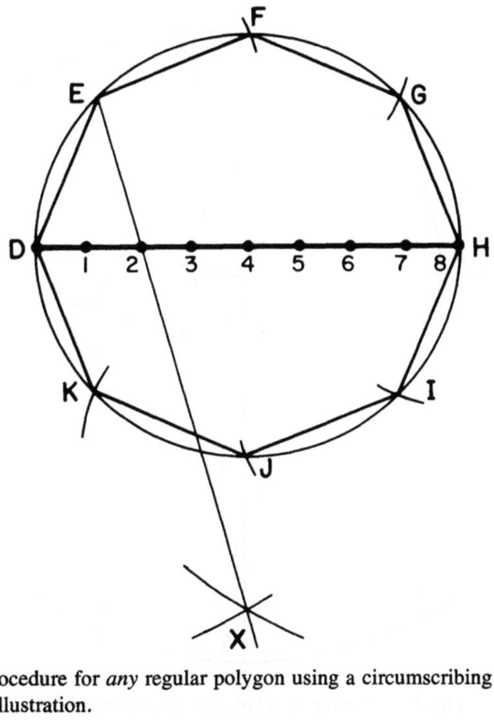

Fig. E-8. Layout procedure for *any* regular polygon using a circumscribing circle as a guide. An octagon is used for illustration.

b. Divide the diameter into as many equal parts as the polygon is to have sides, e.g., 8. (see Procedure 10 in this appendix)
c. With *D*8 as radius, swing arcs from *D* and 8 to locate point *X*.
d. Draw *XE* through point 2. (always use point 2.)
e. Join DE, i.e., one side of the required polygon.
f. With DE as radius locate points *F*, *G*, *H*, *I*, *J*, *K* on the circle.
g. Connect these points by straight lines to form the sides of the polygon.

9. Lay Out a Spiral

Construct: Spiral 1 *ABCDEFGH* by using arcs of circles, center at 2 (see Fig. E-9).

Procedure:

a. Lay out square 1, 2, 3, 4 and extend sides as indicated.
b. Using radius 21, swing arc 1*A* about point 2.
c. Using radius 3*A*, swing arc *AB* about point 3.
d. Using radius 4*B*, swing arc *BC* about point 4.

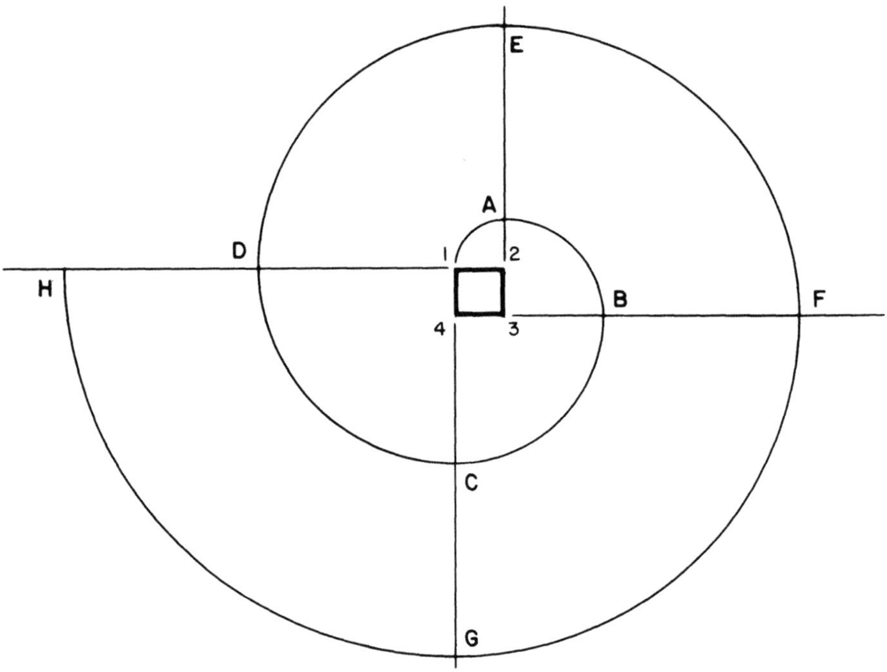

Fig. E-9. Layout of a spiral with center at 2.

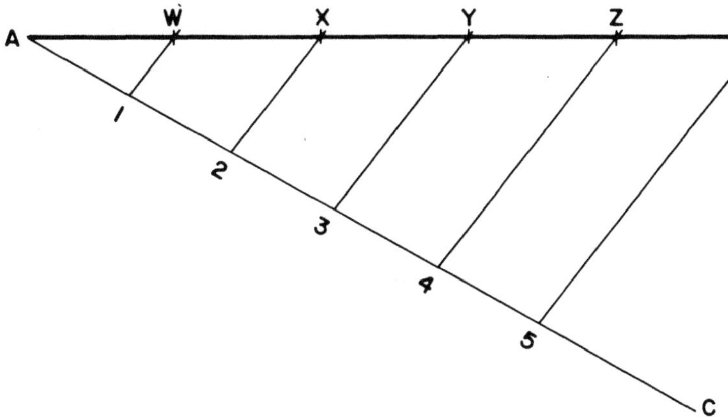

Fig. E-10. Procedure for dividing any line into equal parts.

e. Using radius 1*C*, swing arc *CD* about point 1.
f. Using radius 2*D*, swing arc DE about point 2.
g. Continue as needed.

10. Divide A Line into Equal Parts

Divide: Line *AB* into a number of equal parts, e.g., five (see Fig. E-10).
Procedure:

a. Lay out any diagonal line *AC* from *AB*.
b. Divide *AC* into five equal lengths (use any convenient unit length).
c. Join *B* and 5.
d. Through points 1, 2, 3, and 4 lay out lines 1*W*, 2*X*, 3*Y*, and 4*Z* (all parallel to *B*5).
e. *AB* is divided into 5 equal parts *AQ*, *WX*, *XY*, *YZ*, and *ZB*.

11. Layout a Line Equidistant from a Curved Line

Construct: Line *XY* equidistant from line *AB* (see Fig. E-11).
Procedure:

a. Select a number of points (the greater the number the greater the accuracy) along *AB*.
b. Swing a series of arcs (radius equal to the required separation) with the points selected in step a as center.
c. Tangents to these arcs is line *XY* equidistant from *AB*.

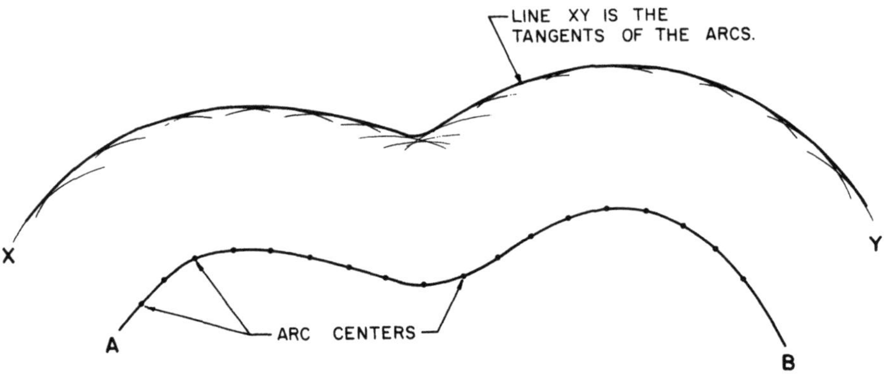

Fig. E-11. Method for laying out a line equidistant from a given irregular line.

SOME ADDITIONAL FIELD TECHNIQUES WITH A TRANSIT

The procedures given below are occasionally used in field work. They are useful in themselves, and they can be used in conjunction with other procedures. Refer to chapter 1 for needed trigonometry.

1. Locating the Intersection of Two Straight Lines

Given: Points *ABC* and *D* (see Fig. E-12).

Find: The intersection (*X*) of *AB* and *CD*.

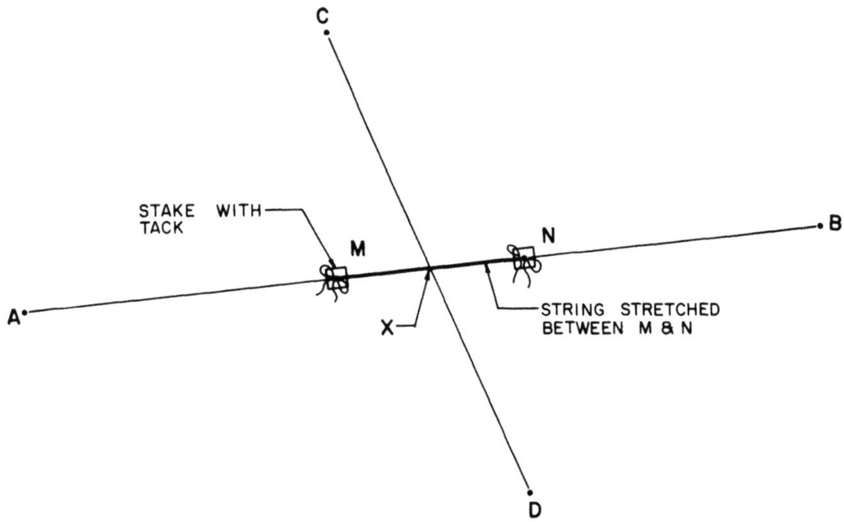

Fig. E-12. A method for locating the intersection of two straight lines.

Procedure:

a. Set transit on point *A*, and sight on *B*.
b. Set stakes (with tacks) at *M* and *N* to straddle line *CD*.
c. Connect *M* and *N* with a string.
d. Set transit at point *C* and sight on *D*.
e. Where line of sight crosses the string is point *X*.

2. Measuring an Angle That Cannot Be Occupied

Method #1

Given: Corner *ABC* (see Fig. E-13)
Find: Angle *ABC* (which cannot be occupied).
Procedure:

a. Establish $A'B'$ parallel to *AB* and $C'B''$ parallel to *CB*.
b. Extend $A'B'$ and $C'B''$ to where the intersect at *X*.
c. Set transit at *X* and measure angle $A'XC'$
d. Angle $A'XC'$ = Angle *ABC*

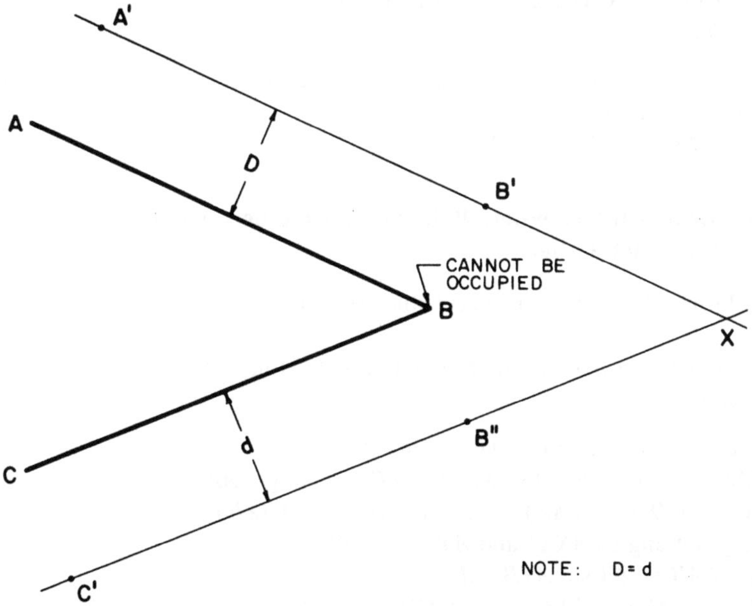

Fig. E-13. Measuring angle *ABC* when *B* cannot be occupied (first method).

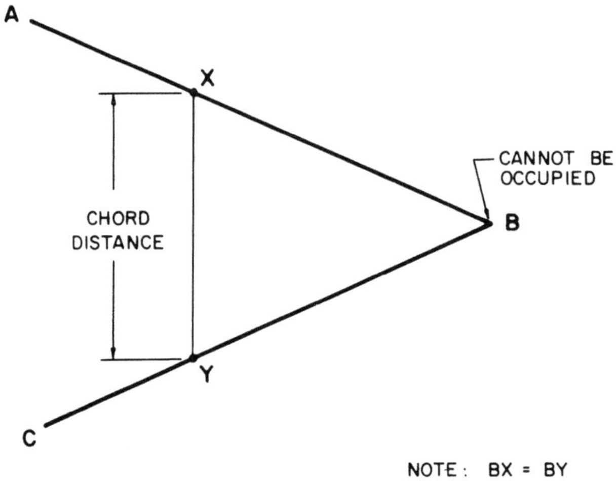

Fig. E-14. Measuring angle *ABC* when *B* cannot be occupied (second method).

Method #2 (more accurate)

Given: Corner *ABC* (see Fig. E-14).

Find: Angle *ABC* (which cannot be occupied).

Procedure:

a. Measure a convenient distance along *AB* and *CB* to *X* and *Y*, respectively.
b. Measure chord distance *XY*.
c. Sin $ABC = XY/XB + YB$.

3. Establishing Intermediate Points on a Line between Two Points Not Visible from Each Other

Given: Points *A* and *B*, not intervisible (see Fig. E-15, and Appendices B and C).

Find: Length of *AB* and intermediate points along *AB*.

Procedure:

a. Lay out and measure random line *AZ*.
b. Measure distances *AX*, *XY*, *YZ*, *ZB*, and angle *AZB*.
c. $AB = (ZB)^2 + (AZ)^2 - 2(ZB)(AZ) \cos (AZB)$
d. Lay out angles AXX' and $AYY' = AZB$
e. $\sin ZAB = ZB \sin AZB/AB$
f. $XX' = AX \sin XAX'/\sin (XAX' + AX'X)$
g. $YY' = AY \sin YAY'/\sin (YAY + AY'Y)$

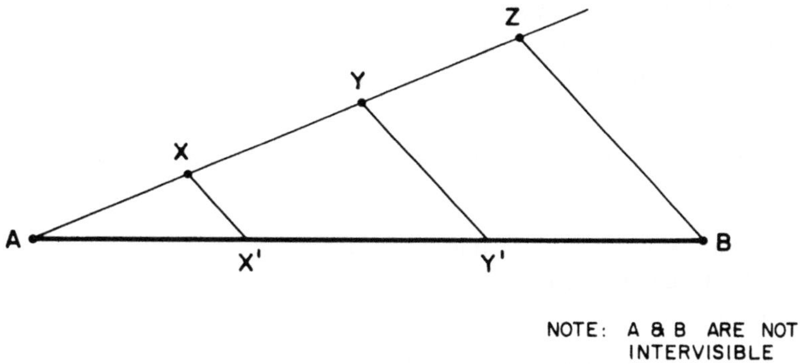

Fig. E-15. Procedure for finding the length of a line and intermediate points on a line when the end points, e.g., *A* and *B*, are not intervisible.

4. Find the Height of an Object with Base Inaccessible

Given: Point *Y* directly above *X* and ground is horizontal (see Fig. E-16, and Appendix B).

Find: Height of *Y* and *X*.

Procedure:

a. Lay out convenient distance *AB*.
b. From transit setup at *A* determine angles *XAY* and *BAX*.

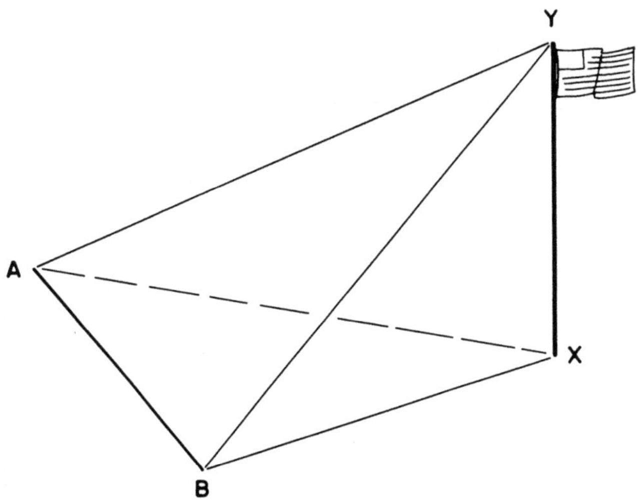

Fig. E-16. One procedure for determining the height of an object, e.g., *XY*, whose base is inaccessible from transit setups.

c. From transit set up at B determine angle ABX.
d. $AX = AB \sin ABX / \sin (ABX + BAX)$.
e. $XY = AX \tan (XAY)$.

5. Find Distance between Two Points Each Inaccessible from Each Other and from Any Third Point

Given: Points X and Y inaccessible from each other (see Fig. E-17).

Find: Distance between X and Y.

Procedure:

a. Establish A on the extension of XY.
b. Lay out AB a convenient length perpendicular to AY.
c. Lay out BC perpendicular to AB (make this line as long as or longer than AY: estimate).
d. Locate transit at P, midway between A and B.
e. Sight on Y and plunge transit to locate Y' on BC.
f. Repeat step e for point X.
g. $XY = X'Y$

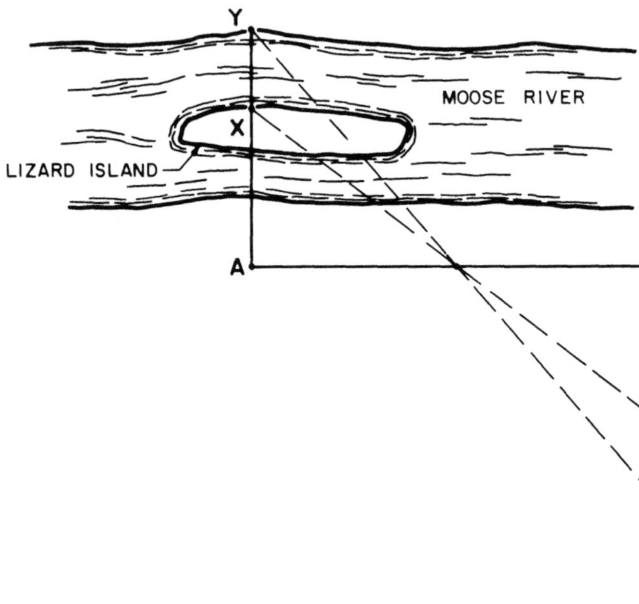

Fig. E-17. Locating two points (inaccessible from each other) from a third point inaccessible from the other two.

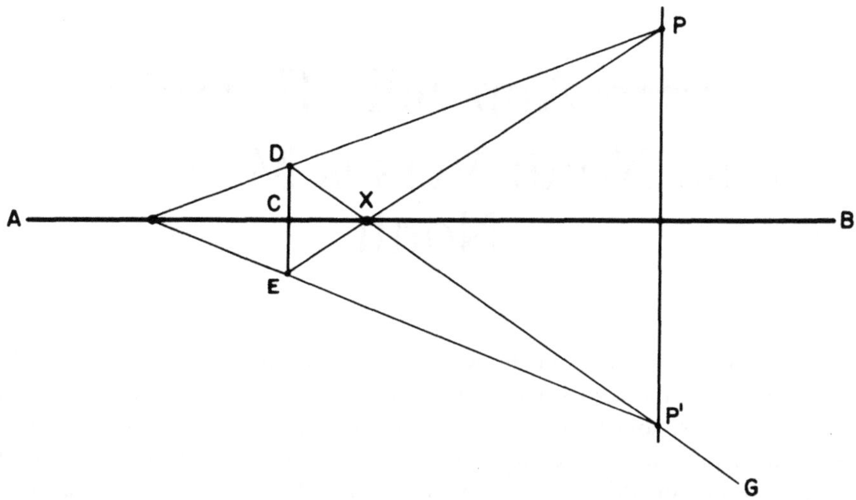

Fig. E-18. Layout procedure for establishing a perpendicular to line *AB* from an inaccessible point *P*.

6. Establish a Perpendicular to a Given Line from an Inaccessible Point

Given: *P* the inaccessible point, and the line *AB* (see Fig. E-18).

Find: Perpendicular *PP'* to line *AB* from inaccessible point *P*.

Procedure:

a. At any point *C* on *AB* and perpendicular to *AB*, layout *CD* at any convenient length.
b. Extend *CD* an *equal distance* on the other side of *AB* to *CE*.
c. Set up transit *A* + *E*, sight on *P*, and mark intersection with *AB* as *X*.
d. With the transit at *D*, sight on *P*, and extend this line from *PD* to (where it intersects *AB*)*F*.
e. From the same setup, extend line *DX* to *G* (a point beyond *P*).
f. From a setup at *F*, sight through *E* to its intersection with *DG*, mark this point *P'*.
g. *PP'* perpendicular to *AB*.

Appendix F
True North Versus Magnetic North

There are two common designations for "north" that are popular on deeds, plot plans, tax maps, municipal maps, etc. They are *true north* (TN) and *magnetic north* (MN). TN, sometimes called *geographic north*, is located at the point of the earth's axial rotation. Reference for magnetic north, i.e., the magnetic north pole, is located on Prince of Wales Island in northern Canada, a distance of roughly a thousand miles from the geographic pole.

Because of this, most large scale maps show two north arrows, one pointing to TN and the other toward MN. The angular separation between the two directions is called the *declination* (see Fig. F-1). The declination varies in different parts of the world depending on the observer's position relative to the two poles (see Fig. F-2). Fig. F-3 is an *isogonic map*, e.g., a map of equal magnetic declination. Isogonic maps can be used to adjust compass readings at various geographic locations.

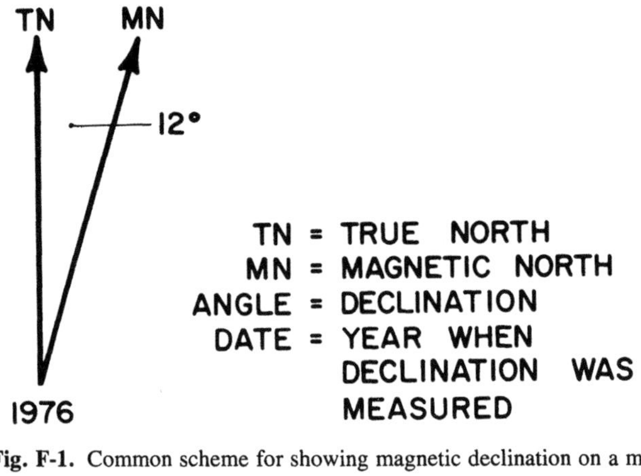

Fig. F-1. Common scheme for showing magnetic declination on a map.

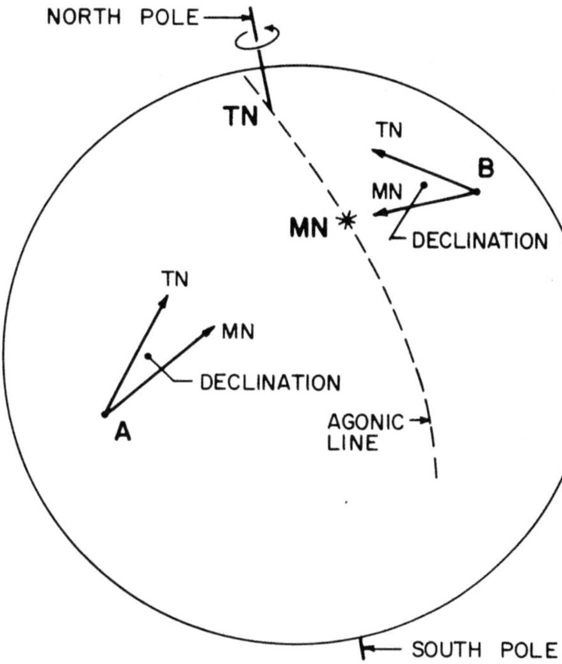

Fig. F-2. Variation in magnetic declination at two different geographic locations.

As a final note, declination will also vary slightly at the same location over a long period of time because of the fact that the magnetic north pole is not stationary. The date of the most recent declination data are usually given along with the two north arrows (see Fig. F-1).

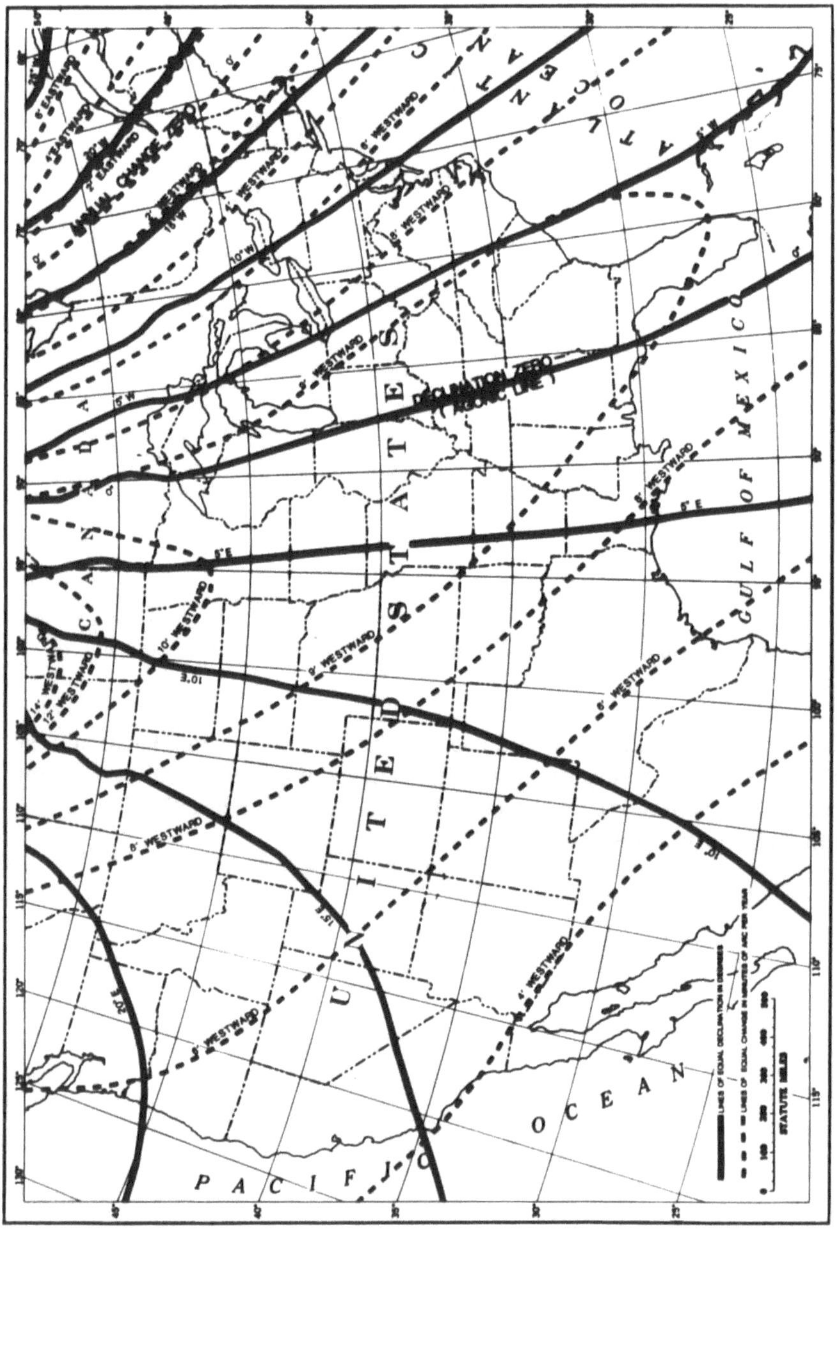

Fig. F-3. Magnetic declination for the contiguous 48 states. (Derived from U.S. Geological Survey Map, I 1283.)

Useful References

American Society of Civil Engineers, Committee on Definition of Surveying and Mapping (1954). *Definitions of Surveying, Mapping, and Related Terms*. American Society of Civil Engineers, New York, 202 p.

Atcheson, Daniel (1986). *Estimating Earthwork Quantities*. Norseman Publishing Company, Lubbock, TX, 221 p.

Barnes, W. M. (1988). *Basic Surveying*. Butterworths Publ. Ltd., London, 196 p.

Barry, Benjamin A. (1981). *Construction Measurements*, 2nd Ed. John Wiley & Sons, New York, 346 p.

Beakley, Geo. C., Autore, Donald D., and Patterson, Terry L. (1984). *Architectural Drawing and Design*. Macmillan Publ. Co., New York, 539 p.

Booth, Norman K. (1983). *Basic Elements of Landscape Architectural Design*. Elsevier Science Publ. Co., New York, 315 p.

Bouchard, Harry and Moffitt, Francis H. (1965). *Surveying*, 5th Ed. International Textbook Co., Scranton, PA, 754 p.

Brock, Gerald C. (1967). *Physical Aspects of Aerial Photography*. Dover Publications, Inc., New York, 267 p.

Burnside, Clifford D. (1979). *Mapping From Aerial Photographs*. Granada Publ. Ltd., London, 304 p.

Ciciarelli, John A. (1984). *Practical Physical Geology, Problems and Solutions*. Gordon and Breach Science Publ., New York, 263 p.

Davis, Raymond E., Foote, Francis S., and Kelly, Joe W. (1953). *Surveying Theory and Practice*, 5th Ed. McGraw-Hill, New York, 1096 p.

Droogan, Irving et al., Editors (1987). *Trigonometry, An Analytical Approach*, 5th Ed. Macmillan Co., London, 396 p.

Feldman, Bernard (1974). *Trigonometry*. Wadsworth Publ. Co., Belmont, CA, 100 p.

Finkl, Charles W., Editor (1984). *The Encyclopedia of Applied Geology* (Vol. 13 of Encyclopedia of Earth Sciences Series: Fairbridge, Rhodes, W., series editor). Van Nostrand Reinhold Co., New York, 644 p.

Fischer, W. A. (1962). "Color aerial photography in geological investigation," *Photogrammetric Engineering*, Vol. 28, No. 1, pp. 133–139.

Frost, R. E. (1946). "Identification of granular deposits by aerial photography, Natl. Res. Council, Hwy. Res. Board, 26th Ann. Mtg. Proc., v. 25, pp. 116–129.

Griffiths, John C. (1967). *Scientific Method in the Analysis of Sediments*. McGraw-Hill Book Co., New York, 508 p.

Hallert, Bertil (1968). *Photogrammetry, Basic Principles and General Survey*. McGraw-Hill Book Co., New York, 348 p.

Hamblin, Wm. K. (1980). *Atlas of Stereoscopic Aerial Photographs and Landsat Imagery of North America*. Tasa Publ. Co., Minneapolis, MN, 208 p.

Heisler, Sanford I. (1984). *The Wiley Engineer's Desk Reference*. John Wiley & Sons, New York, 565 p.

Herubin, Charles H. (1978). *Principles of Surveying*, 2nd Ed. Reston Publishing Co., Reston, VA, 324 p.

Hickerson, Thomas F. (1964). *Route Location and Design*, 5th Ed. McGraw-Hill Book Co., New York, 634 p.

Hornung, Wm. J. (1965). *Reinhold Data Sheets for Architects, Engineers, Designers, Draftsmen*. Reinhold Publ. Corp., New York, 238 p.

Howard, John A. (1970). *Aerial Photo-Ecology*. American Elsevier Publ. Co., New York, 325 p.

Jackson, W. P. (1979). *Building Layout*. Craftsman Book Co., Carlsbad, CA, 238 p.

Kissam, Philip (1971). *Surveying Practice*, 2nd Ed. McGraw-Hill Book Co., New York, 482 p.

Kodak Scientific and Technical Publication (1985). *Kodak Filters for Scientific and Technical Uses*. Eastman Kodak Co., Rochester, NY, 94 p.

Lattman, Laurence H. and Ray, Richard G. (1965). *Aerial Photographs in Field Geology*. Holt, Reinhart, and Winston, New York, 221 p.

Lo, Chor Pang (1986). *Applied Remote Sensing*. Longman Scientific and Technical Publ., Essex, England, 393 p.

Marsh, Wm. M., and Dozier, Jeff (1986). *Landscape Planning, Environmental Applications*. John Wiley & Sons, New York, 356 p.

Miller, Victor C. (1961). *Photogeology*. McGraw-Hill Book Co., New York, 248 p.

Miller, Victor C. and Westerback, Mary (1988). *Interpretation of Topographic Maps*. Merrill Publishing Co., Columbus, OH, 416 p.

Milne, P. H. (1984). *Basic Programs for Land Surveying*. E. and F. N. Spon Publ., London, 442 p.

Milne, P. H. (1987). *Computer Graphics for Surveying*. E. and F. N. Spon Publ., London, 241 p.

Mitchell, Hugh C. (1969). *Definitions of Terms Used in Geodetic and Other Surveys*. U.S. Govt. Printing Office, Washington, DC, 87 p.

Mitchell, R. J. (1983). *Earth Structures Engineering*. Allen and Unwin, Inc., Boston, 265 p.

Monkhouse, Francis J. and Wilkinson, H. R. (1971). *Maps and Diagrams: Their Compilation and Construction*. Methuen Publishers, London, 522 p.

Munson, Albe E. (1975). *Construction Design for Landscape Architects*. McGraw-Hill Book Co., New York, 212 p.

Parker, Harry and MacGuire, John W. (1954). *Simplified Site Engineering for Architects and Builders*. John Wiley & Sons, Inc., New York, 250 p.

Rabenhorst, Thomas D. and McDermott, Paul D. (1989). *Applied Cartography: Source Materials for Map Making*. Merrill Publishing Co., Columbus, OH, 168 p.

Ray, Richard G. (1960). *Aerial Photographs in Geologic Interpretation and Mapping*. U.S. Geological Survey Professional Paper, 373 p.

Rich, John L. (1947). "Reconnaissance mapping from oblique aerial photographs without ground control," *Photogrammetric Engineering*, Vol. 13, No. 4, pp. 600–609.

Root, James B. (1985). *Fundamentals of Landscaping and Site Planning*. AVI Publ. Co., Westport, CT, 158 p.
Schmitz, Michael J. (1981) *Surveying Computations, Mastery or Mystery*. Landmark Enterprises, Inc., Rancho Cordova, CA, 184 p.
Seelye, Elwyn G. (1954). *Field Practice, A Data Book for Civil Engineers*, Vol. 3, 2nd Ed. John Wiley & Sons, New York, 394 p.
Sharp, Henry, Jr. (1958). *Elements of Plane Trigonometry*. Prentice-Hall, Inc., Englewood Cliffs, NJ, 274 p.
Shuster, R. L. and Krizek, R. J., Editors (1978). *Landslides, Analysis and Control, Special Report 176*. Transportation Research Board, National Research Council, National Academy of Sciences, Washington, DC, 234 p.
Siegle, Arthur (1979). *Basic Plane Surveying*. Van Nostrand Reinhold Co., New York, 295 p.
Speak, P., and Carter, A. H. C. (1977). *Map Reading and Interpretation*. Longman Group Publ. Ltd., London, 79 p.
Strandberg, Carl H. (1967). *Aerial Discovery Manual*. John Wiley & Sons, New York, 249 p.
U.S. Department of the Army (1954). *Tactical Interpretation of Air Photos*. U.S. Army Technical Manual 30-246. Headquarters, Dept. of the Army, Washington, DC.
U.S. Department of the Army (1969). *Map Reading*. U.S. Army Field Manual 21-26. Headquarters, Dept. of the Army, Washington, DC.
U.S. Department of the Army (1971). *Elements of Surveying*. U.S. Army Technical Manual 5-232. Headquarters, Dept. of the Army, Washington, DC.
U.S. Department of the Army (1985). *Construction Surveying*. U.S. Army Field Manual 5-233. Headquarters, Dept. of the Army, Washington, DC.
Way, Douglass (1973). *Terrain Analysis, A Guide to Site Selection using Aerial Photographic Interpretation*. Dowden, Hutchinson, and Ross, Publ., Stroudsburg, PA, 392 p.
Wolf, Paul R. (1983). *Elements of Photogrammetry*, 2nd Ed. McGraw-Hill Book Co., New York, 628 p.

Index

Index

MIX
Papier aus verantwortungsvollen Quellen
Paper from responsible sources
FSC® C105338

If you have any concerns about our products,
you can contact us on
ProductSafety@springernature.com

In case Publisher is established outside the EU,
the EU authorized representative is:
Springer Nature Customer Service Center GmbH
Europaplatz 3, 69115 Heidelberg, Germany

Printed by Libri Plureos GmbH
in Hamburg, Germany